COMMUNICATION
AND
RADAR SYSTEMS

COMMUNICATION
AND
RADAR SYSTEMS

NICOLAOS S. TZANNES, Ph.D.
Professor of Electrical Engineering
University of Patras
Greece

PRENTICE-HALL, INC., *Englewood Cliffs, N.J.* *07632*

Library of Congress Cataloging in Publication Data

TZANNES, N. S. (NICOLAOS S.) (date)
    Communication and radar systems.

    Bibliography: p.
    Includes index.
    1. Radar. 2. Telecommunication. I. Title.
TK6575.T95 1985   621.3848   84-2005
ISBN 0-13-153545-5

Editorial/production supervision
and interior design: *Theresa A. Soler*
Manufacturing buyer: *Anthony Caruso*

Printed in the United States of America

10 9 8 7 6 5 4 3 2 1

ISBN 0-13-153545-5 01

Prentice-Hall International, Inc., *London*
Prentice-Hall of Australia Pty. Limited, *Sydney*
Editora Prentice-Hall do Brasil, Ltda., *Rio de Janeiro*
Prentice-Hall Canada Inc., *Toronto*
Prentice-Hall of India Private Limited, *New Delhi*
Prentice-Hall of Japan, Inc., *Tokyo*
Prentice-Hall of Southeast Asia Pte. Ltd., *Singapore*
Whitehall Books Limited, *Wellington, New Zealand*

To Michael Andrew

CONTENTS

= 3 = **AMPLITUDE MODULATION:
        COMMUNICATION SYSTEMS**   71

= 4 = **AMPLITUDE MODULATION:
        RADAR SYSTEMS**   104

= 5 = **ANGLE MODULATION:
        COMMUNICATION SYSTEMS**   141

PREFACE

Most graduating engineers know nothing about radar; it is not part of their undergraduate or, in most cases, graduate programs. This was acceptable when radar was used only by the military, but now radar has crept into many civilian applications, and the time is ripe for a change.

I first toyed with the idea of teaching a course on radar systems to last-year electrical engineering students during the early 1980s. My first problem was a text. Although there was a plethora of books on radars, all of them were written for the specialist or the advanced graduate student. My immediate reaction was to develop my own notes, leading possibly to a new book suitable for the endeavor at hand. Soon, however, I realized that such an effort would be fruitless. The undergraduate electrical engineering curriculum was already packed with courses and there was little if any room for radar systems, even as an elective. There was only one alternative under these conditions, and that was to combine the material with an already existing course.

The present book has been written with the belief that radar systems are a natural part of communication systems. The two fields have enough in common that the final result of their combination is much less than the sum of the two fields. Even so, a new book had to be written on the combined subjects, as the many excellent existing texts on communications have very little or nothing on the subject of radar systems. It is my hope that the way I have combined the two fields makes their joint study comprehensible and easy to grasp.

The first part of the book (chapters 1-10) covers the analysis of communication and radar systems in a noise-free regime; an approach taken for pedagogical reasons. Chapter 1 is an introductory general discussion on communication and radar systems, an effort to point out their similarities and differences, and to lay the foundation for their joint analysis. Chapter 2 gives background material on signal analysis of the type usually found in communication and radar systems books; it is presented here in a manner useful for both. The remainder of the chapters cover communication and radar systems in a parallel fashion (AM communication systems followed by AM radar systems and so on) a method that results in saving time and space, and that enables the reader to consolidate knowledge by comparison. It should be noted that the joint analysis of radar systems had to be presented in a way that violates the usual order of presentation in radar books. In any event, I now feel that this order of presentation serves educational purposes much better than the traditional one.

The second part of the book (chapters 11–15) covers the analysis of both types of systems in the presence of additive noise. Chapters 11 and 12 comprise the usual background material on probability and random processes, which is the foundation for the mathematical treatment of noise. Chapter 13 looks at some typical communication systems with an eye on performance in the presence of noise and emphasis on the method of analysis. Chapter 14 is a parallel effort for radar systems from the point of view of their overall functions and not their specific type. The last chapter in the book is an introduction to the field of electronic warfare, a rather controversial topic, presented here to illustrate the often conflicting goals of design and optimization.

I feel the entire book can be covered in one semester at the senior level, assuming the background materials in Chapters 2, 11, and 12 can be covered rapidly. If the students' backgrounds are not strong, a two-semester course is in order, filling in where necessary with material from the problems. In any event, the first ten chapters of the book are the absolute minimum that should be mastered by graduating Electrical Engineers before they enter the job market.

I have written this book the way I talk and lecture. I could, of course, claim the reason for this style is to make the book readable. But the real reason is that I read the following in the book *Palm Sunday* by Kurt Vonnegut, Jr., (1981): "Newspaper reporters and technical writers are trained to reveal almost nothing about themselves in their writings. This makes them freaks in the world of writers, since almost all the other ink-stained wretches in that world reveal a lot about themselves to readers. We call them revelations, accidental and intentional, elements of literary style."

After that *I knew* I had to write it as I speak. Maybe I resented the word "freak." Maybe deep down I harbored illusions of writing a technical book in literary style. Anyway, it turned out to be quite easy to let the writing flow as it came, and it came as it is presented in this book. At the end I realized that Mr. Vonnegut will always be right, at least about technical writers. There is not much you can put in about yourself when you are analyzing chirp radar—*accidental* or *intentional.* So I just hope the book is simply more readable.

Many thanks to Mrs. Niki Sarantoglu for typing the manuscript.

N.S. Tzannes
Rion, Patras, Greece

COMMUNICATION
AND
RADAR SYSTEMS

1

INTRODUCTION

Well-informed people know that it is impossible to transmit the voice over wires and that, were it possible to do so, the thing would be of no practical value.

A newspaper editor in 1865[1]

1.1 PROLEGOMENA

The oldest communication system is the person-to-person type, the one that operates when we talk to each other, transmitting messages back and forth by conversation. We, of course, had nothing to do with the development of this system; evolutionary forces are responsible for it, and the result is not yet well understood.

Like all systems, the person-to-person type has some difficulties, and many of them still remain uncorrected, as they appear to be caused by psychological factors. There was one difficulty, however—that of its inability to transmit messages over long distances, for which something could be done—and was. Fire, smoke, doves, and so on, quickly moved in to alleviate the problem and extend the range of this natural system. The origins of such primitive systems have been lost in antiquity. But crude as they may appear to us today, they were satisfactory for the human needs of their time. In fact, they had all the elements of a sound communication system, as will be seen during the discussion in the following section.

The first recorded instance of an artificial communication system has been found in the writings of the Greek historian Polivius[2] (201–120 B.C.). This system transmitted messages by using two groups of five torches, each group

---

[1] As reported by Alvin Toffler in *Future Shock* (New York: Random House, Inc., 1970).
[2] See Still (1946).

1

being placed on a wall 2 m high. Various combinations of torches in the two walls resulted in the codification of the 24 Greek letters and the blank that separates words from each other. Such walls were built in a way that messages could be carried over long distances—not very rapidly, of course, but faster than any other known method. It is worthwhile mentioning that this system is a pulse-code modulated system (PCM), and such systems are the latest word in modern communications, as we shall see in due time.

Modern communication systems are based on electricity and magnetism, and as such they probably owe their development in the discoveries of Oersted (1819) and Ampère (1820). This development has been phenomenal, most of it in the last 50 years. We now have transatlantic cables carrying voice messages across continents, satellites transmitting voice and picture messages across stars, and all these transmissions are dissipated in fractions of a second. Advances in basic science promise even faster developments for the future, developments that may change completely our presently known life-styles.

Radar systems, like communication systems, also have their origin in nature. Various animals (e.g., porpoises, bats) are known to have the ability to transmit acoustic waves and to obtain information about the outside world, by sensing their reflections (echoes). We, of course, are not a complete radar system but we certainly are half of one, the receiver half. Human beings cannot emit waves (although some people claim they can), but we can receive the reflection of waves emitted from other sources, and thus obtain information about the outside world. We do so, for example, every time we use our various senses (radar receivers). *Seeing* is accepting and analyzing optical reflections from objects. The only difference is that we do not emit the light waves to begin with, at least we did not, in the early part of our evolution. Nowadays, of course, we do, with a flashlight or other sources of optical waves, so that with a little ingenuity we have turned ourselves into a good radar system.

Extending our natural range as a radar system to greater distances did not start until the discovery of the telescope in the year 1608. But the telescope still did not have its own source of illumination, and depended mainly on the sun or other stars. Modern electromagnetic (or acoustic) radar differs in this important detail. It emits its own waves and then sits back to receive the return echoes and to analyze them. Its ability to sense the environment and obtain information about it has been extended to gigantic distances, sometimes millions of miles away. And it can do so in the dark, under cloud cover, under water, or various other adverse conditions that render other instruments inoperative.

Radar systems are more recent than communication systems. Although some of the principles of their operations were known earlier, the first radar systems appeared just before World War II. Since then, their development has been phenomenal. Despite the fact that their initial purpose was military (detecting enemy aircraft), radar systems are everywhere around us, in police (or passenger) vehicles, in ships, in airports, in weather stations, and in astronomical observatories. Their contributions to human well-being is steadily climbing. Knowledge of their principles of operation will soon become an integral part of the electrical engineering studies of undergraduates, and this is the reason for including them in this book.

Communication and radar systems are different, of course, particularly in the goals they wish to accomplish. Communication systems transmit information from one place to another. Radar systems are sensing devices. They send out a signal and they analyze the echo's changes from the original signal to deduce information about the object that reflected the signal. Despite this difference of purpose, their similarities are greater, and that is the reason they can be studied together in a unified manner. Both systems use signals for transmission, so signal theory is the common background for both. Both must convert this signal to electromagnetic waves by using devices that operate in the same manner. The waves travel through media that are similar in both cases. On the receiver side, both systems must receive a signal, usually contaminated by noise, and the information that it carries must be extracted. It should, therefore, be no surprise that a unified approach for analysis is available in both cases.

This first chapter is quite general in its approach. The first two of the sections that follow are devoted to each system separately, with the discussion from the overall systemic point of view. The third section compares the two, and points out their similarities and differences in some detail. The final section of the chapter reviews the approach and level of presentation of the contents of the entire book.

We should warn the reader that there are a lot of generalizations in the remainder of this chapter. This may be offensive, especially to the experts, as generalizations are *rarely* absolutely correct. Nevertheless, they have some merit, particularly in introducing a subject, a case in point in this, our first chapter. The idea here is to give an overall picture of the topics, a picture that readers must try to keep in the back of their minds or they may loose interest in the details that follow.

1.2 THE GENERAL COMMUNICATION SYSTEM

The purpose of a communication system is to transmit some piece of information from point A (the source) to point B (the destination). This goal is accomplished by a series of devices or systems interconnected in the manner shown in Fig. 1.2.1. This figure represents a very general communication system in block diagram form and as such, it is not very accurate or detailed. Nevertheless, it is satisfactory as an initial encounter with the system.

The signal $f(t)$ represents the information that needs to be transmitted. This $f(t)$ could be any number of things: a speech signal, a landscape photograph, the thoughts of a human being, the variations of temperature in the bottom of the sea, and so on. In most cases, this initial information signal can not be transmitted to the destinations in its "raw" form. For example, the thoughts of a human being cannot be transmitted directly to the brain of another

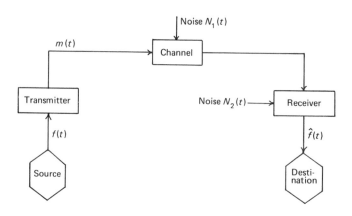

Figure 1.2.1   General communication system.

human being. They are not suitable for transmission as they are; they need some processing that will make them so. This processing is done by the system designated as the **transmitter** in the block diagram of Fig. 1.2.1.

The transmitter is therefore a device (or group of devices) whose function is to accept the raw $f(t)$ and turn it into a signal *that is suitable for transmission through the channel*. That, obviously, means that the transmitter (or the designer of the transmitter) must know the natures of $f(t)$ and channel, so that it makes the proper transformation. In the "human" communication system, the transmitter would include the organs and processes that transform the thoughts into a speech signal uttered by the mouth. The thoughts are transformed into acoustic waves, and these waves can travel through the channel, the air space between speaker and listener. Nature has, therefore, designed an excellent transmitter for this communication system, albeit only for short distances. For large distances, the acoustic waves of speech are not suitable for transmission. We have to come in and augment nature's work with electronic devices. Having discovered that electromagnetic waves (unlike sound waves) can travel great distances without much attenuation, we proceeded to develop an addition to nature's transmitter, which transforms the speech signal to an electromagnetic signal. In a radio communication system, the speech signal would, of course, be the raw $f(t)$, and the devices that transform it to an electromagnetic wave, the transmitter. In a television system, the raw $f(t)$ would be pictures and the transmitter would transform them into electromagnetic waves for wireless transmission or into electrical waves for cable transmission. The key words that obviously characterize the nature of the transmitter are the words "suitable for transmission through the channel."

Let us now consider the block designated as the **channel** in Fig. 1.2.1. This is defined as the transmission medium between transmitter and receiver. The most common type of channel is the atmosphere, that is, the air space between source and destination of information. Wire channels are also very popular, particularly in telephony or in computer-to-computer communication systems. In sonar the channel is water. In outer space communications (earth to other stars, earth to satellites, satellite to satellite, etc.) the channel is the atmosphere, free space, or both. Recently developed optical communication systems utilize channels made of plastic strands (fibers).

The channel is one of the key components of a communication system, because it is usually there and you are stuck with it. True, sometimes you can change it, but not without enormous costs in time and money. Channel changes for long-distance communications are very rare and occur once or twice in a person's lifetime, usually ushering new areas of civilization. Transatlantic cables were an example of a drastic channel change. Satellite communications can also be viewed as an example of such a change, although not a straightforward one. In both cases these new channels resulted in immense increases in the volume of information transmission between human beings. This had a profound effect on human civilization, an effect that is continuously being studied by humanists, as it is still not well understood.

So in most cases the channel is there and it is the transmitter that must be designed as the proper interface between it and $f(t)$. The channel must be studied well, so that the types of signals that it will permit to pass it must be ascertained. In fact, the channel presents us with more problems than that. It is there that we usually find the serious effects of noise [$N(t)$ in Fig. 1.2.1] for the first time. So it is not enough to know what signals can pass through the channel. You must also study the noise in the channel, for often, even though the signal goes through, it could be unrecognizable at the output of the channel.

The noise $N(t)$ usually comes as a package deal with the channel; that is, you are stuck with it as well. Noise can broadly be defined as a collection of signals that are undesirable and which distort the information signal during its passage through the channel. Usually, noise is caused by *unpredictable* natural phenomena, but not always. If you are trying to transmit information with smoke signals, for example, you would not do it in a dark night, for darkness (as noise) completely overshadows the signal. You might, however, try torches (or flashlights), which, in turn, would be useless during the daytime. These are both examples of predictable noise. Channels, however, more often than not contain unpredictable noises (electrical storms, outer space phenomena), and these must be studied well. Interestingly enough, such noises are not as unpredictable as they first appear. We will see toward the end of the book that they can be modeled mathematically in a manner that explains quite well their effects on the information signals that pass through the channel. So much for the channel, at least in this introductory presentation.

Let us now go to the other side of the communication system, the destination. There, we first encounter the block **receiver**. Generally speaking, the function of the receiver is to accept the signal that comes out of the channel and convert it back to its original form $f(t)$, or as is usually the case, $\hat{f}(t)$, which is an approximation of $f(t)$. The better the receiver, the closer this $\hat{f}(t)$ will resemble the original $f(t)$. So we see that the receiver is generally intended to do the inverse operation of the one done by the transmitter. Its real task, however, is much more difficult. The signal that enters is contaminated by channel noise, it is altered by the other channel characteristics, and it is weakened in power due to the traveled distance. Add to that the fact that new and strong noise [$N_2(t)$] is now present in the receiver, and you have a serious mess. No wonder entire books have been written on just this aspect of the communication systems, the receiver. And the end of the problems that it presents is not yet in sight.

To make a block and call it "receiver" is, of course, an oversimplification, as it was in the case of the transmitter. There are many subsystems here (subblocks) and the noise $N_2(t)$ is the cumulative effect of all noises that pop up in these subsystems. The receiving antenna, for instance, may have noise, due to rain. Amplifiers, used to restrengthen the weakened signal, usually have noise in their components (thermal noise in resistors, noise in diodes and transistors, etc.). Generally speaking, all human-made devices have noise, and this includes the devices in the transmitter as well. The only reason that we excluded the effects of noise in the transmitter is because the signal power there is high and the effects of noise are *relatively* minor. This, however, is not always the case. In PCM systems, "quantization noise" is a serious type and it appears in the transmitter. The block system of Fig. 1.2.1 is meant only as a general, common case, not as the only one. When we come to examine specific systems, additional blocks will be introduced that will make the function of the blocks of Fig. 1.2.1 clearer.

The final block in Fig. 1.2.1 is the destination. It is assumed here that the destination "wants" the signal in its original form, and that is why the receiver transformed it to $\hat{f}(t)$. Often, however, the destination "desires" the signal in a different form, not the original one, $f(t)$. For example, $f(t)$ may be speech, but the destination may be printed text, not another human being who will hear it. In this case, the output of the receiver will be $\hat{f}(t)$, but in a form suitable for the destination (a typewriter output, for example).

The discussion above had as its main purpose a description of the overall function of the communication system, the type of work performed by its subsystems, and the problems that may appear. Of course, as in all engineering problems, there are two main aspects that concern us, the analysis and the synthesis. This book is concerned primarily with analysis. We will take most of the existing communications systems and see how they do—what it is that they do. Of course, such an analysis helps a great deal in the design of new systems, especially if they are similar to the existing ones. Bona fide synthesis of communication systems is not, however, the purpose of this book, and will not be discussed in what follows, except in specific cases that are motivated by the analysis.

The reader will do well at this point to think about known communication systems and try to put them in the framework of Fig. 1.2.1. What are, for example, the various blocks in a television system? What about the primitive system using a dove? The telephone? What about the human (person-to-person) communication system? The last one has been discussed somewhat above, but its complete analysis still defies human reasoning. Somehow, $\hat{f}(t)$ does not always come close to the original thoughts of the speaker. Information is also transferred visually (e.g., gestures, facial expressions), not only acoustically. The context plays a key role in what $\hat{f}(t)$ finally is, so that the same $f(t)$ "means" different things in different contexts. There is feedback in this system as well, for often the speaker changes his or her thoughts while watching the listener's reactions. Tough stuff. It should not surprise anyone if we omit the analysis of such a system and stick to the artificial ones from now on. It is, however, intriguing to consider it from time to time and illustrate certain concepts, for we are quite familiar with it intuitively, even though we cannot analyze it in a completely convincing manner.

1.3 THE GENERAL RADAR SYSTEM

Let us now take up the general radar[3] system and discuss it in the same manner as the general communication system. This will be done with the aid of Fig. 1.3.1.

Generally speaking, the radar's basic goal is to extract information about an object (the target) outside of the radar itself. The way it achieves this purpose is by first transmitting a signal from its antenna. This signal, in the form of an electromagnetic wave, bounces off the object, and proceeds next to the receiver of the radar. This bouncing off (reflection) of the signal *changes some of its parameters*. The receiver measures these changes and extracts information about the target.

It should be noted that the components of the radar system are not that different from those of a communications system. There is a **transmitter** that sends out a signal *suitable for passage through the channel*. The **channel** is there

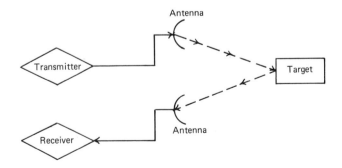

Figure 1.3.1   General radar system in pictorial form.

[3]The word "radar" is formed by the initials of the phrase "Radio detecting and ranging."

again; in fact, the signal traverses it twice, once on the way to the target and
again on the return. A block named **receiver** appears here as well, to receive the
signal and measure its parameter changes, which will provide information
about the target. Figure 1.3.2, a block diagram form of the radar system, points
out a bit more clearly the similarities in the blocks of the two systems.

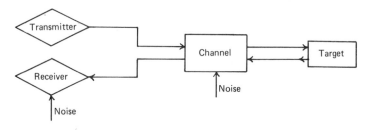

Figure 1.3.2   Block diagram of a radar system.

The fact that there exist similarities in the two systems does not, of course,
make them the same. We have already mentioned that their purposes are different. Other differences will also crop up as we take a closer look at the system
below.

Let us start out by noting that in this system the receiver knows exactly
the nature of the signal sent out by the transmitter. In this respect the radar
system is quite different from a communication system. Information here does
not originate at the transmitter but *at the target*, a point that will become clearer
later.

After the pulse is transmitted through the antenna toward some direction,
the receiver waits for a reflected signal (an echo) from a possible target. If none
appears, he concludes that no target is there and emits a new pulse probably in
a different direction. If, on the other hand, an echo does appear, this echo is
studied carefully for information about the target. By this we mean that the
receiver compares the received signal with the transmitted one, notes the changes
in its parameters, and deduces various types of knowledge about the target from
these changes. The target, in other words, *is the information source* without
wishing it—in fact, often without *realizing it*.

What type of information can actually be deduced about the target in this
manner? An amazing amount, sometimes a complete description of its nature
(shape, size, etc.), location, direction of movement, and speed. We will be a
little more specific about this below.

The most immediate type of information that is deduced from an echo is
the distance of the target. This is done by measuring the total time span from the
instant of transmittal to the instant of reception. This time measurement,
coupled with the fact that the speed of the electromagnetic wave is known, leads
to an immediate evaluation of the distance traveled, and half of that is the
radar-to-target distance. Of course, knowing the distance does not pinpoint

the exact location of the target. This too can be done, however, by noting the angles of the antenna axis as it points its power to the target. And that is not all. The size of the target has a lot to do with the size of the echo (in power); in fact, it is usually directly proportional. And the shape of the echo's spectrum, more specifically, its shift from the original transmitted spectrum, gives out information about the velocity of the target (Doppler phenomenon). All these things will, of course, be taken up and discussed in great detail as the book progresses. The present discussion is only an introduction to the subject, designed to motivate the reader and arouse his or her interest.

It should be noted in Fig. 1.3.2 that the by-now-familiar noise enters this system as well, in the channel and the receiver. Most of these noises are similar to those in communications systems, but there are some differences. One of the most important noises in radar is "clutter," and that does not appear in communication systems. Clutter noise is the sum total of echoes that return to the receiver from terrain objects (hills, trees, sea, etc.), objects that are of no interest to the radar. Luckily, such objects are stationary (or move very slowly), and this leads to the ability to detect them and ignore them. Nevertheless, clutter noise is very bothersome, especially for low-flying objects, and must be dealt with, together with all the other types of noise. This will be done in an introductory manner during the last two chapters of the book.

So far we have discussed the general type of radar, the one most commonly in use. Variations of this general form are in use, but they are minor and do not effect the analysis or the principles behind it. There exist radars, for example, with different locations for their transmitter and receiver, and they are called bistatic. In fact, there are radars with many transmitters and receivers (multistatic), all of them placed in different locations. Such combinations make the measurement of certain parameters more accurate. There are also radars that emit and receive acoustic waves, just as those of porpoises and bats. The most common example of such a radar is **sonar**, used in underwater sensing. But there are acoustic radars even for atmospheric sensing, although their use is relatively restricted, owing to their limited range. In this book all of our attention will be given to the electromagnetic type of radar with only one receiver and transmitter, the one usually called **monostatic** radar.

1.4 COMPARISON OF THE TWO SYSTEMS

The similarities and differences between the two types of systems will become quite apparent as our discussion evolves through the book. The present treatment is only an introduction, in a general context. The reader will do well to return to it after finishing the book so that he or she can again see the overall picture. Some of the concepts used here may not be completely familiar at this stage.

The fundamental difference between the systems is their purpose, as we mentioned in the preceding section. Their main similarity, on the other hand, is that both are subject to the same principles of analysis, since both of them use signals to carry information, and process them similarly to extract the information.

Let us now be a little more specific about the major and minor differences and similarities of the two systems.

Signals.   Signals carry the information or obscure it (noise) in both systems. In this general respect, the mathematical description of signals (signal theory) is the common tool for the analysis of both. The difference here lies in the *variety* of signals used by the two systems. Communication systems carry a larger variety of signals, voice, video, data, and so on, which can be in various forms (continuous, discrete time or sampled, quantized, etc.). In fact, it is usually the type of signal which determines the type of communication system that must be used to transmit it. So we have AM and FM systems, which are used primarily in transmitting *analog* signals: PAM, PDM, PPM systems, which transmit *sampled* signals, and PCM, which transmit *sampled* and *quantized* signals. In radar, the signal variety is restricted by the nature of the purpose of the system. A large number of radars, for example, are designed to operate with one type of signal, a pulse (repeated at specific instants). Another type operates using only a single-frequency cosine. The signals for a radar are picked with a single purpose in mind: to facilitate the measurement of the changes in a parameter. If you want to measure the time delay, what better signal is there than the impulse, which has a precise point-of-time reference? And if you want to measure the shift in the spectrum, you cannot do better than using a cosine, which has a distinct line spectrum. The signals used are limited in number, but they are quite important. In fact, some radars take their name from the signal that they use, as, for example, pulse radar and continuous-wave radar. But these are specific signals and not classes of signals (analog, sampled, digital) as in the case of communication systems. Often, radars are named after the type of antenna that they use (monopulse, conical scan), or the operation that they perform (scanning, tracking, searching), not the signal that they use.

The upshot of the discussion above is that both systems need signal theory in their analysis. The amount that is needed for communication systems is enough for both, since the type of signals used in radar is a small subclass of those used in communication systems. This is a major reason both systems can be analyzed simultaneously.

Processing of signals.   The type of processing that goes on in both systems is very similar. Transmitters modulate the signals in the same manner, and receivers demodulate them and try to eliminate the noise. Of course, even though the operations are the same, that does not mean that the actual practical devices that perform these operations are identical. There exist differences in their design, primarily because of the power and frequency ranges involved.

But such differences are of no concern to us here, since we discuss their operation from a systemic point of view.

Antenna.   Here the differences are much more accentuated. Antennas, are, of course, the devices that act as the coupler (the transducer) between the transmitter and the transmission medium. Their main purpose is to convert the electrical signals to electromagnetic ones, suitable for passage through the channel. This purpose is pretty much their only one in communication systems, and that is the reason they are usually mentioned, and then replaced by a filter, or completely ignored. Unfortunately, this cannot be done in radar systems, where they play a second role just as important as their first one. They are actually used in the main purpose of the radar: in finding the object and in measuring some of its characteristics. That is why a great variety of antennas are used, and why some radars take their name from this device. They vary in construction, in radiation patterns, even in principle. Since they help in measuring the location of the target, they are capable of rotation on three axes. This, of course, requires motors, gears, control systems, and the like—gadgetry that is seldom found in the common types of communication systems. Luckily, the amount of antenna theory required for a satisfactory understanding of radar systems is not very great, and can be covered in simple ways that suit our purposes.

The above constitute, in our opinion, the *key* characteristics of both systems, at least from the point of view of systems analysis, discussed in this work. Secondary similarities or differences between the two systems will be pointed out later, as the material evolves.

=2=

SIGNAL ANALYSIS

The beginning of wisdom is the definition of terms.

Socrates

2.1 INTRODUCTION

Signals, as we have noted, carry the information in both communication and radar systems. It is quite natural, then, to start with them.

Our main task in this chapter is to present various ways of specifying (or modeling) our signals in a mathematical framework. Most of the material is devoted to this task. We will discuss methods of describing signals in the time domain by straightforward ways, or by means of expansions. We will also consider specifying them in the frequency domain, via their amplitude and phase spectra. Both methods are useful and must be well understood. There are cases when the time-domain description facilitates the analysis of a system, and times when analysis is easier with a frequency-domain description. Our ultimate goal is, of course, to be able to trace signal passage through the various blocks (devices) of our systems, and this can be done only if we first learn how to model them mathematically.

Aside from individual signal descriptions, the chapter also includes some mathematical relationships *between* signals, relationships such as inner products, distances, and correlations, which are used in comparing them. Such comparisons are useful in radar detection—for example, where one must compare the transmitted signal with its echo.

This material constitutes background needed for the analysis of both communications and radar systems. Not all that is needed, however, is included in this chapter. Some of it (Doppler effect, sampling theorem, quantization, etc.) is taken up in chapters where it is most appropriate.

It should be mentioned that most of the material in this chapter is considered a review. It is assumed that the reader has been exposed to most of the topics in a prior course with contents such as those covered by McGillem and Cooper (1974) in their textbook. The only difference here may be a matter of approach and emphasis. The examples are also picked in a manner consistent with our unified presentation of both radar and communication systems.

2.2 SOME FUNDAMENTAL NOTIONS ABOUT SIGNALS

A signal[1] is, of course, a function of some parameter. When we write $f(t)$, we mean that we have a function of the parameter t. In mathematical terms, the signal $f(t)$ is defined by giving its *domain of definition* (the set of values that t takes) and its *range* [the set of values that $f(t)$ takes]. A fundamental way, therefore, of specifying a signal is by giving both its domain of definition and its range, and this can be done in two ways.

1. *By a formula.* Often we specify a signal by a mathematical relation of the form $f(t) = at + b$ or $f(t) = 10 \cos 3t$. When we do this, it is assumed that t takes all the values on the real line $(-\infty, +\infty)$. This method is neat and economical, but often not possible in practical problems.
2. *By a table.* This, too, is a well-known method and needs no elaboration. The table has two columns, one with the values of t, and another with the corresponding values of $f(t)$.

Later in this chapter we are going to discuss some alternative ways of describing our signals, ways that will make our eventual analysis of communications and radar systems easier. But first we must do some groundwork.

On the basis of the range and domain of definition type of description, we can distinguish four major categories of signals. If the range is continuous, the signals are called **analog**. Obviously, there are two types of analog signals: **continuous time** and **discrete time**. Similarly, if the range is discrete, the signals are called **digital**. Again, there are two varieties of digital signals, depending on the parameter t: **discrete-time** and **continuous-time**—just as in the case of analog signals. It is clear that the type of range is designated by the terms "analog" and "digital," whereas the type of parameter is spelled out more clearly.

---

[1] For the first nine chapters, signals will be assumed to be of the deterministic type, not the stochastic one. The difference will be explained in Chapter 12.

Example 2.2.1

(a) The signals mentioned in (1) above are of the continuous-time analog variety.

(b) The signal

$$f(t) = \begin{cases} \cos t & \text{for } t \in (0, 1, 2, \ldots) \\ 0 & \text{elsewhere} \end{cases} \tag{2.2.1}$$

is also analog but of the discrete-time variety. (Why?)

(c) The signal

$$f(t) = \begin{cases} 1 & \text{for } 0 \leq t \leq 1 \\ 0 & \text{elsewhere} \end{cases} \tag{2.2.2}$$

is a continuous-time digital signal.

(d) The signal

$$f(t) = \begin{cases} 1 & \text{for } t \text{ a positive integer} \\ -1 & \text{for } t \text{ a negative integer} \\ 0 & \text{elsewhere} \end{cases} \tag{2.2.3}$$

is of the remaining type (i.e., discrete-time digital). These examples also illustrate that when t does not take all the values in the real line, it must be so specified.

Let us now take up another classification of signals, the one that divides them into **power** and **energy** types. To understand the difference, we must first backtrack a bit, to the notions of power and energy.

If we have an ideal voltage source $v(t)$ across a resistor R, as shown in Fig. 2.2.1, then the instantaneous power $p(t)$ delivered to the resistor is given by

$$p(t) = i^2(t)R = \frac{v^2(t)}{R}$$

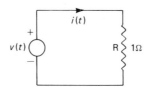

Figure 2.2.1   Illustration of power.

If $R = 1$, then

$$p(t) = i^2(t) = v^2(t)$$

By extension of this idea, given any signal $f(t)$, then $f^2(t)$ is called its **instantaneous power**, meaning that it is the instantaneous power that this signal would deliver to a 1-Ω resistor if it had the chance to do so, whether it was an ideal voltage or current source.

The next steps are now obvious. The expression

$$E = \int_{t_1}^{t_2} f^2(t)\, dt \tag{2.2.4}$$

is called the **energy** of the signal *in the interval* (t_1, t_2), and

$$E = \int_{-\infty}^{+\infty} f^2(t) \, dt \qquad (2.2.5)$$

its *entire* **energy**. They represent, of course, the energy that $f(t)$ would deliver as a source to the $R = 1\text{-}\Omega$ resistor over the specified time intervals.

Our final step is that the expression

$$P_{ave} = \frac{1}{t_2 - t_1} \int_{t_1}^{t_2} f^2(t) \, dt \qquad (2.2.6)$$

is called the **average power** of $f(t)$ in the (t_1, t_2) *interval*, with

$$P_{ave} = \lim_{T \to \infty} \frac{1}{T} \int_{-T/2}^{T/2} f^2(t) \, dt \qquad (2.2.7)$$

its average power *over all time*.

With the foregoing concepts in mind, we can now proceed to our classification of signals into energy and power types.

An **energy signal** $f(t)$ is one whose total energy is finite, that is,

$$\int_{-\infty}^{+\infty} f^2(t) \, dt < \infty \qquad (2.2.8)$$

Signals of this type tend to drop to zero for increasing values of t; otherwise, the area under their $f^2(t)$ curve would not be finite. The signal $f(t) = e^{-|t|}$ is an energy signal, for example, and so is the signal (2.2.2) in Example 2.2.1.

When a signal does not obey (2.2.8), it is usually a power signal. A **power signal** is one whose energy may be infinite, but whose average power given by (2.2.7) is finite. Periodic signals such as $f(t) = \cos \pi t$ are power signals since their total energy is infinite,

$$\int_{-\infty}^{+\infty} \cos^2 \pi t \, dt = \infty \qquad (2.2.9)$$

whereas its *total* average power is finite. It should also be noted that the total average power of an energy signal is always zero (why?) but not, of course, its average power over a specific interval.

The above is all we are going to say about signal classifications. It is by no means all-inclusive. There are other ways of classifying or dividing signals into subgroups (periodic, symmetric, etc.), and some of them we will see later.

We close this section with a few items of notation that seem to cause a lot of trouble to students of the material. Those who have mastered these operations can proceed to the next section.

1. *Parameter shifts.* Let us assume that we have a signal $f(t)$ (see Fig. 2.2.2a) and we are asked to find $f(t - 2)$. This last expression means, of course, that the signal $f(t)$ is shifted 2 units of time to the right, as shown in Fig. 2.2.2b. If we wanted to find $f(t + 2)$, we would shift $f(t)$ to the left by 2 units. One method of remembering the direction of the shift is by finding the value of t

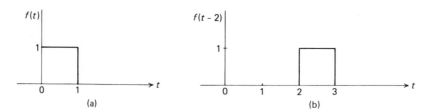

Figure 2.2.2   Illustration of shifts.

that makes the argument of $f(t + a)$ equal to zero. If this value is positive, the shift is to the right, and vice versa.

2. *Negation of the argument.* If we have an $f(t)$ and must find $f(-t)$, we flip the graph of $f(t)$ over the axis $t = 0$. But what if we must find $f(2 - t)$, which is $f(t - 2)$ with the "argument" (the term inside the parentheses) negated?

The answer is that $f(2 - t)$ is the $f(t - 2)$ flipped over the axis $t = 2$, that is, the axis defined by equating the argument to zero (see Fig. 2.2.3). In other words, $f(t - 2)$ and $f(2 - t)$ are symmetric with respect to the axis $t = 2$. This is important, and often missed. For some reason or other, people seem to think that all flips are with respect to $t = 0$.

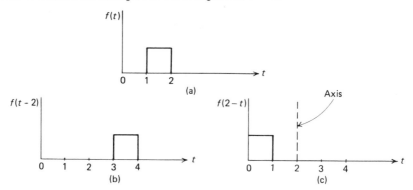

Figure 2.2.3   Illustration of argument negation.

3. *Scaling the parameter.* Another situation that will often crop up in the analysis of our systems is starting with some $f(t)$ and having to find $f(at)$. In general, the best way to do this is to take $f(t)$ and plug in at for t. Often, however, the question that must be answered is whether a *finite-duration* signal is *expanded* or *squeezed* in time by such a scaling of its parameter. In the last case, the answer can easily be found by checking one or two points.

Let us look at an example. Let the $f(t)$ be the one shown in Fig. 2.2.4 and note that $f(10t)$ and $f(t/10)$, differ only in the point where they hit the t axis. To find this point, one equates the scaled argument (at) to the original point $t = 10$ and solves for t. Thus the point where $f(10t)$ hits the axis is $10t = 10$ (i.e., $t = 1$), whereas $f(t/10)$ hits it at $t = 100$. It is a good idea to remember

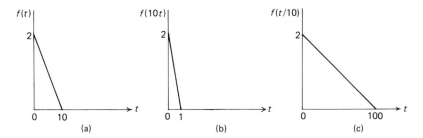

Figure 2.2.4   Parameter scaling.

that in scaling of the form $f(at)$, the curve is expanded when $a < 1$ and squeezed when $a > 1$.

This method works even if the signal $f(t)$ does not start at $t = 0$. In this last case, however, the signal $f(at)$ not only gets distorted, *but also shifted.* If the signal of Fig. 2.2.4 had its start at $t = 1$ rather than $t = 0$, the start of $f(10t)$ would be at $t = 1/10$, whereas its finish is at $t = 1$. This is not only interesting, it is quite useful and it actually helps in explaining the Doppler effect, which we will discuss in a later chapter (Chapter 4).

Example 2.2.2

For the $f(t)$ shown in Fig. 2.2.5, find $f(2 - t/2)$.

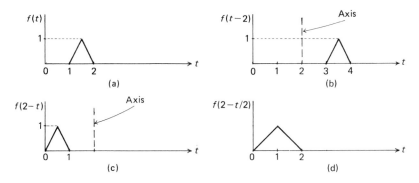

Figure 2.2.5   Multiple operations on an $f(t)$.

Solution.   This, of course, is an example that includes all the operations discussed above. To arrive at the result, we first find $f(t - 2)$, then $f(2 - t)$, and then the desired $f(2 - t/2)$, as shown in Fig. 2.2.5. One must be very careful to ensure that the sequence of operations will achieve the desired result. If, for example, we first find $f(t/2)$, then shifting by 2 gives $f((t - 2)/2)$, and negating the argument gives $f(1 - t/2)$, which is not the desired result. The reader should also note that there are various ways of arriving at the correct result.

All the "tricks" above are given, of course, without proof. It is assumed that the reader can find one if he or she wishes. One thing the reader must not do, however, is to underestimate the importance of these results. Operations of this sort appear repeatedly in the analysis of our system in both the time and frequency domains. We will come across them soon enough. They certainly appear in the convolution integral, which the reader is presumed to have as background (see Problem 2.3).

2.3 GENERALIZED FUNCTIONS

There is a group of well-known signals, including the impulse, the unit step, and the ramp, which have a plethora of uses in systems analysis. This section is devoted to a presentation of the principal notions involving such signals, usually called **generalized functions**. The treatment will be elementary. The interested reader can look up Lighthill (1959) for a more thorough presentation.

The impulse.   One rather unsophisticated way of describing the impulse (or delta) signal is by the expression

$$\int_{-\infty}^{+\infty} \delta(t)\, dt = \int_{0^-}^{0^+} \delta(t)\, dt = 1 \tag{2.3.1}$$

and $\delta(t) = 0$ for $t \neq 0$. Its value at $t = 0$ is not mentioned in this description, but that does not stop us from visualizing this signal and even sketching it, as shown in Fig. 2.3.1, even though mathematicians frown on such activities.

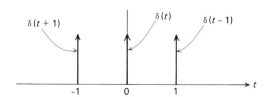

Figure 2.3.1   The impulse.

In this regard, $\delta(t - t_0)$ means that our impulse is *located* at $t = t_0$; that is, that it represents a shifted version of $\delta(t)$ at $t = t_0$. This shifted form is described by

$$\int_{-\infty}^{+\infty} \delta(t - t_0)\, dt = \int_{t_0^-}^{t_0^+} \delta(t - t_0)\, dt = 1 \tag{2.3.2}$$

and $\delta(t) = 0$ for $t \neq t_0$.

It is quite interesting to note that the description of $f(t)$ above is somewhat unique; that is, it does not obey the usual methods described in the preceding section (formula or table). It describes $\delta(t)$ not so much by its values, but by *what it does to an integral*. Before we elaborate on this new way of describing signals, let us first look at this function from a mathematical point of view.

Mathematicians *define* the impulse function $\delta(t)$ by the expression

$$\int_{-\infty}^{+\infty} \varphi(t)\, \delta(t)\, dt = \varphi(t)\Big|_{t=0} = \varphi(0) \tag{2.3.3}$$

where $\varphi(t)$ is a well-behaved (continuous) function at $t = 0$ called the **testing function**. Again we note that what is given is a property of $\delta(t)$ when it appears in the integrand of an integral: namely, the *sampling property*. It should be noted that if one starts with this definition, expression (2.3.1) follows immediately by taking $\varphi(t) = 1$, certainly a testing function. In fact, this definition can lead to a number of properties of $\delta(t)$, as seen below.

(1) $$\int_{-\infty}^{+\infty} \varphi(t)\, \delta(t - t_0)\, dt = \varphi(t_0) \tag{2.3.4}$$

This one can be easily proved by changing variables, $t - t_0 = x$. This leads to

$$\int_{-\infty}^{+\infty} \varphi(x + t_0)\, \delta(x)\, dx = \varphi(x + t_0)\Big|_{x=0} = \varphi(t_0)$$

the last step being taken using the definition (2.3.3). This property can be used to derive expression (2.3.2).

(2) $\quad\quad\quad \delta(t) = \delta(-t) \quad\quad$ [i.e., $\delta(t)$ is an even function] $\quad\quad$ (2.3.5)

To show this, we must first understand what is meant by the equality (2.3.5), a rather subtle point. Equality of delta functions can be tested only by using their definition. This means that *two delta functions are equal if they give the same result in an integral with a testing function $\varphi(t)$*. Thus one must show that

$$\int_{-\infty}^{+\infty} \varphi(t)\, \delta(t)\, dt = \int_{-\infty}^{+\infty} \varphi(t)\, \delta(-t)\, dt$$

which is not very difficult if one makes a change of variable ($x = -t$, etc.).

(3) $\quad\quad\quad f(t)\, \delta(t) = f(0)\, \delta(t) \quad\quad$ for $f(t)$ continuous at $t = 0$ $\quad\quad$ (2.3.6)

Here again we must show an equality of two delta functions, or, more generally, two generalized functions. This means we must show that

$$\int_{-\infty}^{+\infty} \varphi(t) f(t)\, \delta(t)\, dt = \int_{-\infty}^{+\infty} \varphi(t) f(0)\, \delta(t)\, dt \tag{2.3.7}$$

and this easily follows by considering $\varphi(t) f(t)$ as a new testing function. Both integrals will give $\varphi(0) f(0)$ [i.e., (2.3.6) holds].

(4) $$\int_{-\infty}^{+\infty} \varphi(t)\, \delta(at)\, dt = \frac{1}{|a|}\varphi(0) \tag{2.3.8}$$

This, like most of them, can be proved by a change of variable. First take $a > 0$, and let $x = at$. Simple manipulations give the result of the integral above as $(1/a)\varphi(0)$. Repeat the procedure with $a < 0$, and the result is the same, proving that (2.3.8) holds.

(5)
$$\int_{-\infty}^{+\infty} f(t - \tau)\, \delta(\tau)\, d\tau = f(t) \tag{2.3.9}$$

This is a very interesting property, and, in fact, quite important. It deals, of course, with the convolution integral of a function $f(t)$ (assumed to be well behaved everywhere) with a delta function. The proof is immediate, from the definition (2.3.3), since the result of the integral should be $f(t - \tau)|_{\tau=0} = f(t)$. But trivial as the proof may be, the reader should learn the result by heart [*the convolution of a signal $f(t)$ with a delta function gives back the $f(t)$*] and should also note that

$$\int_{-\infty}^{+\infty} f(t - \tau)\, \delta(t - t_0)\, d\tau = f(t - t_0) \tag{2.3.10a}$$

That is, $f(t)*\delta(t - t_0)$ (the asterisk denotes convolution) leads to a shifted $f(t)$, the shift being equal to the point of *location* of the delta function.

This last property was, of course, discussed with the assumption that the reader is familiar with the convolution integral. If this is not the case, he or she should immediately look up a reference such as McGillem and Cooper (1974), and should understand especially the graphical method of evaluating convolutions. In this endeavor, the material at the end of the preceding section is quite useful.

Since we are at this point, we might as well mention that we will also extend Eq. (2.2.10a) to two impulses, even though we would have an awful time trying to prove it. Thus

$$\delta(t - t_0)*\delta(t - t_1) = \delta(t - t_0 - t_1) \tag{2.3.10b}$$

which we will have some occasions to use.

More of the properties of $\delta(t)$ appear in the Problems. The key idea is the mathematical definition (2.3.3), which, as we mentioned above, is unique. In fact, it can be generalized and then used to define many functions, which are then named generalized functions. In this context, a generalized function $g(t)$ is defined by an integral of the form

$$\int_{-\infty}^{+\infty} \varphi(t)g(t)\, dt = N[\varphi(t)] \tag{2.3.11}$$

with $\varphi(t)$ a properly specified testing function and $N[\varphi(t)]$ a number somehow obtained from $\varphi(t)$. In the case of the $\delta(t)$, $N[\varphi(t)] = \varphi(0)$. Other cases will be seen later.

This may seem like a strange way to define (and specify) signals, but a little thought will reveal that it is not. In practical problems—for example, when you measure a signal and specify its values—in reality you are not specifying the signal, but rather, the result of the convolution of the signal with the impulse response of the instrument. And the convolution is, in fact, an integral. So philosophically at least, this method should not surprise us.

Derivatives of the impulse. The impulse has derivatives, as strange as this may sound, and they are denoted by

$$\delta^{(n)}(t) = \frac{d^n \, \delta(t)}{dt^n} \tag{2.3.12}$$

Mathematically, they are defined by

$$\int_{-\infty}^{+\infty} \delta^{(n)}(t)\varphi(t) \, dt = (-1)^n \frac{d^n \varphi(t)}{dt^n}\bigg|_{t=0} = (-1)^n \varphi^{(n)}(0) \tag{2.3.13}$$

with $\varphi^{(n)}(t)$ as a testing function, assumed differentiable at $t = 0$. This definition is, of course, of the (2.3.11) variety. Of all these derivatives, the first one is the most important, and is called the **doublet**. As we shall not have much use for these generalized functions in this book, we will not bother with any of their properties.

The unit step. The unit step $u(t)$ can also be defined as a generalized function by the integral

$$\int_{-\infty}^{+\infty} \varphi(t)u(t) \, dt = \int_{0}^{\infty} \varphi(t) \, dt \tag{2.3.14}$$

where $\varphi(t)$ is a testing function (that behaves well in the negative axis). Of course, the $u(t)$ can also be defined in a straightforward manner by the expression

$$u(t) = \begin{cases} 1 & \text{for } t > 0 \\ 0 & \text{for } t < 0 \\ \text{undefined at } t = 0 \end{cases} \tag{2.3.15}$$

which is the most common method in the field of engineering. This last definition can also lead to a graph of this function as shown in Fig. 2.3.2.

Figure 2.3.2 The unit step.

It can be shown that the impulse is the derivative of the unit step, but we do not have the space to do it here. Problem 2.7 deals with this, so we leave it for the reader to prove.

The unit ramp. The integral of the unit step is called the unit ramp, and is denoted by $r(t)$. It is simply the function $r(t) = t$ for $t > 0$ [$r(t) = 0$ for $t < 0$] and needs no further elaboration here. It is also the last generalized function that we will have to deal with in this work.

We mentioned at the beginning of this section that generalized functions are quite useful. Although this will become apparent as the book progresses, it will not hurt to pursue it a little more at this point.

To begin with, the entire chapter is devoted to signal descriptions. These functions have shown us a new way to describe signals, via integrals—and that is not all. Generalized functions can be used to specify other signals in a simple way, a fact that should be familiar to the reader.

Take, for example, the two signals shown in Fig. 2.3.3. It is obvious that $f(t)$ can be written as

$$f(t) = 3[u(t) - u(t - 1)] \qquad (2.3.16)$$

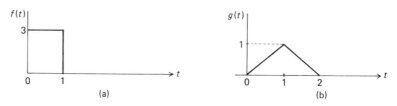

(a)   (b)

Figure 2.3.3 Signals specified by generalized functions.

and that

$$g(t) = r(t) - 2r(t - 1) + r(t - 2) \qquad (2.3.17)$$

although this last expression is not so obvious. Many other signals can be written as shifted, linear combinations of generalized functions, and the reader should work enough of these types of problems to feel comfortable.

In fact, generalized functions are used not only to describe signals, but systems as well. A linear system, for example, is commonly defined by its output to a delta input (see Problem 2.3). So there are certainly plenty of uses. We will keep coming across them in both the time domain and the frequency domain (which will be defined very soon).

2.4 SIGNAL VECTOR SPACES: EXPANSIONS

In our quest to find ways to model signals mathematically, in the preceding section we came across the idea of writing them as linear combinations of other, well-known signals. This idea leads to the formal field of signal expansions, which represents one of the major methods of specifying signals. This section is devoted to this all-important topic.

The reader has, no doubt, met various types of signal expansions in other studies, expansions with names such as *Fourier series*, *Walsh functions*, *Bessel functions*, and *Legendre polynomials*. These are, actually, examples of one and the same theory, a theory whose main points are outlined below. The treatment may seem somewhat abstract at first, but it will eventually make a lot of sense. If the reader bears with us until some of the details are covered, he or she will find that the benefits of this discussion are very worthwhile.

The central idea behind all expansions is the concept of a vector space.

Let us, therefore, first define it and then we will show how signals can actually be thought to form such a space.

Vector space.   A vector space S, defined over the field F of real (or complex) numbers, is a collection of things, $X, Y, Z, \ldots$, called vectors. In addition, there are two operations defined over S, "addition" $(+)$ of two vectors and "multiplication" $(\cdot)$ of a vector and a number in F, both of which result in a new vector which is also in S. The vectors, the numbers, and the operations must also obey the following relations (axioms).

(1) There exists ($\exists$) zero vector $\bar{0}$ such that $X + \bar{0} = X$ for all ($\forall$) $X \in S$.

(2) For every $X \in S$ $\exists$ a vector $-X \in S$ such that $X + (-X) = \bar{0}$.

(3) $X + Y = Y + X, \forall\ X, Y \in S.$

(4) $(X + Y) + Z = X + (Y + Z), \forall\ X, Y, Z \in S.$ (2.4.1)

(5) $1 \cdot X = X, X \in S$, where 1 is the number 1 in the field F.

(6) $c(X + Y) = cX + cY$ for $X, Y \in S$ and $c \in F$.

(7) $(a + b)X = aX + bX$ for $X \in S$ and $a, b \in F$.

(8) $(ab)X = a(bX)$ for $X \in S$ and $a, b \in F$.

The rather lengthy process above defines a vector *space*. There are, of course, many such vector spaces, the most familiar of which is the one with vectors of the three-dimensional space, specified as $X = ai + bj + ck$. (Can you prove that this vector space obeys the definition above?) What is of interest to us, however, is that special classes of signals *can also form a vector space*. All periodic signals with the same period T make up a vector space. The same is true for all continuous (or piecewise continuous) signals in some interval $[a, b]$ of their parameter. To show these assertions, one has to prove that these examples obey the definition, but that is quite easy and is left as an exercise.

Next we need to develop the concepts of inner product, distance, and norm in a vector space, all necessary for the developments we are pursuing.

Inner product.   The inner product (IP) or dot product is a numerically valued function between two vectors in a vector space S. It is denoted by (X, Y), and it obeys the following axioms.

(1) $(X, Y) = (Y, X)^*$ where * denotes the complex   (2.4.2)
     conjugate

(2) $(cX, Y) = c(X, Y)$ (2.4.3)

(3) $(X + Y, Z) = (X, Z) + (Y, Z)$ (2.4.4)

(4) $(X, X) \geq 0$ with equality when $X = \bar{0}$ (2.4.5)

This definition is, of course, quite abstract—things usually are if you take them directly out of the field of mathematics. But we will not let that deter us. The IP is a very useful expression between two vectors in a vector space, and is loaded with meaning. Let us look at it a little more closely.

First, we note that the definition above tells you only what an IP is. It does not tell you how to find one. Luckily for us, in most vector spaces somebody has found one and we can use it. Take, for example, the vector space with $X = a_1 i + a_2 j + a_3 k$ and $Y = b_1 i + b_2 j + b_3 k$. An IP for this space which obeys the definition above is (check it)

$$(X, Y) = \sum_{i=1}^{3} a_i b_i \tag{2.4.6}$$

There can be many IPs in a vector space, as long as they obey the definition. In each case, no matter how they are defined, the meaning they carry is always the same. The IP gives you the projection of one vector onto the other (i.e., how much of X is in Y, and vice versa). This is, in fact, quite remarkable. It gives you, in mathematical terms, an expression of "similarity" between two vectors. Note, for example, that if $X = i$ and $Y = j$, their IP of (2.4.6) is zero. These two are quite dissimilar; in fact, they are *orthogonal* to each other.

Let us now look at a vector space of interest to us, one that deals with signals. Let it be the one whose vectors are all the continuous energy signals in the interval $[0, \infty)$. A legitimate IP for this space between any two of its vectors, $f(t)$ and $g(t)$, has been found to be the one expressed by

$$(f(t), g(t)) = \int_0^\infty f(t)g(t)\, dt \tag{2.4.7}$$

That it obeys axioms (2.4.2) through (2.4.5) is left for the reader to ascertain. In any event, we now have an IP for signals in the vector space above, and similar ones exist for other signal vector spaces. And with this expression, we have its meaning. The IP of (2.4.7) designates the projection of $f(t)$ on $g(t)$ (or vice versa) —in other words, how similar they are. In fact, *if the IP between two signals is zero, the signals are called orthogonal*. They are as unlike each other as possible. This last fact, as well as the meaning of the IP, are quite important and will be used elsewhere, not only in the expansion theory that we are developing (see, for example, Section 2.10).

Our next order of business is the concept of the norm (or absolute value).

Norm. The norm (N) is a numerically valued function of individual vectors of a vector space S. It is denoted by the expression $N(X)$ and obeys

(1) $N(X) \geq 0$ with equality if $X = \bar{0}$ (2.4.8)

(2) $N(aX) = |a| N(X)$ (2.4.9)

(3) $N(X + Y) \leq N(X) + N(Y)$ (2.4.10)

The norm is a more familiar quantity to engineers. It represents, of course, the magnitude of the vector, a measure of how big it is. In three-dimensional space the norm of $X = a_1 i + a_2 j + a_3 k$ is usually taken to be the expression

$$N(X) = \sqrt{\sum_{i=1}^{3} a_i^2} \qquad (2.4.11)$$

but there is nothing special about this expression. It is acceptable because it obeys the definition. Other expressions can have these properties and can serve as norms, just as well.

The search for norms is not as hard as it is for IPs. This is due to the fact that if you have found an IP, this IP can give you a norm by the expression

$$N(X) = \sqrt{(X, X)} \qquad (2.4.12)$$

In other words, $N(X)$ (a possible one, not the only one) is given as the square root of the IP of X with itself. We will not prove this useful result here, but we will certainly use it to find a norm for our signal vector space with IP (2.4.7). This norm is

$$N(f(t)) = \sqrt{\int_0^{\infty} f^2(t)\, dt} \qquad (2.4.13)$$

and it gives a number that represents the magnitude of the signal. In fact, the reader should note that this norm is nothing but the *square root of the energy of the signal*. This strengthens its interpretation, since the energy is certainly indicative of the magnitude of the signal. It also gives us a bridge between the abstract notions of mathematics and the more concrete notions of electrical engineering.

Distance.   Distance (or metric) is a numerically valued function of two vectors in a vector space S. It is denoted by $d(X, Y)$ and obeys the following axioms:

(1) $d(X, Y) \geq 0$ with equality iff $X = Y$ (2.4.14)

(2) $d(X, Y) \leq d(Z, X) + d(Z, Y)$ (2.4.15)

Any function of two vectors that obeys the definition above qualifies as a distance and many have been found in various vector spaces. Actually, one need not look very far for one. It can be shown that the function

$$d(X, Y) = N(X - Y) \qquad (2.4.16)$$

fills the void quite nicely, so that if one has found a norm, he or she also has a distance. In the three-dimensional space with Eq. (2.4.11) as a norm, the distance is usually taken as

$$d(X, Y) = \sqrt{\sum_{i=1}^{3} (a_i - b_i)^2} \qquad (2.4.17)$$

which is obtained by using (2.4.16). If we use the same route, a distance for our signal space with norm (2.4.13) is

$$d(f(t), g(t)) = \sqrt{\int_0^{\infty} [f(t) - g(t)]^2\, dt} \qquad (2.4.18)$$

but, of course, it is not the only one that can be found (see Problem 2.12).

The distance (D) is a measure of similarity between two vectors. If it is zero, the vectors are identical. This is extremely important, and we will see an application of it immediately below. In fact, we now have two expressions that provide us with a measure of similarity between two signals in some vector space: the inner product and the distance. They do so, of course, in opposite ways. The distance is zero when the signals are identical, whereas the IP is zero when they are completely unlike each other. The importance of such facts will become apparent as the material develops.

We are now ready to discuss the main goal of this section, expansions of signals in terms of linear combinations of other signals. We will actually cover two results, one general and one specific. We will see their difference below.

2.4.1 General Expansions: The Orthogonality Criterion

Let us assume that we have a vector space S of signals, in which an IP, an N, and a D have been found. Let it be, in fact, the space of signals with finite energy (or norm) in $(0, \infty)$ for which expressions (2.4.7), (2.4.13), and (2.4.18) can serve us, so that the results can be obtained in a concrete way. This, we hope, will not distract us from appreciating its wide applicability.

Next we assume that we wish to expand a signal $f(t)$ as a linear combination of a group of other signals $\phi_i(t)$ $(i = 1, 2, 3, \ldots, n)$. By this we mean that we seek constants C_i $(i = 1, 2, \ldots, n)$ that will result in the "best" approximation of $f(t)$ in terms of the $\phi_i(t)$, of the form

$$f(t) \approx \sum_{i=1}^{n} C_i \phi_i(t) \tag{2.4.19}$$

The first thing we must clear up is what we mean by *best*. That is where the distance comes in. Ideally, we would like Eq. (2.4.19) to be an equality. This would require the distance between $f(t)$, and its approximation by the sum, to be zero. If we cannot hit this ideal, it is natural to strive for the C's that will make the distance minimum. That is what we mean by best. In mathematical jargon, we seek *the optimum C's, with optimality criterion the minimum distance*. The distance, as we noted, measures the degree of similarity, so it represents the *error* in the expansion. In fact, the reader should recognize the expression (2.4.18) as resembling those expressions usually referred to as "mean-squared errors." The way to find the C's so that the distance in question is minimum is dictated by the following theorem, often called the orthogonality criterion.

Theorem 2.4.1 (The Orthogonality Criterion). In the framework above, the C's that minimize the distance

$$d\left(f(t), \sum_{i=1}^{n} C_i \phi_i(t)\right) = \sqrt{\int_0^{\infty} \left[f(t) - \sum_{i=1}^{n} C_i \phi_i(t)\right]^2 dt} \tag{2.4.20}$$

are the ones that make the difference signal vector

$$f(t) - \sum_{i=1}^{n} C_i \phi_i(t)$$

orthogonal to each one of the $\phi_i(t)$, that is, the C's that satisfy the IP equations

$$\left(f(t) - \sum_{i=1}^{n} C_i \phi_i(t), \phi_j(t) \right) = \int_0^{\infty} \left(f(t) - \sum_{i=1}^{n} C_i \phi_i(t) \right) \phi_j(t) \, dt = 0 \quad (2.4.21)$$

for all $j = 1, 2, \ldots, n$. This theorem holds both ways. If the C's satisfy (2.4.21), the D is minimum, and if D is minimum, the C's satisfy (2.4.21).

Proof. The proof of the theorem is quite easy. We will prove the straight statement and leave the converse as an exercise for the reader (Problem 2.13).

To minimize the distance (2.4.20), it suffices to minimize its square, so that our expressions are easier. The minimization is done with respect to the C's. Differentiating with respect to C_1 and equating the result to zero, we obtain

$$\int_0^{\infty} \left(f(t) - \sum_{i=1}^{n} C_i \phi_i(t) \right) \phi_1(t) \, dt = 0 \quad (2.4.22)$$

which gives us immediately the first equation ($j = 1$) of Eq. (2.4.21). The others are obtained by differentiating with respect to the remaining C's. It should be emphasized that the theorem holds for all appropriately defined vector spaces and not only for the one with the IP, N, and D that we considered.

So now we have a new systematic way of describing signals in terms of other signals. Even though this description is often an approximation, it is quite important. It is especially helpful in the transmission (or processing) of signals. If the $\phi(t)$'s are known to the receiver of a communication system, for example, one need only transmit the C's, and this may be fast and economical. This idea will become clearer when we cover the second theorem of this section, after the following example.

Example 2.4.1

Expand the signal $f(t) = e^{-3t}u(t)$ as a linear combination of the signals $\phi_1(t) = e^{-t}u(t)$ and $\phi_2(t) = e^{-2t}u(t)$. Also find the "error" of the expansion.

Solution.   The signals are defined for $t \geq 0$, so they can be thought of as belonging to the vector space discussed above. We seek C_1 and C_2 so that the expansion

$$f(t) \approx C_1 \phi_1(t) + C_2 \phi_2(t) \quad (2.4.23)$$

is optimum. The C's are found from immediate application of the equations (2.4.21). Thus

$$C_1 \int_0^{\infty} e^{-t} e^{-t} \, dt + C_2 \int_0^{\infty} e^{-2t} e^{-t} \, dt = \int_0^{\infty} e^{-3t} e^{-t} \, dt \quad (2.4.24)$$

$$C_1 \int_0^{\infty} e^{-t} e^{-2t} \, dt + C_2 \int_0^{\infty} e^{-2t} e^{-2t} \, dt = \int_0^{\infty} e^{-3t} e^{-2t} \, dt \quad (2.4.25)$$

are two equations in the unknown C_1 and C_2 which can easily be solved after the evaluation of the integrals.

The "error" of the expansion is taken to be the distance between $f(t)$ and its expansion. Once the C's are found, the error is evaluated from

$$\text{error} = \sqrt{\int_0^\infty [f(t) - C_1\phi_1(t) - C_2\phi_2(t)]^2 \, dt} \qquad (2.4.26)$$

strictly a matter of calculus.

2.4.2 Expansions on an Orthonormal Set of Signals

It is often the case that the $\phi_i(t)$ of the expansion are *orthonormal*. This means that they are orthogonal to each other (their IPs are all zero) and that their individual norms are all unity. In mathematical terms the ϕ's are orthonormal if and only if

$$(\phi_i(t), \phi_j(t)) = \delta_{ij} \qquad (2.4.27)$$

where δ_{ij} (the Kronecker delta) is 1 for $i = j$ and zero when $i \neq j$. It should be noted that when $i = j$ the IPs become the norms, or rather the square of the norms, which is equivalent since their value is unity.

Under these conditions Theorem 2.4.1 simplifies, and becomes Theorem 2.4.2.

Theorem 2.4.2.   In the framework of Theorem 2.4.1 and when the $\phi_i(t)$ are orthonormal, the C_i are given by

$$C_i = (f(t), \phi_i(t)) = \int_0^\infty f(t)\phi_i(t) \, dt \qquad (2.4.28)$$

for each value of i [i.e., by the IP of $f(t)$ with each member of the group of $\phi(t)$'s].

Proof. The proof is immediate from the result of Theorem 2.4.1. If the condition of orthonormality is used in Eq. (2.4.21), the terms involving the IPs of the ϕ's are zero and each equation reduces to the form of Eq. (2.4.28). The statement and proof of the converse of the theorem are omitted.

Theorem 2.4.2 is, of course, much easier to apply since the C's are found immediately, not from the solution of n simultaneous equations as in the case of Theorem 2.4.1. This simplification is not without a penalty, however. One must have a set of orthonormal signals. Such a set is called a **basis** or a **coordinate system**, and is not easy to find. In fact, most of those who have discovered one have been immortalized by having their name attached to their find (Fourier, Legendre, Bessel, etc.).

The result above becomes even more important when the *basis is complete*. A **complete basis** is a group of orthonormal signals such that no other can be found with the same property. When this is the case (and it often is), the approximation in (2.4.19) is an equality, a result that will not be proven here for lack of space. Not only that—the error in this case will be zero and this immediately leads to the expression (prove it)

$$\int_0^\infty f^2(t)\, dt = \sum_{i=1}^n C_2^2 \qquad (2.4.29)$$

which is called the *Parseval identity*, an interesting result that gives the energy of the signal in terms of coefficients. Most of the orthonormal bases used in expansions have been shown to be complete, a nice fact which makes (2.4.29) valid.

Example 2.4.2

Consider the signal $f(t) = e^{-t}[u(t) - u(t-1)]$ and the two $\phi(t)$'s (see Fig. 2.4.1)

$$\phi_1(t) = u(t) - u(t-1)$$
$$\phi_2(t) = u(t) - 2u(t-0.5) + u(t-1)$$

of the well-known group of Walsh functions (a complete basis). Expand $f(t)$ in terms of the other two.

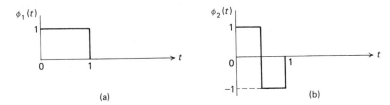

Figure 2.4.1   First two members of the Walsh basis.

Solution.   In solving a problem like this, one must first make sure that all matters pertaining to the vector space formulation are in order. In this case, the vector space can be the one of finite energy signals in [0, 1], so that the previously defined **IP**, *N*, and *D* are valid with limits (0, 1) in the integrals.

Next we note that $\phi_1(t)$ and $\phi_2(t)$ are orthonormal since

$$\int_0^1 \phi_1^2(t)\, dt = \int_0^1 \phi_2^2(t)\, dt = 1$$

and

$$\int_0^1 \phi_1(t)\phi_2(t)\, dt = 0$$

This means that Theorem 2.4.2 applies. Therefore,

$$C_1 = \int_0^1 e^{-t}\, dt$$

$$C_2 = \int_0^{0.5} e^{-t}\, dt - \int_{0.5}^1 e^{-t}\, dt$$

It should be noted that the expansion is an approximation. There is an error that can be found. The Walsh basis is complete, but in this problem we have not used all its members.

In closing, we should emphasize that the results are general, but that each case depends on the IP and the N and D that are found from it. We will see in the next section, for example, that the vector space of periodic functions with period T will have a slightly different IP when it includes signals that are complex functions of a real variable. In any event, this section has introduced us to ways of modeling signals in terms of other signals in a formal way. The methods of doing so with the use of generalized signals (steps, ramps, etc.) were not well organized but rather ad hoc.

2.5 FOURIER SERIES: DISCRETE SPECTRA

Fourier series expansions can be viewed as an example of the general theory of expansions. There are other ways to present this expansion, but the approach we are taking gives a better overall perspective of signal descriptions.

Let us consider the vector space of piecewise continuous signals of time, in some interval $(0, T)$. If we denote such signals as $f(t)$, $g(t)$, and so on, the following expressions are legitimate IP, N, and D, as we noted in the preceding section.

1. *Inner product:*

$$(f(t), g(t)) = \int_0^T f(t)g(t)\, dt \tag{2.5.1}$$

2. *Norm:*

$$N(f(t)) = \sqrt{\int_0^T f^2(t)\, dt} \tag{2.5.2}$$

3. *Distance:*

$$d(f(t), g(t)) = \sqrt{\int_0^T [f(t) - g(t)]^2\, dt} \tag{2.5.3}$$

Now let us consider the set of signals

$$\phi_{10} = \frac{1}{\sqrt{T}}, \qquad \phi_{1n} = \sqrt{\frac{2}{T}} \cos n\omega_0 t, \qquad \phi_{2n} = \sqrt{\frac{2}{T}} \sin n\omega_0 t \tag{2.5.4}$$

where $\omega_0 = 2\pi/T$ and $n = 1, 2, 3, \ldots$.

This set is actually a basis, that is, all these signals have unity norms and they are orthogonal to each other, a fact that can be easily proved by the reader. The basis is also complete (the proof is omitted). The upshot of all this is that Theorem 2.4.2 holds, and any $f(t)$ in the space can be written as

$$f(t) = C_{10}\frac{1}{\sqrt{T}} + \sum_{n=1}^{\infty} C_{1n}\phi_{1n} + \sum_{n=1}^{\infty} C_{2n}\phi_{2n} \tag{2.5.5}$$

where the C's are given by

$$C_{10} = (f(t), \phi_{10}) = \int_0^T f(t) \frac{1}{\sqrt{T}} \, dt \qquad (2.5.6)$$

$$C_{1n} = (f(t), \phi_{1n}) = \sqrt{\frac{2}{T}} \int_0^T f(t) \cos n\omega_0 t \, dt \qquad (2.5.7)$$

$$C_{2n} = (f(t), \phi_{2n}) = \sqrt{\frac{2}{T}} \int_0^T f(t) \cos n\omega_0 t \, dt \qquad (2.5.8)$$

This expansion is called the *trigonometric Fourier series*. In most books it has the form

$$f(t) = a_0 + \sum_{n=1}^{\infty} a_n \cos n\omega_0 t + \sum_{n=1}^{\infty} b_n \sin n\omega_0 t \qquad (2.5.9)$$

a form that can be obtained easily from the above by defining the ϕ's without the constants $\sqrt{2/T}$ and $\sqrt{1/T}$, and by lumping these constants with the new coefficients a_0, a_n, and b_n. If the form (2.5.9) is used, the coefficients are given by

$$a_0 = \frac{1}{T} \int_0^T f(t) \, dt \qquad (2.5.10)$$

$$a_n = \frac{2}{T} \int_0^T f(t) \cos n\omega_0 t \, dt, \qquad b_n = \frac{2}{T} \int_0^T f(t) \sin n\omega_0 t \, dt \qquad (2.5.11)$$

It should be noted that they are still the IPs of $f(t)$ with the signals of the Fourier basis, although they have been scaled by new constants.

Let us look next at a second "form" of this series, the **exponential** form. To do this, we must enlarge our vector space a bit, and introduce a new IP, N, and D. We enlarge the space by accepting signals that are also functions of a complex variable. Under these conditions, expressions (2.5.1), (2.5.2), and (2.5.3) become:

4. *Inner product:*

$$(f(t), g(t)) = \int_0^T f(t) g^*(t) \, dt \qquad (2.5.12)$$

5. *Norm:*

$$N(f(t)) = \sqrt{\int_0^T |f(t)|^2 \, dt} \qquad (2.5.13)$$

6. *Distance:*

$$d(f(t), g(t)) = \sqrt{\int_0^T |f(t) - g(t)|^2 \, dt} \qquad (2.5.14)$$

where the asterisk (*) denotes a complex conjugate. The changes above were made so that these expressions meet their axioms in the new, enlarged space. (Can you prove this?)

Consider now the set of ϕ's defined by

$$\phi_n = \frac{e^{jn\omega_0 t}}{\sqrt{T}} \qquad \text{for } n = -\infty, \ldots, -1, 0, 1, 2, \ldots \qquad (2.5.15)$$

This set is again a complete basis. Therefore, any $f(t)$ in the space can be written as

$$f(t) = \sum_{n=-\infty}^{+\infty} C_n \phi_n = \sum_{-\infty}^{+\infty} C_n \frac{e^{jn\omega_0 t}}{\sqrt{T}} \qquad (2.5.16)$$

where

$$C_n = (f(t), \phi_n) = \int_0^T f(t) \frac{e^{-jn\omega_0 t}}{\sqrt{T}} \, dt \qquad (2.5.17)$$

The negative sign of the exponent of the integrand of (2.5.17) is, of course, due to the complex conjugate needed in the IP.

The exponential Fourier series is often written as

$$f(t) = \sum_{-\infty}^{+\infty} F_n e^{jn\omega_0 t} \qquad (2.5.18)$$

where the coefficients F_n are now

$$F_n = \frac{1}{T} \int_0^T f(t) e^{-jn\omega_0 t} \, dt \qquad (2.5.19)$$

This form is equivalent to (2.5.16). Simply note that

$$F_n = \frac{C_n}{\sqrt{T}} \qquad (2.5.20)$$

It is the versions (2.5.9) and (2.5.18) that we will use primarily in the remainder of the book, in order to conform to the accepted convention. We do feel, however, that the forms obtained directly from expansion theory are more illustrative of the concepts.

The two forms of the Fourier series are related to each other, and one can easily be obtained from the other. Table 2.5.1 has both forms and the conversion formulas for the coefficients. These formulas can be derived from the known Euler relationships

$$\cos \omega t = \frac{e^{j\omega t} + e^{-j\omega t}}{2}, \qquad \sin \omega t = \frac{e^{j\omega t} - e^{-j\omega t}}{2j} \qquad (2.5.21)$$

Table 2.5.1 also includes a third form of the Fourier series (the polar form), easily obtained from the other two, which will not be discussed here. In fact, we will mainly use the form (2.5.18) in what follows, and it is with this form in mind that we later develop discrete spectra.

TABLE 2.5.1   Various Forms of the Fourier Series

Type	Expansion	Coefficients	Conversion Formulas
Exponential	$f(t) = \sum\limits_{n=-\infty}^{+\infty} F_n e^{jn\omega_0 t}$	$F_n = \dfrac{1}{T}\int_0^T f(t)e^{-jn\omega_0 t}\,dt$	$F_0 = a_0$ $F_n = \tfrac{1}{2}(a_n - jb_n)$
Trigonometric	$f(t) = a_0 + \sum\limits_{n=1}^{\infty} a_n \cos n\omega_0 t$ $+ \sum\limits_{n=1}^{\infty} b_n \sin n\omega_0 t$	$a_0 = \dfrac{1}{T}\int_0^T f(t)\,dt$ $a_n = \dfrac{2}{T}\int_0^T f(t)\cos n\omega_0 t\,dt$ $b_n = \dfrac{2}{T}\int_0^T f(t)\sin n\omega_0 t\,dt$	$a_n = F_n + F_{-n}$ $b_n = j(F_n - F_{-n})$ $a_0 = F_0$
Polar	$f(t) = a_0 + \sum\limits_{n=1}^{\infty} A_n$ $\cos(n\omega_0 t + \phi_n)$		$A_n = \sqrt{a_n^2 + b_n^2} = 2|F_n|$ $\phi_n = -\tan^{-1}\dfrac{b_n}{a_n}$

A Fourier series can be used to describe many signals in a specified interval $(0, T)$, but it is used primarily for a special class of signals, periodic ones with period T. This is due to the fact that if we make the expansion for the interval $(0, T)$ of such a periodic function, the expansion holds outside of it as well, for all t. To prove this, we proceed as follows.

A **periodic function** $f(t)$ is, by definition, one for which

$$f(t) = f(t + T) \tag{2.5.22}$$

where T is its period. Let us assume that we have an expansion of $f(t)$ obtained by using the interval $(0, T)$, and it is of the form (2.5.18). If we evaluate the expansion at $t = t + T$, we obtain

$$f(t + T) = \sum_{n=-\infty}^{+\infty} F_n e^{jn\omega_0(t+T)} = e^{jn\omega_0 t}\sum F_n e^{jn\omega_0 t} \tag{2.5.23}$$

but

$$e^{jn\omega_0 T} = \cos n\omega_0 T + j\sin \omega_0 T = \cos 2n\pi + j\sin 2n\pi = 1$$

since $\omega_0 = 2\pi/T$. Thus (2.5.22) holds for the expansion as well, and the assertion is proved. A Fourier series expansion is, then, a complete description of a periodic signal, valid for the entire time axis.

With this last note we complete our presentation of Fourier series, except for the notions of discrete spectra, which will be discussed after the examples. Our discussion has not been complete because it is assumed that the reader has seen the material before. Such questions as those involving convergence, symmetry, differentiation, and integration of the series are omitted, since no use will be made of them in the remainder of the book.

Example 2.5.1

Find the two Fourier series for the periodic signal $f(t)$, described in the interval of its period by

$$f(t) = A \sin \pi t \qquad \text{for } 0 < t \leq 1$$

Solution. This signal is, of course, a rectified sine wave. Let us first find the trigonometric form and then arrive at the exponential form by the conversion formulas of Table 2.5.1. We note that when $\omega_0 = 2\pi$,

$$a_0 = \frac{1}{T} \int_0^T A \sin \pi t \, dt = \frac{2A}{\pi} \tag{2.5.24}$$

$$a_n = 2 \int_0^1 \sin \pi t \cos n2\pi t \, dt = \frac{4A}{\pi(1 - 4n^2)} \tag{2.5.25}$$

$$b_n = 0 \tag{2.5.26}$$

Therefore,

$$f(t) = \frac{2A}{\pi} + \sum_{n=1}^{\infty} \frac{4A}{\pi(1 - 4n^2)} \cos 2\pi n t \tag{2.5.27}$$

From the above and since

$$F_n = \tfrac{1}{2}(a_n - jb_n) = F_{-n} \qquad \text{(since } b_n = 0\text{)}$$

we have that

$$F_n = \tfrac{1}{2}a_n$$

and

$$f(t) = \sum_{n=-\infty}^{+\infty} \frac{2A}{\pi(1 - 4n^2)} e^{j2\pi n t} \tag{2.5.28}$$

Example 2.5.2

Find the exponential Fourier series of the periodic pulse shown in Fig. 2.5.1.

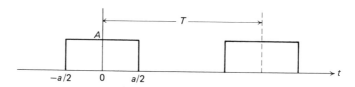

Figure 2.5.1 Periodic pulse.

Solution. Let us find the expansion in the interval $-T/2$ to $T/2$.

$$F_n = \frac{A}{T} \int_{-a/2}^{a/2} e^{-jn\omega_0 t} \, dt = \frac{aA}{T} \frac{\sin (n\omega_0 a/2)t}{(n\omega_0 a/2)t} \tag{2.5.29}$$

a result obtained by using Euler's relationship (2.5.21) after integration.

This example is very important and we shall return to it toward the end of the section. A periodic pulse is a very common signal in radar systems, and its Fourier series expansion is useful in the analysis of such systems. Of course, in practice the pulse may start at $t = 0$, but that is simply a time-shifted version of the signal of Fig. 2.5.1.

Example 2.5.3

Find the exponential Fourier series of the signal

$$f(t) = 6 \cos 3t - 8 \sin 3t \qquad (2.5.30)$$

Solution. This signal is already in a trigonometric series form and use of the conversion formulas of Table 2.5.1 gives the result immediately. Assuming that

$$a_1 = 6, \qquad b_1 = -8 \qquad (2.5.31)$$

then

$$F_1 = \tfrac{1}{2}(6 + 8j) \qquad \text{and} \qquad F_{-1} = \tfrac{1}{2}(6 - 8j) \qquad (2.5.32)$$

It should be noted that the F's are complex numbers. In the previous two examples they turned out to be real, but in general they are complex, and in fact, for real $f(t)$'s, F_n and F_{-n} are conjugates of each other. These facts are important in understanding the meaning of discrete spectra, which come next in our presentation.

We mentioned earlier (at the beginning of Section 2.4) that expansions are an alternative way of specifying signals. All one has to do is to give the *basis* (Fourier, Bessel, etc.) and then the coefficients. Often this method is easier than specifying the signal in some other way. Computers, for instance, usually have the elements of the basis stored in them, and a few coefficients can lead to an adequate description of the signal. This specification of the coefficients, when used in conjuction with the exponential Fourier series, leads to the notions of spectra, which we will cover next in this section. Spectra associated with the other Fourier series forms will be omitted.

Let us consider an $f(t)$ that has been expanded as

$$f(t) = \sum_{n=-\infty}^{+\infty} F_n e^{jn\omega_0 t} \qquad (2.5.33)$$

with $\omega_0 = 2\pi/T$ and $(0, T)$ the interval of the expansion. The coefficients F_n *completely specify* this $f(t)$ in this interval [in fact, everywhere if $f(t)$ is periodic]. To transmit this $f(t)$, all one has to do is to transmit these F_n's and the nature of the series (the basis).

Now these F_n's are complex numbers. A common way of specifying such numbers is by giving their *amplitude* and their *phase*. When these two are given in the form of a graph, the **amplitude** and **phase spectra** of the signal $f(t)$ result.

But what sort of graphs are we to make? To understand this we must elaborate further on what we mean by the term *complete description of the signal $f(t)$ by its Fourier coefficients*.

Each coefficient multiplies an element of the basis of the series. A complete description of the signal requires not only the F_n's, but also the element exp $(jn\omega_0 t)$, which each of them multiplies. If, for example, someone tell us that the coefficients of some $f(t)$ are

$$F_1 = 3 + 2j \quad \text{and} \quad F_{-1} = 3 - 2j$$

we still do not know how to write this signal $f(t)$. A complete description requires knowledge of the elements that F_1 and F_{-1} multiply. If, for example, we are told that F_1 multiplies the element e^{j3t}, then we know everything, and we can write

$$f(t) = F_1 e^{j3t} + F_{-1} e^{-j3t}$$

The above can be accomplished by giving only the part of the exponent $n\omega_0$, since we all know the form of the elements. If we use the axis ω, then all we have to give are the F_n's and the points $n\omega_0$ for each F_n. In the example above, the information F_1 at $\omega = 3$ and F_{-1} at $\omega = -3$ is sufficient. In fact, just the information F_1 at $\omega = +3$ is sufficient, since we know the rest from the nature of the series. The **amplitude spectrum** of $f(t)$ is a plot of its $|F_n|$ versus the values $\omega = n\omega_0$, which correspond to the exponents of the Fourier elements. The **phase spectrum** of $f(t)$ is a plot of the phase angle [angle (F_n)] versus the same ω. These two spectra are called *discrete* (or *line*) *frequency spectra of* $f(t)$—*discrete*, because they are defined only at specific values of $\omega = n\omega_0$ and *frequency* because the axis ω represents the frequency of the basis elements.

Although the spectra are basically the specification of the F_n's in the form of the two graphs, they provide us with more meaning, because of the nature of the series. They give us at a glance the various harmonics present and their relative amplitudes and phases. Let us not forget that this expansion, despite its complex exponential form, is still an expansion on sines and cosines (harmonics), and that such a specification is very useful, since filters are also specified in a way (transfer functions) that shows *the range of frequencies which are allowed to pass them*.

Example 2.5.4

Plot the amplitude and phase spectra of the $f(t)$ of Example 2.5.3.

Solution. We have that

$$|F_1| = \sqrt{9 + 16} = 5 = |F_{-1}|$$

and since

$$F_1 \longrightarrow \omega = 3 \quad \text{and} \quad F_{-1} \longrightarrow \omega = -3$$

we have the amplitude spectrum of Fig. 2.5.2a. Similarly, angle $(F_1) = \tan^{-1}\left(\frac{4}{3}\right) \approx 53°$, angle $(F_{-1}) = -$angle $(F_1) \approx -53°$ and the phase spectrum is shown in Fig. 2.5.2b.

It should be noted that since F_n and F_{-n} are complex conjugates, the amplitude spectrum is an even function of ω, and the phase spectrum, an odd one. With these facts in mind, *we need to show the plots for* $\omega \geq 0$ *only*.[2]

The notion of harmonics and their relative amplitudes is further refined by the definition of the power spectrum, the last type of spectrum associated with

---

[2]These symmetry relations hold only when $f(t)$ is a real function of t (see Problem 2.17).

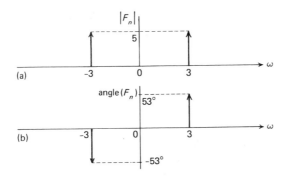

Figure 2.5.2   Spectra of $f(t)$ of Example 2.5.2.

Fourier series expansion. The whole idea is based on the Parseval identity, expression (2.4.29). In the case of the exponential Fourier series this identity becomes

$$\int_0^T f^2(t)\, dt = \int_0^T f(t) \sum_{n=-\infty}^{+\infty} F_n e^{jn\omega_0 t}\, dt$$

$$= \sum_{-\infty}^{+\infty} F_n T\left(\frac{1}{T} \int_0^T f(t) e^{jn\omega_0 t}\, dt\right) = T \sum_{-\infty}^{+\infty} F_n F_{-n}$$

or

$$\frac{1}{T} \int_0^T f^2(t)\, dt = \sum_{n=-\infty}^{+\infty} |F_n|^2 \qquad (2.5.34)$$

This last expression shows that the average power of $f(t)$ over T equals the sum of the $|F_n|^2$, and as such, it gives us a way of finding this power using the coefficients. Furthermore, each pair $|F_n|^2$ and $|F_{-n}|^2$ represents the average power of a $F_n e^{jn\omega_0 t} + F_{-n} e^{-jn\omega_0 t}$ term, which represents a harmonic. For this reason, we usually plot $|F_n|^2$ versus $\omega = n\omega_0$, and call the result the **power spectrum** of $f(t)$. Using this plot, we can also estimate the relative average power of a harmonic (or harmonics) by dividing its power by the total average power.

Example 2.5.5
Sketch the amplitude, phase, and power spectra of the signal in Example 2.5.2. Find the percent average power in the direct-current (dc) term. Examine the changes in the spectra caused by the changes in the period T.

Solution.   The coefficients have already been found for this signal and are given in (2.5.29).

The signal $(\sin x)/x$ comes up very often in signal analysis and has been named the *sampling function*. It is denoted by

$$\text{Sa}(x) = \frac{\sin x}{x} \qquad (2.5.35)$$

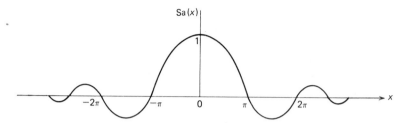

Figure 2.5.3   Signal Sa (x).

and is given in graph form in Fig. 2.5.3. Note that at $t = 0$ its value is Sa $(x) = 1$ (by L'Hospital's rule) and that it is zero at $x = \pm n\pi$.

Using this last signal, the coefficients of the periodic pulse can be written as

$$F_n = \frac{aA}{T} \text{Sa} \left(\frac{n\omega_0 a}{2}\right) \qquad (2.5.36)$$

This means that these F_n are determined by the values of the signal Sa (x) at the points $x = (n\omega_0 a)/2$.

Let us now see how changes in T affect the spectra. Let us first take $A = 1$, $a = 50$ ms, and $T = 0.25$ s. With these values,

$$F_n = \tfrac{1}{5}\text{Sa}\left(\frac{n\pi}{5}\right) \qquad (2.5.37)$$

They are, of course, real, although they can be negative (when they are negative, their phase angle is $\pm\pi$), and their values are at $\omega = \pm n\omega_0 = \pm n8\pi$.

Figure 2.5.4 shows the amplitude spectrum of the periodic pulse for the values of A, a, and T given above. The phase spectrum is simple and will not be given. The power spectrum is like Fig. 2.5.4, with the values $|F_n|^2$ rather than $|F_n|$. The average power in the dc term is

$$|F_0|^2 = |\tfrac{1}{5}|^2 = 0.04 \text{ W}$$

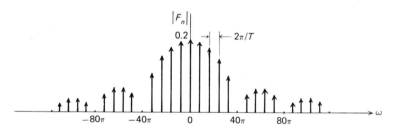

Figure 2.5.4   Amplitude spectrum of a periodic pulse for $T = 0.25$ s, $a = 50$ ms.

The total average power can be found by expression (2.5.34). It is easier to find it by using its left-hand side (i.e., the time-domain expression). The result is

$$\frac{1}{T} \int_{-a/2}^{a/2} f^2(t)\, dt = 4(a) = 0.20 \text{ W}$$

The percent average power in the dc term is, therefore,

$$\frac{0.04}{0.20} = 20\%$$

Let us now change the period to $T = 0.5$ s and see the effect on the amplitude spectrum. Now we will have

$$F_n = 0.01 \, \text{Sa} \left(\frac{n\pi}{10}\right)$$

with the $|F_n|$ versus ω spectrum shown in Fig. 2.5.5. The percent average power in the dc term is now 10%, as the reader can easily ascertain.

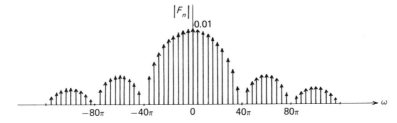

Figure 2.5.5   Amplitude spectrum of a periodic pulse for $T = 0.5$ s, $a = 50$ ms.

What is the principal difference between Figs. 2.5.4 and 2.5.5? Well, as T increases, the lines get closer together (more frequencies), and vice versa. In fact, for $T \longrightarrow \infty$, the curve becomes continuous, as we shall see in the next section. This effect, which can be produced in the spectrum by changing the period, is important. It will be very useful later in the study of radar systems.

This last remark completes our discussion of Fourier series expansions and the resulting spectra. It is worth restating that this expansion is best when the signal is periodic, since it is then valid for all t. If we want a Fourier series for a nonperiodic signal, we can have one only for a specific interval, not for all t. A harmonic *expansion* of a nonperiodic function for all t is possible only with the use of the Fourier transform, and that is our next topic.

2.6 THE FOURIER TRANSFORM: SPECTRA

We noted in the preceding section that expansions on a Fourier series basis were more than just an alternative way of representing a signal. They led to the notions of spectra (harmonic content), notions that will prove quite useful in tracing signals as they go through communications and radar systems. This property of Fourier series is so useful that we would like to extend it to aperiodic signals, that is, signals that are not periodic and are defined over the entire time axis. Such signals can be expanded on other types of bases (Laguerre functions, for example), but such expansions, although useful in other applications, do not aid us much in analyzing our systems, for they do not have the frequency interpretation we seek.

The extension of the exponential Fourier series to an aperiodic function $f(t)$ leads to the **Fourier transform**. This name is given to the old Fourier series coefficients (F_n), which are now denoted by $F(\omega)$. The new formula for these coefficients (the Fourier transform) is

$$F(\omega) = \int_{-\infty}^{+\infty} f(t)e^{-j\omega t}\, dt \qquad (2.6.1)$$

The difference between this expression and the old expression for the coefficients (2.5.19) is in the exponent of the exponential term, where we now have ω instead of $n\omega_0$. This means that the inner product is now between $f(t)$ and the functions $e^{j\omega t}$ defined for all values of ω, not only the values $n\omega_0$. In other words, the expansion is on a new set with an uncountable infinity of elements. The result is, of course, that the new coefficients are defined for each value of ω; that is, they are a continuous function of ω, and they have a new name, the Fourier transform of $f(t)$.

In terms of these coefficients, the original $f(t)$ can be written as

$$f(t) = \frac{1}{2\pi} \int_{-\infty}^{+\infty} F(\omega)e^{j\omega t}\, d\omega \qquad (2.6.2)$$

an expression that corresponds to equation (2.5.18) and gives us the $f(t)$ in terms of an expansion. The integral (rather than a summation) is necessary in (2.6.2), since both the elements of the expansion and $F(\omega)$ are no longer discrete functions of ω. Expression (2.6.2) is called the **inverse Fourier transform** of $F(\omega)$. The signal $f(t)$ and its Fourier transform $F(\omega)$ are called a **Fourier transform pair**, *and they are unique.*

We viewed the Fourier transform above as an extension of the Fourier series for aperiodic functions, without proving that this is so from a mathematical point of view. Such a proof actually exists (see Stark and Tuteur, 1979), but it will be omitted here. Study of Example 2.5.5 will give the reader the idea behind such a proof. As the period of the periodic pulse tends to infinity, the result is more and more harmonics, until the spectrum becomes continuous.

An interesting mathematical question involves the type of condition that signals must obey in order for their Fourier transform to exist. It has been proven that a sufficient condition for the existence of the Fourier transform is that of Dirichlet.[3] This is obeyed when $f(t)$ is single valued, with a finite number of maxima, minima, and discontinuities in any finite interval, and satisfies

$$\int_{-\infty}^{+\infty} |f(t)|\, dt < \infty \qquad (2.6.3)$$

---

[3]This condition also guarantees the existence of the Fourier series. For more on conditions for the existence of Fourier transforms and series, see the excellent text by Papoulis (1962).

The Dirichlet condition is sufficient but not necessary. Our energy signals obey it (prove it). In fact, all useful signals in the analysis of communication and radar systems have a Fourier transform, even the power signals, although sometimes one must revert to limit operations to find them (transform in the limit), as we shall see later.

The Fourier transform (FT) of a signal $f(t)$ is generally a complex function of ω. As such, it can be written as

$$F(\omega) = |F(\omega)|\, \text{angle}\, (F(\omega)) \qquad (2.6.4)$$

To specify $F(\omega)$ completely in terms of plots, one must give $|F(\omega)|$ and angle $(F(\omega))$ separately. In analogy to our discussion of Fourier series spectra, we can now define $|F(\omega)|$ and angle $(F(\omega))$ as the **amplitude** and **phase spectrum**, respectively. These two spectra of a signal $f(t)$ are called **continuous spectra**, since $F(\omega)$ is defined for all ω, and not at the specific points $n\omega_0$ as was the case with periodic $f(t)$'s. These spectra carry the same meaning as they did in the case of the Fourier series [i.e., an indication of the harmonic content of the signal $f(t)$]. It is hoped that the reader comprehends this well. In fact, it was for this reason that we presented the Fourier transform as the extension of the Fourier series.

The idea of harmonic content becomes more concrete with the introduction of some sort of power spectrum. This was done in the preceding section and will be repeated below for aperiodic energy signals. Power signals present some mathematical difficulties and will be treated after the section on Fourier transform properties.

Energy Signals.   The energy of such a signal $f(t)$ is given by

$$E = \int_{-\infty}^{+\infty} f^2(t)\, dt \qquad (2.6.5)$$

This $f(t)$, like all energy signals, has a Fourier transform $F(\omega)$ given by expression (2.6.1), and can be written in terms of $F(\omega)$ as in expression (2.6.2). Using this last expression to replace one of the $f(t)$'s in the integrand of the energy above, we have

$$E = \frac{1}{2\pi} \int_{-\infty}^{+\infty} F(\omega) \left[\int_{-\infty}^{+\infty} f(t) e^{j\omega t}\, d\omega \right] dt \qquad (2.6.6)$$

Interchanging the order of integration (can you do this?), we obtain

$$E = \frac{1}{2\pi} \int_{-\infty}^{+\infty} f(t) \int_{-\infty}^{+\infty} F(\omega) e^{j\omega t}\, d\omega\, dt \qquad (2.6.7)$$

A glance at the definition of the inverse FT (2.6.2) reveals that the term in parentheses on the right-hand side of (2.6.7) is $F(-\omega)$ [i.e., the complex conjugate of

$F(\omega)$]. Finally, then,

$$E = \int_{-\infty}^{+\infty} f^2(t)\, dt = \frac{1}{2\pi} \int_{-\infty}^{+\infty} |F(\omega)|^2\, d\omega \qquad (2.6.8)$$

a relation that is analogous to the Parseval identity encountered in expansions, strengthening the notion that the FT represents the coefficients of an expansion. Of course, the relation gives the *energy* (not the average power) of the signal in both domains (time and frequency), but this was to be expected. We are dealing here with energy signals, whose average power over all time is zero.

The term $|F(\omega)|^2$ is now denoted by $S(\omega)$ and is called the **energy density spectrum** (or spectral density) of $f(t)$. It can, of course, do more than just provide us with another way of finding the energy. It can be used to find the amount of energy (in percent, if we like) in specific bands of frequencies. For example, the energy in the interval of harmonics (ω_1, ω_2) can be obtained by

$$E_{(\omega_1, \omega_2)} = \frac{1}{2\pi} \int_{-\omega_1}^{-\omega_2} S(\omega)\, d\omega + \frac{1}{2\pi} \int_{\omega_1}^{\omega_2} S(\omega)\, d\omega = \frac{1}{\pi} \int_{\omega_1}^{\omega_2} S(\omega)\, d\omega \qquad (2.6.9)$$

Since a harmonic is made up of both the positive and negative frequencies, Equation (2.6.9) reduces to its final form for real signals $f(t)$, for which $|F(\omega)|$ [and $S(\omega)$] prossesses even symmetry.

Let us now illustrate these notions with some examples. The results of both of the following examples are given in the Table 2.6.1, but we work them out any way, just to get a flavor of the math involved.

Example 2.6.1 (One-Sided Exponential)

For the signal $f(t) = Ke^{-at}u(t)$, find and plot all three spectra. Also find the percent energy in the harmonic band [0, 10] rad/s.

Solution. First we find the FT $F(\omega)$. From definition (2.6.1) we have

$$F(\omega) = \int_{-\infty}^{+\infty} Ke^{-at}u(t)e^{-j\omega t}\, dt = \int_{0}^{\infty} e^{-(a+j\omega)t}\, dt = \frac{K}{a + j\omega} \qquad (2.6.10)$$

The three spectra desired are

$$|F(\omega)| = \frac{K}{\sqrt{a^2 + \omega^2}} \qquad (2.6.11)$$

$$\text{angle}\,(F(\omega)) = -\tan^{-1}\frac{\omega}{a} \qquad (5.6.12)$$

$$|F(\omega)|^2 = S(\omega) = \frac{K^2}{a^2 + \omega^2} \qquad (2.6.13)$$

Their plots are given in Fig. 2.6.1.

To find the percentage of the energy in the [0, 10] band, we first find the total energy E. Using the time-domain formula, we have

$$E = \int_{-\infty}^{+\infty} f^2(t)\, dt = \int_{0}^{\infty} K^2 e^{-2t}\, dt = \frac{K^2}{2} \qquad \text{joules}$$

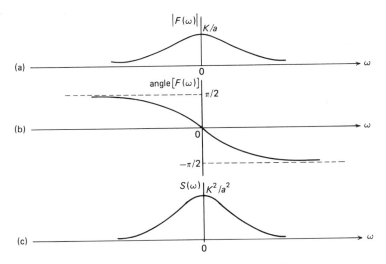

Figure 2.6.1   Spectra of $f(t) = Ke^{-at}u(t)$: (a) amplitude; (b) phase; (c) energy density.

The energy in the band [0, 10] can only be found by using expression (2.6.9). This yields

$$E_{(0,10)} = \frac{K^2}{\pi} \int_0^{10} \frac{1}{a^2 + \omega^2}\, d\omega = \text{(find it)}$$

and the desired percentage is easily obtainable. It should be noted that $f(t)$ has most of its energy in this low-frequency band (low-frequency signal).

Example 2.6.2 (Rectangular Pulse)
Find all three spectra of the signal

$$f(t) = \begin{cases} 1 & \text{for } -\frac{T}{2} \le t \le \frac{T}{2} \\ 0 & \text{elsewhere} \end{cases} \tag{2.6.14}$$

Solution.   Direct application of the definition of $F(\omega)$ gives

$$F(\omega) = \int_{-T/2}^{T/2} e^{-j\omega t}\, dt = \frac{e^{-j\omega t}}{-j\omega} \bigg|_{-T/2}^{T/2} = \frac{T \sin(\omega T/2)}{\omega T/2} = T\,\text{Sa}\!\left(\frac{\omega T}{2}\right) \tag{2.6.15}$$

This is a real function of ω. The amplitude spectrum of $f(t)$ is sketched in Table 2.6.1. The phase spectrum is easily obtained (find it). The energy density spectrum is the square of the amplitude spectrum. Study these spectra well. They will be very useful later in our study of communication and radar systems.

Table 2.6.1 lists a few more energy signals, their FTs and their amplitude spectra. We shall make repeated use of this table in examples and problems.

TABLE 2.6.1   Fourier Transforms of Energy Signals

$f(t)$		$F(\omega)$	$|F(\omega)|$
Rectangular pulse	$\mathrm{rect}\left(\dfrac{t}{T}\right)$	$T\,\dfrac{\sin(\omega T/2)}{\omega T/2}$	
	$T\,\mathrm{Sa}\left(\dfrac{T}{2}t\right)$	$2\pi\,\mathrm{rect}\left(\dfrac{\omega}{T}\right)$	
Triangular	$1-2\dfrac{|t|}{T},\ |t|<\dfrac{T}{2}$ $0\quad$ elsewhere	$\dfrac{T}{2}\left[\dfrac{\sin(\omega T/4)}{\omega T/4}\right]^2$	
Exponential	$e^{-at}u(t)$	$\dfrac{1}{j\omega+a}$	
Double exponential	$e^{-a|t|}$	$\dfrac{2a}{a^2+\omega^2}$	
Gaussian	$e^{-a^2 t^2}$	$\dfrac{\sqrt{\pi}}{a}\,e^{-(\omega^2/4a^2)}$	
Damped sine	$e^{-at}\sin(\omega_0 t)u(t)$	$\dfrac{\omega_0}{(a+j\omega)^2+\omega_0^2}$	
Damped cosine	$e^{-at}\cos(\omega_0 t)u(t)$	$\dfrac{a+j\omega}{(a+j\omega)^2+\omega_0^2}$	
Cosine pulse	$\cos\omega_0 t\,\mathrm{rect}\left(\dfrac{t}{T}\right)$	$\dfrac{T}{2}\left[\dfrac{\sin(\omega-\omega_0)T/2}{(\omega-\omega_0)T/2}\right.$ $\left.+\dfrac{\sin(\omega+\omega_0)T/2}{(\omega+\omega_0)T/2}\right]$	
	$\dfrac{1}{\beta-a}(e^{-at}-e^{-\beta t})u(t)$	$\dfrac{1}{(j\omega+a)(j\omega+\beta)}$	

2.7 PROPERTIES OF THE FOURIER TRANSFORM

Up to now our main concern has been with mathematical description of signals. Generalized functions, expansions, the Fourier transform—all these have been for the most part alternative ways of describing signals by means of other signals or in other domains.

The analysis of communication and radar systems demands more than the ability to describe signals by a mathematical model. One must also be able to keep track of a model while going through the various devices (blocks) of the system. A given signal starts at some point (the source, for example), and its model may be known there quite well. Soon, however, it enters the transmitter, amplifiers, filters, antennas, and the like, and everytime it passes through, its model has been changed. How does one keep track of it so that he or she can perform a thorough analysis of the system?

The Fourier transform properties play a vital role in the ability to track the model of a signal as it traverses the various devices of a system. This section is devoted to a discussion of these properties and their usefulness. Not all of them will be proven here, but nearly all of them will be useful in the remainder of the book.

In discussing the FT properties we shall work in conjunction with Table 2.7.1, which includes the most important ones. By the way, the reader would do well to restudy the end of Section 2.2 at this point, as many of the properties deal with shifts, scalings, and the like.

TABLE 2.7.1 Properties of the Fourier Transform

Property	Time Domain	Frequency Domain
1. Linearity	$af(t) + bg(t)$	$aF(\omega) + bG(\omega)$
2. Duality	$F(t)$	$2\pi f(-\omega)$
3. Scaling	$f(at)$	$\dfrac{1}{\|a\|} F\left(\dfrac{\omega}{a}\right)$
4. Delay	$f(t \pm t_0)$	$F(\omega)e^{\pm j\omega t_0}$
5. Frequency shifting	$f(t)e^{\pm j\omega_0 t}$	$F(\omega \mp \omega_0)$
6. Convolution	$\displaystyle\int_{-\infty}^{+\infty} f(\tau)g(t - \tau)\, d\tau$	$F(\omega)G(\omega)$
7. Time product	$f(t)g(t)$	$\dfrac{1}{2\pi} \displaystyle\int_{-\infty}^{+\infty} F(\xi)G(\omega - \xi)\, d\xi$
8. Differentiation	$\dfrac{d^n f(t)}{dt^n}$	$(j\omega)^n F(\omega)$
9. Integration	$\displaystyle\int_{-\infty}^{t} f(\tau)\, d\tau$	$\dfrac{F(\omega)}{j\omega} + \pi F(0)\delta(\omega)$

Property 1 (Linearity).   This property says basically that the FT of $af(t) + bg(t)$ is equal to $aF(\omega) + bG(\omega)$, where $F(\omega)$ and $G(\omega)$ are the FTs of $f(t)$ and $g(t)$. We stated this property in words so that the reader can understand the use of the table.

The proof of the linearity property is quite easy. The FT of the sum is, by definition,

$$\int_{-\infty}^{+\infty} [af(t) + bg(t)]e^{-j\omega t}\, dt = aF(\omega) + bG(\omega) \qquad (2.7.1)$$

So if any of our signals are multiplied by a constant (amplification), their Fourier transforms are also. Or if we have a sum of signals, the FT is the sum of their FTs—a very nice little property with a lot of usefulness.

Property 2 (Duality). This is a difficult property to visualize. The main problem with it is the mathematical symbolism. The property states that if $f(t)$ has a FT of the form $F(\omega)$, then a time signal of the form $F(t)$ has a FT that looks like $2\pi f(-\omega)$. The primary idea here relates to the symbols f and F. To see this, let us look at an example.

Suppose that

$$f(t) = e^{-t}u(t)$$

Then its $F(\omega)$ is (from Table 2.6.1)

$$F(\omega) = \frac{1}{1 + j\omega} \qquad (2.7.2)$$

Now the $F(t)$ of the property is

$$F(t) = \frac{1}{jt + 1}$$

that is, it is a time function of the form (2.7.2). Its FT $f(\omega)$ would then be

$$2\pi f(-\omega) = 2\pi e^{\omega}u(-\omega) \qquad (2.7.3)$$

that is, it would be related by (2.7.3) to the original signal $f(t)$.

The usefulness of this property is that for every $F(\omega)$ that you find, you actually have two. You also have the FT of the time function $F(t)$, or rather, you can find it. Thus Table 2.6.1 actually has 20 FTs rather than the 10 that are listed.

We will not prove this property here, as we do not feel that it adds anything to our understanding. It is just a mess of symbolism and change of variables.

Property 3 (Scaling). The proof of this property is quite typical of all proofs of FT properties. Let us prove it first, and discuss its importance afterward.

Let us consider $a > 0$. The FT of $f(at)$ is, by definition,

$$\int_{-\infty}^{+\infty} f(at)e^{-j\omega t}\, dt \qquad (2.7.4)$$

Let now $at = x$. This means that $dt = (1/a)\, dx$ and that there is no change in the limits of the integral. Thus Eq. (2.7.4) becomes

$$\frac{1}{a} \int_{-\infty}^{+\infty} f(x)e^{-j(\omega/a)x}\, dx = \frac{1}{a}F\left(\frac{\omega}{a}\right) \qquad (2.7.5)$$

If the procedure is repeated with $a < 0$, the result will be such that in general the FT of $f(at)$ will be $(1/|a|)F(\omega/a)$, as is claimed by the property.

The scaling property is quite important and very useful. Its practicality is, of course, the fact that you can trace the changes in the FT when the signal $f(t)$ passes through a device that speeds it up $(a > 1)$ or slows it down $(a < 1)$ (see the end of Section 2.2). But its importance is also conceptual, for it tells us what happens to the frequency content of a signal that is subjected to such time changes. The property clearly states that if the duration in time of $f(t)$ is decreased $(a > 1)$, the duration of its $F(\omega)$ increases (i.e., its frequency content increases), and vice versa. Short-duration signals have wide-frequency contents, and signals that go on in time have very few frequencies. In fact, the relation of time duration and frequency duration (bandwidth) can be made even more precise by the uncertainty principle, which will be discussed in a later section.

We should remind the reader that scaling changes not only distort the signal but also shift it. (Where did we first mention this?) Thus a scale change in time (and possibly a shift there) can easily cause a shift in $F(\omega)$ as well as a distortion. This remark will be used in the explanation of the Doppler effect in a later chapter.

Property 4 (Delay).   To prove this property, we apply the definition of the FT to the signal $f(t \pm t_0)$. This yields

$$\int_{-\infty}^{+\infty} f(t \pm t_0)e^{-j\omega t}\, dt$$

Changing variables, $t \pm t_0 = x$, the result is

$$\int_{-\infty}^{+\infty} f(x)e^{\pm j\omega t_0 - j\omega t}\, dt = e^{\pm j\omega t_0}F(\omega) \qquad (2.7.6)$$

as claimed by the property.

The reader should now glance at (2.7.6) to see the meaning of the property. Delays in time add the term $e^{\pm j\omega t_0}$ to the FT of a signal; that is, they add a phase angle of value ωt_0 at each point ω. They do not, however, change the amplitude spectrum at all, since the magnitude of $e^{j\omega t_0}$ is unity. This result is intuitively pleasing. Shifting a signal in the time domain does not change its shape, and therefore should not change its frequency content.

Property 5 (Frequency Shifting).   This is perhaps the most important property of all, and most communication and radar systems cannot be analyzed without it. Let us first prove it, and then we will see what it really tells us. The proof is quite easy and involves a judicious change of variables, as in the proof of most of the properties.

Starting with the definition of the FT of $f(t)e^{j\omega_0 t}$, we have

$$\int_{-\infty}^{+\infty} f(t)e^{-j(\omega - \omega_0)t}\, dt = F(\omega - \omega_0) \qquad (2.7.7)$$

and if the exponent is negative, we will have $F(\omega + \omega_0)$. So multiplication of signal $f(t)$ by a term of the form $e^{j\omega_0 t}$ shifts the FT (and thus the spectra) to the left by ω_0, or to the right if the exponent is negative.

Of course, terms like $e^{\pm j\omega_0 t}$ do not exist in practice. But sines and cosines do, and they are related to exponentials by the well-known Euler relationships. Thus, if we multiply $f(t)$ by $\cos \omega_0 t$, this will be equal to

$$f(t) \cos \omega_0 t = \tfrac{1}{2}[f(t)e^{j\omega_0 t} + f(t)e^{-j\omega_0 t}] \tag{2.7.8}$$

and combining Property 5 with Property 1, we will have the following result.

If $f(t)$ has a FT $F(\omega)$, then

$$f(t) \cos \omega_0 t \longleftrightarrow \tfrac{1}{2}[F(\omega - \omega_0) + F(\omega + \omega_0)] \tag{2.7.9}$$

It is this last result (often called the **modulation theorem**) that makes this property so important. It tells us in simple words that multiplication of a signal by a cosine of frequency ω_0 shifts the spectra to the right and to the left by ω_0 and halves the amplitude. Conversely, it tells us that if we wish to shift the spectrum of a signal by a certain amount ω_0, all we have to do is multiply it by $\cos \omega_0 t$. Such shifts of the spectra are so necessary in communications and radar systems that these systems would not exist without this property.

The reader can easily verify that

$$f(t) \sin \omega_0 t \longleftrightarrow \frac{1}{2j}[F(\omega - \omega_0) - F(\omega + \omega_0)] \tag{2.7.10}$$

which tells us that similar shifts occur with a multiplication by a sine, but with different phase angles. This is not surprising since sines and cosines differ only in the phase angle.

This property and the modulation theorem that it contains are so important that they merit an example all their own.

Example 2.7.1

Find the amplitude spectrum of $g(t) = f(t) \cos 10^4 t$ if

$$f(t) = 4[u(t + 1) - u(t - 1)] \tag{2.7.11}$$

Solution.   First, let us find the FT of $f(t)$. Using Table 2.6.1, we have

$$F(\omega) = 8\frac{\sin \omega}{\omega}$$

with an amplitude spectrum $|F(\omega)|$ of the form shown in the table. Now use of the Property 5 gives the FT of $f(t) \cos 10^4 t$ as

$$G(\omega) = \frac{8}{2}\left[\frac{\sin (\omega - 10^4)}{\omega - 10^4} + \frac{\sin (\omega + 10^4)}{\omega + 10^4}\right] \tag{2.7.12}$$

with $|G(\omega)|$ as shown in Fig. 2.7.1. It should be noted that $g(t)$ now has its frequencies around $\omega = 10^4$ rather than around $\omega = 0$ as is the case with $f(t)$. This example is quite important in practice. Communications and radar systems alike use time pulses to carry information, and it is a common case that such pulses are multiplied by cosines to shift their spectra to desirable locations, as we shall see in later chapters.

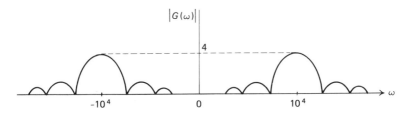

Figure 2.7.1   Amplitude spectrum of $g(t)$.

Property 6 (Convolution).   The proof of this property follows immediately by finding the FT of the convolution integral. Indeed, starting with this,

$$\int_{-\infty}^{+\infty} \int_{-\infty}^{+\infty} f(\tau)g(t - \tau)e^{-j\omega t}\, d\tau\, dt \tag{2.7.13}$$

and changing variables $x = t - \tau$ (the limits of the integrals do not change), we have

$$\int_{-\infty}^{+\infty} f(\tau)e^{-j\omega\tau}\, d\tau \int_{-\infty}^{+\infty} f(x)e^{-j\omega x}\, dx = F(\omega)G(\omega) \tag{2.7.14}$$

which establishes the correctness of the property.

This property is very useful because convolution is a common operation in the analysis of systems. It is assumed here that the reader is familiar with the fact that the output of a linear system $c(t)$ is the convolution of the input $f(t)$ and the impulse response of the system $h(t)$ (those readers who are not should check Problem 2.3).

Performing a convolution is usually tedious, as it must often be done graphically, as the reader presumably knows. Property 6 gives an alternative method of doing it, by a product operation in the frequency domain. We will see in the analysis of our systems later that the use of this property often simplifies the tracing of the changes in the signals that enter and exit the various devices of a system.

Property 7 (Time Product).   Property 7 is the dual of 6, and its proof will be omitted. Its usefulness appears again in the analysis of systems, when our purpose is to keep track of the signals that traverse the devices. Strange as it may sound, there are times when convolution in frequency is easier to do than a product in time (see Problem 2.25, for example).

Properties 8 and 9 (Differentiation and Integration).   We will not bother proving these properties, as we will make only very limitted use of them later. The reader should try doing it as an exercise.

This does it for the FT properties, at least for now. But do not underestimate them. This book (and many others) would not exist without them.

It is time that we did an example, one that needs more than one of these properties for its solution. Let us pick one that has application in both communication and radar systems.

Example 2.7.2

Consider the $f(t)$ defined in Eq. (2.7.11). Then assume that this $f(t)$ goes through a bunch of devices and exits as

$$g(t) = \frac{d[f(3t - 5) \cos 10^5 t]}{dt} \tag{2.7.15}$$

Find the FT $G(\omega)$.

Solution.   There are four operations involved in (2.7.15); let us take them one at a time. The first one is a time shift. So the FT of $f(t - 5)$ will be (Property 4)

$$F(\omega)e^{-j5\omega} \tag{2.7.16}$$

Then we have a scale change (note that if you first take the scale change and then the shift, you run into problems). Using Property 3, we have

$$f(3t - 5) \longleftrightarrow \frac{1}{3} F\left(\frac{\omega}{3}\right) e^{-j(5/3)\omega} \tag{2.7.17}$$

Next, we have a multiplication by a $\cos 10^5 t$. Thus the modulation theorem (which is not in Table 2.7.1, but presumably represented there by Property 5) gives

$$f(3t - 5) \cos 10^5 t \longleftrightarrow \frac{1}{6}\left[F\left(\frac{\omega - 10^5}{3}\right) e^{-j(5/3)(\omega - 10^5)} \right.$$
$$\left. + F\left(\frac{\omega + 10^5}{3}\right) e^{-j(5/3)(\omega + 10^5)} \right] \tag{2.7.18}$$

The desired $G(\omega)$ is simply the product of this last expression with $j\omega$, by Property 8. Pretty nifty. Can you sketch $|G(\omega)|$?

2.8 FOURIER TRANSFORMS OF POWER SIGNALS: FILTERS

We can now use the FT properties and some additional mathematical acrobatics to come up with the FT of some power signals that are very useful. Now, power signals do not satisfy the Dirichlet condition, but this does not mean that they cannot have a FT. (Why?) They often do, even though such a FT is sometimes called **in the limit**.

Let us start with the delta function $\delta(t)$. (Is it a power signal, an energy signal, or neither?)

By brute force (i.e., the definitions of the delta function and of the FT) we get that

$$\int_{-\infty}^{+\infty} \delta(t)e^{-j\omega t}\, dt = e^{-j\omega t}\Big|_{t=0} = 1 \tag{2.8.1}$$

that is,

$$\delta(t) \longleftrightarrow 1 \tag{2.8.2}$$

make up a transform pair. Now let us use Property 2 (duality) and come up with the FT of a constant $f(t) = A$. The result is

$$A \longleftrightarrow 2\pi\delta(\omega) \tag{2.8.3}$$

[actually, we also used the fact that $\delta(\omega) = \delta(-\omega)$, proven under Eq. (2.3.5)]. A constant A is definitely a power signal. Now, what about a signal like

$$f(t) = Ae^{-j\omega_c t} \tag{2.8.4}$$

That is just a constant multiplied by an exponential. It should have a FT. (Give an example.)

$$F(\omega) = 2\pi A\, \delta(\omega - \omega_c) \tag{2.8.5}$$

But why should we care about (2.8.5), since no $e^{-j\omega_c t}$ exists? Because cosines and sines do, and we need their FTs. It is therefore easy to show that

$$\cos \omega_c t \longleftrightarrow 2\pi[\delta(\omega - \omega_c) + \delta(\omega + \omega_c)] \tag{2.8.6}$$

since a cosine is a sum of terms like (2.8.4) (Euler relation) and the linearity property can then be applied.

Table 2.8.1 includes these FTs, and some others obtained in a similar way. But most of the signals encountered in analysis of systems are not in this table (or in Table 2.6.1). That is where high dexterity in the use of the properties really counts. If, for example, we tried finding the FT of the signal $g(t)$ in Example 2.7.2 without the use of the properties, we would not get very far. For it is not in any table, and the brute-force method of trying the Fourier integral would not lead to much. To develop such dexterity one needs to practice a lot. A few of the problems at the end of the chapter provide this opportunity.

Some of the transform pairs in Table 2.8.1 are quite useful and rather interesting. If the reader would stop and think a little, he or she would see an interesting similarity between the FT and the coefficients of the Fourier series, in the case of periodic (and even aperiodic) functions. Take, for example, the cosine, sine, or the general periodic function in Table 2.8.1.

The Fourier series (FS) coefficients of such functions (complex numbers in general) end up as delta functions in the FT domain at the same ω points, with an amplitude change, which does not effect the overall picture. That is comforting. After all, we argued at the beginning of Section 2.6 that the FT is nothing but a generalization of the FS to aperiodic functions.

The reader would do well to learn many of the FTs of Table 2.8.1 by heart, at least their overall picture. Of particular importance for our future analyses is the pair involving the impulse trains in time and frequency, which are often spoken as combs and symbolized as shown in the table (from Woodward 1953).

TABLE 2.8.1   Fourier Transforms of Power Signals

$f(t)$	$F(j\omega)$	$\|F(j\omega)\|$
$\delta(t)$ Unit impulse	1	
A Constant	$2\pi A\,\delta(\omega)$	
$\cos \omega_0 t$ Cosine	$\pi[\delta(\omega - \omega_0) + \delta(\omega + \omega_0)]$	
$\sin \omega_0 t$ Sine	$-j\pi[\delta(\omega - \omega_0) - \delta(\omega + \omega_0)]$	
$u(t)$ Unit step	$\pi\delta(\omega) + \dfrac{1}{j\omega}$	
$tu(t)$ Unit ramp	$j\pi\delta'(\omega) - \dfrac{1}{\omega^2}$	
$\mathrm{sgn}\, t$ Signum signal	$\dfrac{2}{j\omega}$	
$\displaystyle\sum_{n=-\infty}^{\infty} F_n\, e^{j\,(2\pi nt/T)}$ Periodic signal	$\displaystyle 2\pi\sum_{n=-\infty}^{\infty} F_n\,\delta\left(\omega - \dfrac{2\pi n}{T}\right)$	
$\mathrm{comb}_T = \displaystyle\sum \delta(t - nT)$ Impulse train	$\dfrac{2\pi}{T} \displaystyle\sum \delta\left(\omega - \dfrac{2\pi n}{T}\right)$	
$e^{j\omega_0 t}$ Complex exponential	$2\pi\,\delta(\omega - \omega_0)$	
$\dfrac{j}{\pi t}$	$\mathrm{sgn}(\omega)$	

Example 2.8.1

Consider the signal

$$f(t) = \text{Sa}\left(\frac{t}{2 \times 10^3}\right) \tag{2.8.7}$$

and assume that it is ideally sampled every T seconds. This corresponds to multiplying this signal with a comb_T; that is, the result is

$$g(t) = \text{comb}_T \text{Sa}\left(\frac{t}{2 \times 10^3}\right) \tag{2.8.8}$$

Find the FT of $g(t)$.

Solution.   This is an example in combining the FTs of Tables 2.6.1 and 2.8.1 and the FT properties.

From Table 2.6.1 we have that

$$F(\omega) = 2\pi 10^3 \text{ rect} (10^3 \omega) \tag{2.8.9}$$

and in view of Property 7 of Table 2.7.1, and Table 2.8.1 where we can find the FT of comb_T

$$G(\omega) = \frac{1}{T} F(\omega) * \text{comb}_{\omega_0} \tag{2.8.10}$$

where $\omega_0 = \dfrac{2\pi}{T}$.

Our final step is the use of the property of a delta function (or functions) in convolution(s). The result is

$$G(\omega) = \frac{1}{T} \sum F(\omega - n\omega_0) \tag{2.8.11}$$

a useful expression that we will find when we discuss sampling in a later chapter.

Now that we know how to find the FT of various signals, we should give some time to these signals' spectra. The amplitude and phase spectra present no serious problem, other than the fact that when we have delta functions, they are a little strange. But what about something like the energy density spectrum, which we defined for energy signals?

We can define here the **power spectral density** $S(\omega)$ by the equation

$$S(\omega) = \lim_{T \to \infty} \frac{|F_T(\omega)|^2}{2T} \tag{2.8.12}$$

where $F_T(\omega)$ is the FT of the power signal not over all time t, but for $-T < t < T$. We will not use this expression very much in this book, so will not discuss it here. There is another way to look at this anyway, as we will see in Section 2.10.

Now that we have finished with the main discussion of the FT and its properties, it is time we took up an interesting issue that pertains to linear systems.

We have remarked in the past (see the last paragraph of Section 2.3) that signals are also useful in describing systems. One variety of systems, the linear, are uniquely defined by their **impulse response,** that is, by their output $h(t)$ when their input is a delta function $\delta(t)$. Now, after studying the FT, we can easily appreciate that such systems can be described in the frequency domain as well, by the FT of $h(t)$, $H(\omega)$, usually called the **system function.** There is one minor difference in interpretation, however. When we think of $H(\omega)$, we usually interpret it not as the frequency content of the system, but as *the range of frequencies that the system allows to pass it.* This interpretation comes from the fact that if $f(t)$ is the input to such a system and $o(t)$ its output, then by the convolution property we have that

$$O(\omega) = H(\omega)F(\omega) \tag{2.8.13}$$

and so wherever $H(\omega)$ is zero, $O(\omega)$ will be zero as well, even though $F(\omega)$ might not be zero there. Thus the frequencies that go through to the output are dictated by $H(\omega)$.

In this respect it is interesting to speculate about the possibility of a linear system (often called a *filter*), whose impulse response is such that all frequencies are passed unchanged; that is, the $f(t)$ goes through completely undistorted. We could allow a delay, however, and so the output of this ideal filter would be

$$o(t) = f(t - t_0) \tag{2.8.14}$$

In view of Eq. (2.8.13) and the time-delay FT property,

$$O(\omega) = H(\omega)F(\omega) = F(\omega)e^{-j\omega t_0} \tag{2.8.15}$$

which solved for $H(\omega)$ gives

$$H(\omega) = e^{-j\omega t_0} = 1\underline{/-\omega t_0} \tag{2.8.16}$$

whose spectra $|H(\omega)|$ and angle $(H(\omega)) = \theta(\omega)$ are sketched in Fig. 2.8.1a.

Although there exist some practical filters that can match such characteristics pretty closely (transmission lines, waveguides, etc.), most practical devices (amplifiers, for example) can approximate this filter only in certain frequency ranges. This leads to the definition of the **ideal low-pass filter** (LPF), **narrowband filter** (NBF) and **high-pass filter** (HPF), with $H(\omega)$ characteristics as shown in Fig. 2.8.1b, c, and d, respectively. We emphasize that such filters are still ideal, and cannot be matched exactly by filters designed with lumped elements. In fact, they are theoretically unrealizable with *causal* filters, as the reader can easily ascertain by finding their inverse FT, which produces their impulse response $h(t)$ (try the NBF, which is the easiest, as it can be obtained from the tables). Anyway, practical filters that approximate these ideal ones also use the names LPF, NBF, and HPF but without the adjective *ideal.*

The ideal filters were derived starting with the assumption that it is desirable to have a filter that allows a signal to pass unperturbed. Now this is not always desirable, even though it may sound like it at first sight. Often we want to perturb the signal to change it, with a specific reason in mind. Let us look

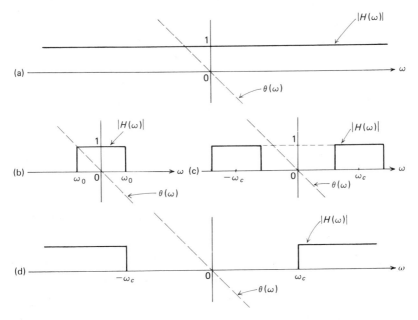

Figure 2.8.1   (a) Ideal distortionless filter; (b) ideal LPF; (c) ideal NBF; (d) ideal HPF.

next at such a case, a case that leads to a result most useful in the analysis of communication and radar systems.

Let us assume that we receive a known signal $s(t)$, which, however, is hard to recognize because of noise effects. To help us recognize it, we can conceive of the possibility of passing it through a linear, causal filter which will *increase its values, enhance it* somehow, even though it may not look the same after passing through the filter.

If the filter's impulse response is $h(t)$, we basically wish to have the output of the filter $s_0(t)$,

$$s_0(t) = \int_0^\infty h(\tau)s(t - \tau) \, d\tau \qquad (2.8.17)$$

as large as possible for each value of t. What would be the form of this "optimum" filter, optimum in the sense that *it maximizes the output at every instant of time* t?

Well, instead of maximizing $s_0(t)$ we can maximize $s_0^2(t)$, which is, of course, the instantaneous power of $s_0(t)$. And this can be done quite simply by using the well-known Cauchy–Schwartz inequality given as Problem 2.10 at the end of the chapter. The result is

$$s_0^2(t) \leq \int_0^\infty h^2(\tau) \, d\tau \int_0^\infty s^2(t - \tau)d\tau \qquad (2.8.18)$$

with equality when

$$h(\tau) = s(t - \tau) \qquad (2.8.19)$$

Naturally, it is this last equality that defines the optimum filter, since it is our only chance to have the maximum for every value of t.

To make more sense out of Eq. (2.8.19), let us assume that $s(t)$ is a signal of finite duration T (say a pulse), a case most typical in digital communications and radar systems. Not only that; let us also try to find the maximum value of $s_0(t)$ with respect to t, assuming of course, that $h(\tau)$ has the optimum form of Eq. (2.8.19). Contemplating Eq. (2.8.18), we note that each integral at its right-hand side is the energy of the input $s(t)$ from $-\infty$ to t (show it by a change of variable), which, of course, will be maximum at $t = T$ (or after), that is, after the $s(t)$ has ended. With this new constraint, the optimum filter becomes

$$h(\tau) = s(T - \tau) \tag{2.8.20}$$

which is called the **matched filter** to $s(t)$. If this filter is used, we are guaranteed a maximum value of $s_0(t)$ at $t = T$, a value that will help us identify the existence of $s(t)$ in the received data. Even though we derived this filter in a rather pedestrian way (talking around the noise, so to speak), its importance is quite celebrated. We will have plenty of occasions to see it again, and even rederive it under different conditions, following Chapter 12.

Example 2.8.2

Find the matched filter for the signal

$$s(t) = u(t) - u(t - T) \tag{2.8.21}$$

Also find $s_0(t)$ and observe its value at $t = T$.

Solution. We have that the matched filter $h(t)$ will be [Eq. (2.8.20)]

$$h(t) = s(T - t)$$

But $s(T - t) = s(t)$, as the reader can easily verify. Therefore,

$$h(t) = s(t) \tag{2.8.22}$$

and now we can begin to see why it is called matched. In the case of a pulse it is not just matched—it is identical.

The output of the matched filter can be easily found (by graphical convolution or the use of FTs) to be a triangle of duration $2T$, and its maximum at $t = T$. The enhancement did take place, but the signal's form was changed from a rectangle to a triangle.

Equation (2.8.22) is not a general result. It holds only when $s(t)$ exhibits certain symmetry which the reader may try to discover. The matched filter is defined by Eq. (2.8.20).

2.9 DURATION, BANDWIDTH, AND THE UNCERTAINTY RELATION

We have come up against the notion of the *width* of a signal a couple of times already in the past. The first time was when we were discussing the effects of time scaling of a signal $f(t)$ (to $f(at)$), right at the beginning of this chapter.

Then we saw this notion again when we were discussing the effects of time scaling on the shape of the FT of the signal, scaling property 3 in Table 2.7.1. In this last instance, in fact, we remarked that if a signal is compressed in time, its FT expands, and vice versa.

The concept of duration or bandwidth is an attempt at putting this notion of width into a more precise mathematical framework. Since signals $f(t)$ are uniquely represented by their FT $F(\omega)$, also a signal (in ω, of course), we ought to be able to find an expression for bandwidth that holds in both domains (or in any domain). Since linear systems (filters) are also represented by signals (their impulse responses), our definition of bandwidth should hold for them as well, even though it might enjoy a slightly different interpretation.

First, let us clear up the following. When we use the term **duration**, we are speaking about the width of the signal in time, and when we say **bandwidth**, we are referring to its width in frequency. That is tradition, and no one should go against it. Unfortunately, tradition (and scientific need) also has it that there is not a single definition of duration and bandwidth. There are quite a few. Of course, that does not really make much difference in results. We get pretty much the same conclusions no matter which definitions we use. Still, the confusion is there, and we shall endeavor to clear up part of it in what follows.

This whole business actually started in the frequency domain, and that is where we will start also. Things are a little neater there, at least for practical signals (not for complex ones), since the amplitude spectrum $|F(\omega)|$ [of a signal $f(t)$] is a positive even function of ω. Of course, this neatness was not the motive for the definition. Need was the motive. We were looking for a quick way to specify the width of the system function of a filter, so that we could get an idea of the width of the band of frequencies that would pass through it. That is where the term "bandwidth" came from.

Let us start with a simple case, an $|F(\omega)|$ that is rectangular around $\omega = 0$, like the FT of a Sa (t) (see Fig. 2.9.1a). Now the actual width of $|F(\omega)|$ is $2\omega_0$. However, in view of the symmetry of $|F(\omega)|$ around $\omega = 0$, we can define our bandwidth B as only $B = \omega_0$, and keep in the back of our mind that we are only referring to the part of $|F(\omega)|$ for $\omega > 0$. So in this case,

$$B = \omega_0 \tag{2.9.1}$$

and we certainly should not have any objections that this represents the spread of $|F(\omega)|$. So far, so good.

Now let us go to Fig. 2.9.1b, which shows a rectangular $|F(\omega)|$, but sitting away from $\omega = 0$, an $|F(\omega)|$ that would be produced if the $f(t)$ of Fig. 2.9.1a was multiplied by $\cos \omega_c t$. Again, we feel that there should be no objection to stating that

$$B = \omega_c + \omega_0 - (\omega_c - \omega_0) = 2\omega_0 \tag{2.9.2}$$

since this is consistent with the notion above of the bandwidth of $|F(\omega)|$ for $\omega > 0$.

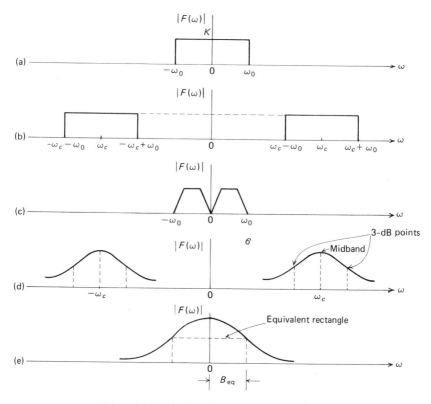

Figure 2.9.1   Various definitions of bandwidth.

This idea of bandwidth can be extended to any $|F(\omega)|$ that has a clear beginning and end, even though it may not be a rectangle, but some other shape. Thus we often consider the bandwidth of the $|F(\omega)|$ of Fig. 2.9.1c [human speech has this kind of $|F(\omega)|$] as $B = \omega_0$, and not many people seem to object to this.

Some do, however, remarking that the triangular portion of the curve attenuates the frequencies of the incoming signal there [if $|F(\omega)|$ belongs to a filter's system function], and this should be taken into consideration.

To account for this, a second definition of bandwidth exists, called **equivalent bandwidth**. This is defined as

$$B_{eq} = \frac{\int_{-\infty}^{+\infty} |F(\omega)|\, d\omega}{2\,|F(\omega)|_{max}} = \frac{\int_{0}^{+\infty} |F(\omega)|\, d\omega}{|F(\omega)|_{max}} \qquad (2.9.3)$$

which is a little more complicated than the one before—not much, though. If the reader would think a little about (2.9.3), he or she will see the meaning easily. The numerator of the last equality is the area under $|F(\omega)|$ for $\omega > 0$. Dividing that by $|F(\omega)|_{max}$, you end up with the width of a rectangle with the same area. So, all this definition says is that the bandwidth of an $|F(\omega)|$ is equal to the width of a rectangle that has the same area; it is taking the $|F(\omega)|$ and *rectanglizing* it, so to speak (see Fig. 2.9.1e). That is why it is given the name "equivalent," meaning equivalent to a rectangle. It is a good way to satisfy the objection. Furthermore, it can be used even in cases when there is no clear beginning or end in $|F(\omega)|$, such as the one shown of Fig. 2.9.1e. That and any other shape can be "rectangularized" [if $|F(\omega)|$ is integrable, anyway] and the definition is fairly acceptable.

There is a third way to define bandwidth, which is quite popular, particularly in the field of filter theory. This one defines the width as the frequency interval over which the $|F(\omega)|$ stays above $1/\sqrt{2}$ times its value at midband. We assume that the reader is familiar with this one from courses in electronics and omit further discussion here. This bandwidth is often called **3-dB bandwidth**, for reasons with which the reader should be quite familiar.

Does the discussion above complete the subject? Hardly. There are still quite a few definitions left. Two of them, the **root-mean-square** (rms) and the **effective** bandwidth, are actually quite important in the study of radar systems. Nevertheless, we will omit them here as we feel that they are better understood after the coverage of probability theory. So the above will have to do us for the time being.

What about time duration? Do the definitions above extend to the time domain as well? Yes, but without the restriction of looking at $f(t)$ for $t > 0$ only, since $f(t)$'s are not often symmetric about $t = 0$. This means that in the case of a rect (t/T), its duration is T and not $T/2$. Equivalent duration of an $f(t)$ is defined as

$$\tau_{\text{eq}} = \frac{\int_{-\infty}^{+\infty} f(t)\, dt}{[f(t)]_{\text{max}}} \qquad (2.9.4)$$

and has exactly the same meaning, but in the time domain. The other two (rms and effective) are also applicable in time, as we shall see when their turn arrives.

Now we are ready to prove an important result that connects duration and bandwidth, called the **uncertainty principle**. It is a result that can be proved using all the definitions of bandwidth and duration,[4] but which we will prove here only for B_{eq} and τ_{eq}. It states that

$$B_{\text{eq}}\tau_{\text{eq}} \geq \pi \qquad (2.9.5)$$

that is, there is a lower bound to the product of the two, meaning that you cannot decrease both of them at the same time beyond a certain point. After this point, decreasing one causes an increase in the other so that their product remains constant. We had actually suspected as much when we were discussing the scaling property of the FT, for $f(at)$ was causing an $F(\omega/a)$ effect in the frequency domain. But now we see it in an official expression with a fancy name attached to it.

[4]Other definitions of B and τ lead to different expressions of the principle, although with the same meaning.

Let us consider proving the form of the uncertainty principle defined by (2.9.5). We must prove that

$$B_{eq}\tau_{eq} = \frac{\int_{-\infty}^{+\infty} |F(\omega)|\, d\omega \int_{-\infty}^{+\infty} f(t)\, dt}{2|F(\omega)|_{max} \qquad [f(t)]_{max}} \geq \pi \qquad (2.9.6)$$

To simplify matters a little, we assume that the $|F(\omega)|$ is such that its maximum value is $\omega = 0$. If it is not, it can be shifted there (by multiplying it with the appropriate exponential) without greatly affecting the proof.

With this assumption and in view of the FT integral,

$$F(0) = \int_{-\infty}^{+\infty} f(t)\, dt = |F(\omega)|_{max} \qquad (2.9.7)$$

Now the inverse FT integral is

$$f(t) = \frac{1}{2\pi} \int_{-\infty}^{+\infty} F(\omega) e^{j\omega t}\, d\omega \qquad (2.9.8)$$

and using the well-known property from calculus that

$$\left| \int f(t)\, dt \right| \leq \int |f(t)|\, dt \qquad (2.9.9)$$

(which makes sense if we recall that such integrals represent the areas under the signals) on both sides of Eq. (2.9.8), we obtain

$$|f(t)| \leq \frac{1}{2\pi} \int_{-\infty}^{+\infty} |F(\omega)|\, d\omega \qquad (2.9.10)$$

since the absolute value of the exponential is unity. Equation (2.9.10) holds for all t, and therefore holds for the t that makes the $f(t)$ maximum. Assuming that this maximum is positive (it usually is), we have

$$[f(t)]_{max} \leq \frac{1}{2\pi} \int_{-\infty}^{+\infty} |F(\omega)|\, d\omega \qquad (2.9.11)$$

If we now use Eqs. (2.9.7) and (2.9.11) on (2.9.6), we easily obtain the uncertainty principle of Eq. (2.9.5).

The reader might wonder at this point why we bother so much about definitions of bandwidth and duration, as well as their relationship expressed by the uncertainty principle. Why, because we are going to need them, of course. Only mathematicians define and prove things so that they can sit back in an armchair and stare at the result with admiration (may we be forgiven for this remark). Here, in a background chapter, everything we do will be needed later. The uncertainty principle is very fundamental in understanding the functions of a radar system and will be seen again in the appropriate chapters. In fact, even the next section is better understood with the concepts of this section in mind.

Example 2.9.1

Find the equivalent duration of the signal $f(t) = e^{-|t|}$.

Solution. Since

$$\int_{-\infty}^{+\infty} e^{-|t|} \, dt = 2 \int_{0}^{\infty} e^{-t} \, dt = 2$$

and in view of the fact that the maximum of $f(t)$ is unity, $\tau_{eq} = 2$ (in time units).

2.10 CORRELATION (AMBIGUITY) FUNCTIONS

We went to a lot of trouble in Section 2.4 to define signal vector spaces with IP, N, and D, and we have a feeling that their full weight has not been appreciated by the reader. In fact, we suspect that he or she may have skipped that section, figuring that its only use is in deriving the Fourier series, which he undoubtedly had seen before, without the need of vector spaces and their abstractions.

Well, such notions should be dispelled quite quickly. The concepts we shall develop here are much better understood if Section 2.4 is well studied. Furthermore, without a good understanding of correlation (or ambiguity) functions, modern radar and communication systems cannot be fully analyzed.

We remarked in Section 2.4 [just below Eq. (2.4.18)] that the distance and the inner product between two signals in a vector space are measures of their similarity, although in a numerically opposite way. This was not idle talk. It can actually be shown mathematically. If we use the distance of Eq. (2.4.18), square, and expand the integral out, we obtain

$$d^2(f(t), g(t)) = \int f^2(t) \, dt + \int g^2(t) \, dt - 2 \int f(t)g(t) \, dt \qquad (2.10.1)$$

A good look at this last expression will show what we are talking about. For the distance squared (and therefore the distance) to be zero, the last term (the IP) must be maximum, something that happens when $f(t) = g(t)$. So there is no doubt about the accuracy of the statement, as far as it goes, anyway. Distance and inner product measure the similarity between two signals. You can call this similarity correlation if you like, or even ambiguity, except that in this last case you reverse the meaning (ambiguity is dissimilarity).

A minute ago, we used the phrase "as far as it goes, anyway." What do we mean by that? We mean that (2.10.1) does not go far enough. Consider, for example, the two signals,

$$f(t) = \text{rect}\left(\frac{t}{2}\right) \qquad (2.10.2)$$

$$g(t) = f(t - 10) \qquad (2.10.3)$$

which are exactly the same in shape, but differ in location; $g(t)$ is $f(t)$ shifted to around $t = 10$. Now if we put them in a vector space with the usual expressions for IP and D, we evaluate their IP to be zero, and their distance the maximum possible value. Conclusion? The $f(t)$ and $g(t)$ are orthogonal (completely dissimilar) to each other. A hard conclusion to swallow, when we know that they are identical, except for the shift.

So the usual D and IP do not go far enough; they cannot account for shifts. What can we do about it? Extend them, of course. Instead of using them only once, use them for all values of shift τ of $g(t)$ with respect to $f(t)$. Define, in other words, a distance that is a function of the shift τ, and end up with something that tells you how similar they are for every possible shift of one of them, with respect to the other. Now if you get zero (for the distance) for some τ, you know that they are equal to each other for such a value of the shift. Needless to say, when the distance is zero for a certain τ, the IP will be maximum for the same τ, and vice versa. Mathematically, this means that our new distance will be

$$D(\tau) = d(f(t), g(t - \tau)) = \int [f(t) - g(t - \tau)]^2 \, dt \qquad (2.10.4)$$

and the inner product,

$$(f(t), g(t - \tau)) = \int f(t)g(t - \tau) \, dt \qquad (2.10.5)$$

If we now take the new distance and expand it out, we will still get an expression like (2.10.1), with the same interpretation, except that the shift τ enters into it, of course.

We have now arrived at the point we wanted. We will have no more to say about distance in this section; we will stick to the shifted inner product entirely. Now that we understand how it came about and what it really means, let us follow tradition and rename it, and then study some of its properties. It is of utmost importance in future chapters.

We define the **time cross-correlation function** $R_{fg}(\tau)$ between two energy signals (finite norm) $f(t)$ and $g(t)$ by the expression

$$R_{fg}(\tau) = \int_{-\infty}^{+\infty} f(t)g(t - \tau) \, dt \qquad (2.10.6)$$

For each value of shift τ [of $g(t)$ versus $f(t)$], $R_{fg}(\tau)$ is an inner product and therefore it tells us how much $f(t)$ resembles $g(t)$ for such a shift. To make this meaning even more concrete, we can even normalize it so that its maximum value is unity (perfect match). This is done by dividing it with the product of the norms of $f(t)$ and $g(t)$, and renaming it into a *cross-correlation coefficient*. But we will not bother with this at this point, as it is not necessary for our future needs.

If $f(t)$ and $g(t)$ are power signals, this expression has to change a bit to take into account that their energy is infinite. For such signals, then,

$$R_{fg}(\tau) = \lim_{T \to \infty} \frac{1}{T} \int_{-T/2}^{T/2} f(t)g(t - \tau) dt \qquad (2.10.7)$$

and if the power signals are periodic, so will be the $R_{fg}(\tau)$, and the limit operation can be skipped by integrating only over its period (check the details on this). By the way, if $R_{fg}(\tau)$ is everywhere zero, the signals are called **uncorrelated** (in time, of course).

Example 2.10.1

Find and interpret the time cross-correlation function of the signals defined by Eqs. (2.10.2) and (2.10.3).

Solution.   Let us shift $g(t)$ with respect to $f(t)$, that is, find

$$R_{fg}(\tau) = \int_{-\infty}^{+\infty} f(t)g(t - \tau)\, dt \qquad (2.10.8)$$

For $\tau > 0$ (shifts to the right), $R_{fg}(\tau) = 0$; so it is for shifts to the left up to $\tau = -8$. Then the IP begins to give values (plotted in Fig. 2.10.1c) until at $\tau = -10$ we get the maximum value. After that it tapers off to zero at $\tau = -12$, staying that way for all values of $\tau < -12$.

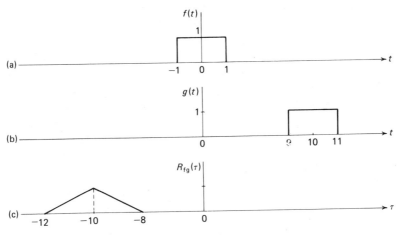

Figure 2.10.1   Cross-correlation.

The interpretation of Fig. 2.10.1c is as follows. The two signals are like each other if $g(t)$ is shifted to the left $8 - 12$ units of time. They resemble each other most at $\tau = -10$. We cannot possibly have many objections about these conclusions, since they agree with our common sense.

Incidentally, it can be easily shown by a change of variable that

$$R_{fg}(\tau) = \int_{-\infty}^{+\infty} f(t + \tau)g(t)\, dt \qquad (2.10.9)$$

which means that you get the same thing by shifting $f(t)$ with respect to $g(t)$ but in the opposite direction.

It should also be noted that the integral (2.10.6) is somewhat *reminiscent* of a convolution, but it *really is not* (integration here is on t, not on τ). Nevertheless, there is a relationship between the two, given by

$$R_{fg}(\tau) = f(t)*g(-t)\Big|_{t=\tau} \tag{2.10.10}$$

which we will have occasion to use later. (Prove it as part of Problem 2.36.) It simply states that you can obtain the $R_{fg}(\tau)$ by filtration of $f(t)$ with a filter whose impulse response is $g(-t)$. For each value of shift, you look at the output at the same value of time. So much for the time cross-correlation function between two signals.

The time has come to pose an innocent little question, which leads us to some unusually interesting results. What happens in the $R_{fg}(\tau)$ if $f(t) = g(t)$? Mathematically, of course, it is easy to answer. The result will be

$$R_{ff}(\tau) = \int_{-\infty}^{+\infty} f(t)f(t-\tau)\, dt \tag{2.10.11}$$

with similar expressions for power or periodic signals. But mathematics is one thing, and meaning and application are another. This is where the worthwhile results enter the picture.

Expression $R_{ff}(\tau)$ [which from now on we will denote by $R_f(\tau)$] is called the **time autocorrelation** (or **ambiguity**) **function** of $f(t)$ and it represents one of the big notions in the field. In view of this, we will devote most of the rest of the section to studying it. Full appreciation of its importance, however, will not come in this section, but later in the applications, starting with Chapter 4.

We start with a few of the mathematical properties of $R_f(\tau)$ which bear on its meaning and applications. We will discuss them with reference to energy signals, but they also hold for power signals as well, with the appropriate changes dictated by the definitions. Figure 2.10.2 illustrates the results in the case of a rectangular pulse.

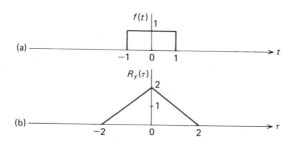

Figure 2.10.2   A pulse and its $R_f(\tau)$.

(1)
$$R_f(0) = \int_{-\infty}^{+\infty} f^2(t)\, dt \tag{2.10.12}$$

This property (easily proved by setting $\tau = 0$) means that its value at $\tau = 0$ provides us with the energy of the signal (the square root of the norm)—a good thing to know.

(2) $$R_f(\tau) = R_f(-\tau) \tag{2.10.13}$$

This property can also be easily proved by simple algebra. It also makes good sense. It should not make much difference which way you shift a signal with respect to itself. The resulting inner product should be the same. Convince yourself of this by working some examples. The property is quite important for many reasons, one of which is that it forces its FT always to be real. (Why?)

(3) $$R_f(0) \geq R_f(\tau) \qquad \text{for every value of } \tau \tag{2.10.14}$$

This property, which is tricky to prove, tells us simply that a signal resembles itself maximally, when there is no shift. If this one did not hold, we might just as well pack up our things and go home. The reason that we have the equal sign is to take care of cases [such as periodic $f(t)$'s] when $f(t)$ can be shifted a certain amount and look like itself again. Try proving it as an exercise, but we warned you—it is tricky.

(4) $$\mathfrak{F}(R_f(\tau)) = |F(\omega)|^2 = S_f(\omega) \tag{2.10.15}$$

This property, which tells us that its FT is the energy density spectrum $S_f(\omega)$ of $f(t)$, is often called the **Wiener–Khintchine theorem**, and it is quite important. It gives us another way to find $S_f(\omega)$ (always real and positive), and if the signal is a power signal, the result is its power spectral density, which is hard to find. To prove it, just go at it with brute force.

$$\overbrace{\mathfrak{F}(R_f(\tau)) = \int_{-\infty}^{+\infty}}^{R_f(\tau)} \int_{-\infty}^{+\infty} f(t)f(t - \tau)\, dt \; e^{-j\omega\tau}\, d\tau \tag{2.10.16}$$

Now change the variable $t - \tau = x$, make sure that no mathematician is looking at you and you get

$$\mathfrak{F}(R_f(\tau)) = \int_{-\infty}^{+\infty} f(t)e^{-j\omega t}\, dt \int_{-\infty}^{+\infty} f(x)e^{j\omega x}\, dx$$

$$= F(\omega)F(-\omega) = |F(\omega)|^2 \tag{2.10.17}$$

(5) $$R_f(\tau) = f(t)*f(-t)\Big|_{t=\tau} \tag{2.10.18}$$

This property is a simple extension of Eq. (2.10.10) and gives us a way to get $R_f(\tau)$ by passing $f(t)$ through a filter with impulse response $h(t) = f(-t)$. Such a filter is often a shifted version of the *matched filter* of $f(t)$ that we derived in Section 2.8. Thus finding $R_f(\tau)$ will result in enhancing the signal in the presence of additive noise as we will see again in later chapters. We can see this approximately even now. Finding the $R_f(\tau)$ provides us with a neat maximum, which is higher in value than any of the values of $f(t)$ (see Fig. 2.10.2) and easier to see in the presence of noise. Of course, one need not use Eq. (2.10.18) to find $R_f(\tau)$. There exist electronic correlators which shift, multiply, and integrate quite rapidly for all shifts and give it to you in a jiffy. Still, whatever you do, *it is equivalent to Eq. (2.10.18)*.

All these things are very interesting and quite important, and the reader should pause and ponder the consequences. The time ambiguity function defines for us the way that an $f(t)$ resembles itself when shifted. This also tells us a bit about how narrow the signal is. A narrow signal (an impulse, for example) will have a narrow $R_f(\tau)$, with a sharp maximum. A smooth signal of long duration will have a wide $R_f(\tau)$ with a barely defined maximum. In fact, $R_f(\tau)$ is usually twice as long in duration as the signal itself (the start–end type of duration of the preceding section), and therefore it *accentuates* the notion of duration of the signal. If $R_f(\tau)$ is *narrow*, who can disagree that $f(t)$ is *very narrow*? And vice versa?

Everything we have said so far generalizes to signals of other parameters, such as space coordinates and the like. Of particular importance is their generalization to functions of frequency ω (i.e., Fourier transforms, spectral densities, etc.). Consider the FT $F(\omega)$ and $G(\omega)$ of two signals $f(t)$ and $g(t)$, respectively. The **frequency cross-correlation function** $\mathcal{H}_{fg}(v)$ is defined as

$$\mathcal{H}_{fg}(v) = \int_{-\infty}^{+\infty} F(\omega)G(\omega-v)\, d\omega \qquad (2.10.19)$$

and carries the same meaning as before, in a frequency-shifting way.

The **frequency ambiguity function** (or **autocorrelation function**) of a signal $f(t)$ is now

$$\mathcal{H}_f(v) = \int_{-\infty}^{+\infty} F(\omega)F(\omega - v)\, d\omega \qquad (2.10.20)$$

and has the same properties as $R_f(\tau)$, except that now they manifest themselves in frequency shifts. Naturally, the uncertainty principle reminds us that if $R_f(\tau)$ is wide, expect $\mathcal{H}_f(v)$ to be narrow, and vice versa.

So now we can study shifts in time and frequency by using the functions we defined in this section. We should also mention that the two expressions can actually be combined into one general expression (called general ambiguity function) which gives $R_f(\tau)$ and $\mathcal{H}_f(v)$ as subcases for $v = 0$ and $\tau = 0$, respectively. We will come across this later.

Example 2.10.2

Find the ambiguity function of a delta function $\delta(t)$.

Solution.   We are including this example since we mentioned the case a little earlier. How do we prove that $R_f(\tau) = \delta(\tau)$?

We can use the convolution property of the impulse [see (2.3.10b)] in conjunction with Eq. (2.10.8), and the result is immediate. The same result holds for an impulse in ω, $\delta(\omega)$. The math bears no discrimination against parameters.

Now it is the reader's turn to do some work, by trying out some of the problems at the end of the chapter. He or she will soon discover that except for those involving pulses, the rest are not at all easy. But after all, neither is convolution.

PROBLEMS, QUESTIONS, AND EXTENSIONS

2.1. Consider the signal $f(t) = e^{-t}u(t)$ where $u(t)$ a step function.
  (a) Find $f(t + 1), f(3 - t)$, and $f(5t + 2)$ and sketch them.
  (b) Find the total energy and average power for $f(t)$ and each signal in part (a). What do you observe about the energy of the first three signals?
  (c) Now let us generalize somewhat. Assume that we start with $f(t)$ and convert it to $g(t) = af(bt)$. If $f(t)$ is an energy signal, find the total energy of $g(t)$. If $f(t)$ is a power signal, find the total average power of $g(t)$.

2.2. Sketch the signal $f(t) = 2u(t + 1) - 3u(t - 1) + u(t - 2)$. Then sketch $f(5 - t)$, $f(3t - 2)$, and $f(2 - 5t)$.

2.3. Recall that the convolution of two signals $f(t)$ and $h(t)$ is generally given by

$$o(t) = f(t)*h(t) = \int_{-\infty}^{+\infty} f(\tau)h(t - \tau)\, d\tau = \int_{-\infty}^{+\infty} h(\tau)f(t - \tau)\, d\tau$$

Recall also that if $f(t)$ is the input to a linear system with impulse response $h(t)$, then $o(t)$ is the output. Now assume that $f(t) = 0$ for $t < 0$ and that $h(t)$ is causal [i.e., $h(t) = 0$ for $t < 0$]. Find the new limits on the convolution integral.

2.4. Consider the signals $f(t) = u(t - 1) - u(t - 2)$ and $f(t - 5)$ (i.e., two rectangular pulses). Find their convolution. You should end up with a triangle. Where is it located? Devise a quick scheme for knowing the location of the triangle.

2.5. As mentioned in Section 2.3, there are a plethora of generalized function properties. Try proving the ones below.
  (a) Can you prove Eq. (2.3.10b)? Where lies the difficulty? (If you cannot prove it, accept it as a definition, as many people do.)
  (b) $\int_{t_1}^{t_2} \delta(\lambda - t)\, \delta(\lambda - t_0)\, d\lambda = \delta(t - t_0), \qquad t_1 < t_0 < t_2$
  (c) $f(t)\, \delta'(t) = f(0)\, \delta'(t) - f'(0)\, \delta(t)$

2.6. Use the properties of generalized functions to evaluate the following integrals.
  (a) $\int_0^\infty (t^2 - 3t - 1)\, \delta'(t - 1)\, dt$
  (b) $\int_0^\infty \delta(t - 3) \cos(\omega(t - 3))\, dt$
  (c) $f(t) * (\delta(t - 1) + \delta(t) + \delta(t + 1))$
  (d) $f(t) * \sum_{n=-\infty}^{+\infty} \delta(t - nT) = f(t)*\text{comb}_T$

2.7. Prove that the impulse is the derivative of a unit step and that the unit step is the derivative of a ramp.

2.8. Show that all continuous signals in $(-\infty, +\infty)$ form a vector space.

2.9. Show that Eq. (2.4.6) is an inner product.

2.10. Prove that Eq. (2.4.12) gives a norm in any vector space, thereby proving that Eq. (2.4.11) is acceptable. Then try to prove the celebrated Cauchy–Schwartz inequality, which is in general

$$(X, Y)^2 \leq \sqrt{N(X)}\sqrt{N(Y)}$$

or, for the signal space,

$$\left[\int f(t)g(t)\,dt\right]^2 \le \int f^2(t)\,dt \int g^2(t)\,dt$$

The proof is sort of tricky. If you cannot do it, look it up in a reference book [Javid and Brenner (1963, p. 130), for example]. We will have occasions to use this inequality in the future.

2.11. Prove that the expression (2.4.16) always provides us with a distance.

2.12. Can the expression

$$\int_0^\infty |f(t) - g(t)|\,dt$$

serve as a distance instead of the one given in Eq. (2.4.18)? Prove it.

2.13. Prove the converse of Theorem 2.4.1 (the orthogonality criterion).

2.14. Prove the Parseval identity, Eq. (2.4.29), under the conditions stated above the equation.

2.15. Two signals, $f(t)$ and $g(t)$, defined in $(0, \infty)$, have expansions

$$f(t) = \sum_{i=1}^n C_i\phi_i(t)$$

$$g(t) = \sum_{i=1}^n B_i\phi_i(t)$$

Prove that under certain conditions (which you must find)

$$\int_0^\infty f(t)g(t)\,dt = \sum_{i=1}^n C_iB_i$$

an equation usually called the *generalized Parseval identity* [note that if $g(t) = f(t)$, you have Eq. (2.4.29)].

2.16. Find the basis called the Walsh functions (in a book, of course), and ponder them for a while. They are quite important.

2.17. Show that if $f(t)$ is a real-valued function of t, and periodic with period T, its amplitude spectrum is an even function of ω and its phase spectrum an odd one.

2.18. Assume that a periodic function $f(t)$ is even [i.e., $f(t) = f(-t)$]. Find the effect that this has on the coefficients of its trigonometric and exponential FS expansion. Repeat the problem with an $f(t)$ that possesses odd symmetry [$f(t) = -f(-t)$].

2.19. The signal

$$f(t) = \begin{cases} 1 - \dfrac{|t|}{T_p} & \text{for } |t| \le T_p \\ 0 & \text{elsewhere} \end{cases}$$

is a triangle (sketch it). Now assume that you have

$$g(t) = f(t)*\text{comb}_T \qquad (T = 10T_p)$$

where the asterisk denotes convolution. Find the F_n FS coefficients of $g(t)$. How are they related to the F_n's of a periodic pulse with half the duration, but with the same period, also centered around $t = 0$?

2.20. The FS coefficients of a periodic signal $f(t)$ are given by

$$F_n = \frac{1}{|2n|} e^{-j \tan^{-1}(n/2)}$$

Find and sketch $|F_n|$, angle (F_n), and the power spectrum. Find the percent of total power in the third and fourth harmonics.

2.21. Let $f(t) = \sum_{n=-\infty}^{+\infty} F_n e^{jn\omega_0 t}$. This signal enters a linear system with transfer function $H(\omega)$. The output is denoted by $g(t)$. Find an expression for the FS coefficients of $g(t)$ in terms of F_n and $H(\omega)$.

2.22. Prove Properties 8 and 9 of Table 2.7.1.

2.23. Find the FT of $f(t) = s(t) + s(t - 4)$, where $s(t) = \text{rect}(t)$. Sketch all three spectra.

2.24. Find the FT $F(\omega)$ of the periodic pulse train defined in Example 2.5.2. Sketch $|F(\omega)|$ for the same values of period and duration as in the example. What do you conclude?

2.25. Show that if the FT of $f(t)$ is such that $F(\omega) = 0$ for $\omega > \omega_c$ and $\omega < -\omega_c$ [such an $f(t)$ is called *bandlimited up to* $\omega = \omega_c$], then $f^2(t)$ is bandlimited up to $2\omega_c$.

2.26. If $f(t) = \text{Sa}^2(at)$, show by using the symmetry property of the FT that $F(\omega)$ is triangular in shape. Add this signal to Table 2.6.1 if you so wish.

2.27. For $f(t)$ as given in Problem 2.26, find the FT of $g(t) = f(10 - 3t) \sin 10^6 t$, and sketch $|G(\omega)|$.

2.28. A signal $f(t)$ with FT

$$F(\omega) = u(\omega + 1) - u(\omega - 1) = \text{rect}\left(\frac{\omega}{2}\right)$$

goes through a series of devices. The first device has as output

$$G(\omega) = \frac{1}{2j}[F(\omega - \omega_c) - F(\omega + \omega_c)]$$

the second

$$Y(\omega) = G(\omega)e^{-j10\omega}$$

and the third

$$X(\omega) = G(10\omega)e^{-j100\omega}$$

Identify what the devices do in the time domain (scaling, shifting, etc.).

2.29. Derive the FT of $f(t) = \text{comb}_T$.

2.30. Find the time cross-correlation function of $f(t) = e^{-3t}u(t)$ and $g(t) = e^{-2t}u(t)$.

2.31. Prove that the time ambiguity function obeys $R_f(0) \geq R_f(\tau)$ for all τ.

2.32. Find $R_f(\tau)$ if $f(t) = e^{-t}u(t)$.

2.33. Find the $R_f(\tau)$ of the signal in Problems 2.23 (two pulses) and 2.24 (infinite pulse train). They will be useful in our discussion of radar systems.

2.34. Find the energy density spectrum of a rect (t/T) by the use of the Wiener–Khintchine theorem. Now repeat for a rect $((t - 5)/T)$. What does this tell you about time ambiguity functions as regards shifts in the signal?

2.35. To strengthen your belief in the usefulness of correlation functions, consider the following two detection problems. Data are coming in and have the form $O(t)$

$= S(t) + N(t)$, where $S(t)$ is periodic with period T, and $N(t)$ is noise, uncorrelated with periodic signals [i.e., $R_{SN}(\tau) = 0$ for all τ]. However, we are not sure if $S(t)$ is present in $O(t)$, or we are receiving only noise, as the noise is high and completely covers $S(t)$ when it is there.

(a) Find the cross-correlation function of $O(t)$ with a locally generated periodic signal $C(t)$ [of the same period as $S(t)$]. What do you conclude from the result? How can we tell whether $S(t)$ is present in $O(t)$?

(b) Assume now that $N(t)$ is such that its $R_N(\tau)$ is zero for $\tau > 10$ (and $\tau < -10$). Now try to devise a detection scheme based on finding the time ambiguity function of the incoming data $O(t)$. For this scheme you do not need to know the period of $S(t)$ as you do in part (a).

2.36. Prepare a small lecture on the matched filter and its relation to autocorrelation functions. Note the point at which the maxima occur in each case. Give some examples that will familiarize you with the finding of matched filters and time ambiguity functions.

=3=
AMPLITUDE MODULATION: COMMUNICATION SYSTEMS

3.1 INTRODUCTION

A primary function of most communication systems is to transmit the human voice over long distances. The human voice (including music) can easily be converted into an electrical signal by using a microphone. The resulting signal appears to have an amplitude spectrum with a maximum frequency of about 15 kHz, and even if it is higher, the human ear seems incapable of detecting it.

Let us assume that we have such a signal and wish to send it over the air to a faraway location. To do so we would have to convert it again, this time to an electromagnetic wave that is capable of traveling through the channel of space. This would be done by an antenna, and therefore the antenna would *match* the signal to the channel. However, things are not as simple as that. A direct adaptation of the human voice signal to the space channel is not possible, unless the antenna is many miles long, a rather impractical situation. The antenna size turns out to be related to the smallest frequency of the signal in an inversely proportional way. A reasonable-size antenna could adapt the signal to the channel only if the signal's lowest frequency were higher than 800 kHz, a very high value compared to the actual lowest frequency of human voice, which is approximately 30 Hz. Are we stuck?

Not quite. True, we cannot (for obvious practical reasons) make a huge antenna. But we have another alternative. We can raise the frequency content of the speech signal to the vicinity of the 800 kHz, or, in more technical terms, we can translate the frequency spectrum of the signal to the desired vicinity.

Such an operation is possible, as we already know, by multiplying the voice signal with a sinusoid (see Section 2.7). If the frequency of the sinusoid is higher than 800 kHz, the resultant signal has a frequency spectrum in this region. This way, the signal is well adapted to the channel, and proper transmission can be accomplished.

Frequency translation of a signal spectrum by multiplication (in the time domain) of the signal with a sinusoid is called **amplitude modulation**. The term comes from the following reasoning, which involves a rather cheerful explanation of the process of matching the voice signal to the space channel.

Let $f(t)$ denote the speech signal whose frequency content is in the region 30 to 15,000 Hz. We noted above that $f(t)$ is incapable of going through the (reasonable-size) antenna, and then the channel. Well, if you cannot cross something yourself, the only reasonable course to follow is to find someone who can, and latch onto this person (if you cannot swim through a river, climb on the back of someone who can, or hire a boat etc.). Such a candidate is a sinusoid, with frequency of say 850 kHz, since its frequency content (an impulse at this frequency) is just right for both the antenna and the air channel. Now that we have found the signal (the *carrier*) that can carry our $f(t)$ through, the next order of business is to find a way for $f(t)$ to latch onto it, so that they can travel the course together. Let us see about that next.

A sinusoid has the form $A \cos(\omega_c t + \theta)$, a fact which means that the three parameters A, ω_c, and θ define it uniquely. These three parameters are also the ones on which we can place our $f(t)$. If we multiply $f(t)$ with the sinusoid, the result is $A f(t) \cos(\omega_c t + \theta)$, and this can be interpreted as the old sinusoid with a new amplitude $[A f(t)]$. In other words, the upshot of the multiplication is to latch our $f(t)$ onto the amplitude of the sinusoid, or to *modulate* the original amplitude of the sinusoid with our signal. This procedure is called *modulating the amplitude of a sinusoid*, and the various systems that use it to cross a channel are called **amplitude modulation systems**. Of course, it is fairly obvious that $f(t)$ can also get placed on ω_c or even θ, and this, when done, gives rise to **frequency** or **phase modulation systems**, respectively. But this is not the right place to discuss these issues, which we postpone for a later chapter.

This rather human-centered explanation tells us the reason for the name "amplitude modulation," and even the term "carrier" for the sinusoid. But, of course, it does not prove anything. After all, if you try to climb on the back of a good swimmer in order to cross a river, you may both sink and drown. Proof that such systems work will come later using established mathematical procedures. Furthermore, there are variations of the scenario above which are not easily explainable in terms of such stories, as we shall soon see.

This chapter is devoted to communication systems that make use of frequency translation or amplitude modulation. Radar systems that use the same principle will be covered in Chapter 4. We should emphasize that all our systems will be analyzed without taking noise into consideration, a plan chosen for pedagogical reasons. Noise and its effects on communication and radar systems will be taken up after such systems are first well understood.

Our presentation in this chapter and the rest of the text is from the *systems approach* point of view. Wave propagation and its effects on the systems is mostly ignored, on the assumption that it is covered elsewhere in an electrical engineer's educational program. The same assumption has caused us to de-emphasize the aspect of communication circuits and devices, that is, the practical units of a system that perform multiplication, addition, envelope detection, synchronization, and the like. Some important such components are presented throughout this text in the form of problems at the end of most chapters, but mostly on a subsystem point of view, not on the basis of its actual circuit configuration.

3.2 THE DOUBLE-SIDEBAND SUPPRESSED CARRIER SYSTEM

The first communication system we shall consider goes by the fancy name "double-sideband suppressed carrier system" (DSB). The reasons for this name will become apparent during the analysis.

A block diagram of such a system is shown in Fig. 3.2.1. This block diagram represents the key functions of the system, the functions that make it the system that it is. Amplifiers, filters, transmission lines to and from the antennas, and the antennas themselves are all considered ideal and omitted from the diagram. The channel is also considered ideal (no noise, no distortion, some loss of amplitude), and as such it is modeled as an ideal bandpass filter (BPF) with a band $[\omega_1, \omega_2]$ (see also Fig. 3.2.2).

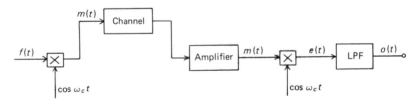

Figure 3.2.1   Block diagram of DSB communication system.

The signal $f(t)$ represents the information signal (usually, human voice with music) that must be transmitted. Our analysis here will take the following form. We wish to show that this system works. This means that we must start with $f(t)$ and follow it through the various devices (blocks) until it finally reaches the destination and becomes $o(t)$. The signal $o(t)$ must be shown to be equal to $f(t)$ or an acceptable facsimile of it. The tracing of $f(t)$ through the blocks will be accomplished by the known methods of signal analysis, in the time or frequency domains, whichever suits our purpose.

We are now ready to plunge into the analysis.

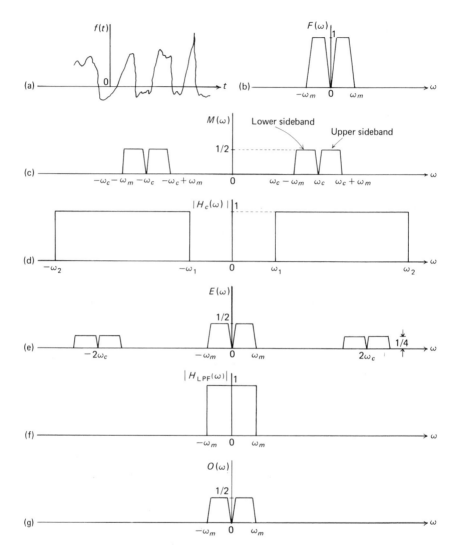

Figure 3.2.2 Pictorial analysis of DSB system: (a) $f(t)$; (b) $F(\omega)$; (c) $M(\omega)$; (d) the channel's $|H(\omega)|$; (e) $E(\omega)$; (f) the LPF's $|H(\omega)|$; (g) the final output $O(\omega)$.

The signal $f(t)$, we repeat, is the information-carrying electrical signal of the human voice, the output of a microphone. Its time representation is something like the signal shown in Fig. 3.2.2a. Its Fourier transform $F(\omega)$ will be assumed real (see Fig. 3.2.2b) so that we can sketch it and follow the analysis with FT sketches as well as mathematics. Its maximum frequency ω_m can be taken as 15,000 Hz, although it can be reduced to as low as 3000 Hz if only speech (no singing) is being transmitted.

The first operation is the multiplication of $f(t)$, with the carrier $\cos \omega_c t$, whose frequency ω_c lies well within the channel's band. The result in time is

$$m(t) = f(t) \cos \omega_c t \qquad (3.2.1)$$

and it is called the amplitude modulation of the carrier. The FT of $m(t)$ can easily be found with the use of Eq. (2.7.9):

$$M(\omega) = \tfrac{1}{2}[F(\omega - \omega_c) + F(\omega + \omega_c)] \qquad (3.2.2)$$

and it is sketched in Fig. 3.2.2c. Expression (3.2.2) can also be found by using the frequency convolution property of the FT (Property 6, Table 2.7.1).

A careful look at Fig. 3.2.2c and d shows that $m(t)$ is indeed in a form suitable for transmission through the reasonable-size antenna, and the space channel. The original low-frequency content of $f(t)$ has been shifted inside the channel's band, and therefore $m(t)$ will go through it untouched, except for a time delay that we choose to ignore. The act of multiplication of $f(t)$ with a carrier has done the trick at the transmitter, and we can now move on to the side of the receiver.

The signal that arrives at the output of the channel has the time shape of $m(t)$, but it may have suffered some attenuation. The first block is an amplifier which returns it to its original amplitude (or higher, if desired). We now come to the next two operations that make up the **demodulation** process, the recovering of the original $f(t)$ from $m(t)$.

The first one is multiplication by a cosine whose frequency ω_c is the same as the carrier's (i.e., the receiver knows this value). Again using the frequency-shifting property, we obtain

$$E(\omega) = \tfrac{1}{2}[M(\omega - \omega_c) + M(\omega + \omega_c)] \qquad (3.2.3)$$

The two terms inside the brackets can easily be written in terms of $F(\omega)$ by using Eq. (3.2.2). Replacing these expressions into (3.2.3), we find (see Fig. 3.2.2e)

$$E(\omega) = \tfrac{1}{2}F(\omega) + \tfrac{1}{4}[F(\omega - 2\omega_c) + F(\omega + 2\omega_c)] \qquad (3.2.4)$$

The final operation is the passage of $E(\omega)$ through an ideal low-pass filter (LPF), whose cutoff frequency is ω_m, so that only the first term of (3.2.4) can cross it. So

$$O(\omega) = H_{\text{LPF}}(\omega)E(\omega) = \tfrac{1}{2}F(\omega) \qquad (3.2.5)$$

or

$$o(t) = \tfrac{1}{2}f(t) \qquad (3.2.6)$$

and there is no doubt that the $f(t)$ made it to the destination. The fact that it made it with half its original amplitude is of no interest, since there are all sorts of amplifiers in the practical system that can correct it.

So we managed to go through our first communication system's analysis effortlessly. In fact, we feel that the reader could have done it alone, since the only two things needed were a FT property, used twice, and the well-known input/output relationship of a linear system, also used twice (where?).

It is interesting to note that our entire analysis was done in the FT domain.

It is quite valid, of course, since the FT of a signal is unique. Validity, however, is not the real reason. This reason is the ease of obtaining the results and then sketching them for illustration. If the reader is unconvinced, he or she should try sketching $m(t)$ [worse yet $e(t)$] to verify what a thankless task it really is.

Talking about sketching, it should be noted that the sketches of Fig. 3.2.2 pretty well tell the story of the whole analysis, and even though mathematicians may not accept it as a complete proof, most engineers would.[1] Furthermore, these sketches can be obtained easily by making use of the FT properties in a direct manner. For example, $M(\omega)$ is the original $F(\omega)$, shifted to ω_c and $-\omega_c$ (right and left by ω_c) and halved in amplitude. Similarly, $E(\omega)$ is a shifted version of the same kind, but now of $M(\omega)$. When $M(\omega)$ is shifted to the right, it gives a term around $\omega = 0$ and one around $\omega = 2\omega_c$. When it is shifted to the left, it gives another term around $\omega = 0$ and one at $\omega = -2\omega_c$. The two terms around $\omega = 0$ add and produce the $\frac{1}{2}F(\omega)$ term in (3.2.4), the only term that passes through the LPF to give the $O(\omega)$ of (3.2.5). The reader should go through this entire shifting scheme with a different $f(t)$ (say, a pulse that will be useful later in the analysis of radar systems) to ensure that he can shift right, left, and so on, without omitting any of the terms.

Now we come to the explanation of the name of the system. The term "double sideband" is owed to the nature of the FT of $m(t)$, the signal that enters the channel. The result of the first shift produces two terms around ω_c (and $-\omega_c$). The term that sits higher than ω_c is called the **upper sideband**, and that which sits before ω_c the **lower sideband**. Both of them contribute to the term "double sideband." The term "suppressed carrier" is based on the fact that $M(\omega)$ has no cosine term (the carrier sort of swims under the water in crossing the river). If it did, it would have an impulse at $\omega = \omega_c$ (and $\omega = -\omega_c$), and no such impulse appears in the sketch of $M(\omega)$ or in Eq. (3.2.3). So much for the name. Now let us take a look at some subtle points of the analysis which will illuminate a bit more the operation of the system and will be useful in future analyses and comparisons between systems.

The analysis is, of course, valid if $\omega_c \gg \omega_m$ (Why? How much bigger?), but that presents no problems in practice in view of the numbers we have already mentioned.

Another important observation is that the two cosines (at the transmitter and the receiver) must be of exactly the same frequency and with no phase difference (i.e., in complete synchronization). Because of this the DSB receiver is often called a **synchronous** detector. Another name for it is a **homodyne** detector, the term used to differentiate it from the operation of **heterodyning**, which means multiplication by a cosine (or frequency translation) that is not of the same frequency. In any event, it is of interest to see what happens when the receiver cosine is not in phase with the incoming one (phase error), or its fre-

---

[1]A mathematician at The Johns Hopkins University was once asked in class to explain what a proof is. "Why," he answered, "anything that the other fellow will accept."

quency is in error, or both. Since the results are useful and we will refer to them in later sections, let us do them in the form of examples for easier identification and reference.

Example 3.2.1

Find the output of the DSB system if the input is $f(t)$ and the synchronous detector makes an error in phase.

Solution. Let us take a simpler mathematical approach this time (there is more than one way to skin a cat), which we purposely avoided before in order to emphasize the shift-to-the-right-shift-to-the-left-and-half routine.[2]

If the receiver cosine is now $\cos(\omega_c t + \varphi)$, then $e(t)$ is obviously

$$e(t) = f(t) \cos \omega_c t \cos(\omega_c t + \varphi) \tag{3.2.7}$$

Using the known trigonometric expression for the term with the phase error, we have

$$e(t) = f(t) \cos^2 \omega_c t \cos \varphi - f(t) \sin \omega_c t \sin \varphi \tag{3.2.8}$$

(note that when there is no phase error, $\sin \varphi = 0$ and $\cos \varphi = 1$). Now the second term of Eq. (3.2.8) will not pass through the LPF. (Why?) Furthermore, the first term can be written as

$$f(t) \cos^2 \omega_c t \cos \varphi = \cos \varphi \, f(t)(\tfrac{1}{2} + \cos 2\omega_c t) \tag{3.2.9}$$

by using the well-known trigonometric identity for $\cos^2 \omega_c t$. The only term to pass through the LPF is the one sitting around $\omega = 0$ in the FT domain, so

$$o(t) = \tfrac{1}{2} f(t) \cos \varphi \tag{3.2.10}$$

The obvious result is that $o(t)$ has suffered a loss of amplitude equal to $\cos \varphi$. If φ is small, this does not greatly affect the operation of the system. Note, however, that if $\varphi \rightarrow 90°$, the output rapidly becomes zero, which is bad news. But no bad news is without some goodness in it, so let us consider the following train of thought.

What Eq. (3.2.10) tells us is that you cannot detect a DSB wave if the phase difference is 90°. In other words, if the transmitter uses a cosine carrier, you cannot use a sine to demodulate it. This is fine, but what if the transmitter uses a sine carrier? Can you demodulate it with a sine? The answer is yes (do it, and prove it). Finally, what if the pairing is sine in the transmitter, cosine in the receiver? This time the answer is no [no proof needed since $e(t)$ is the same as in the pairing we proved above].

The good news comes directly from the question-and-answer period above. It is obvious (to us anyway) that you can send *two signals* $f_1(t)$ and $f_2(t)$ over the same channel with carriers of exactly the same frequency, as long as you pair them sine–sine and cosine–cosine, respectively. The modulated waves $f_1(t) \cos \omega_c t$ and $f_2(t) \sin \omega_c t$ have amplitude spectra that are sitting on top of each other, yet these two waves can be separated out at the receiver with proper demodulation. To appreciate this good news even further, read the material after the next example (and Section 3.5).

Example 3.2.2

Find and discuss what happens to the output of a DSB system if the synchronous detector makes an error in frequency.

---

[2]This routine also constitutes a basic step in square dancing, we are told, the halving corresponding to bowing low.

Solution. Let us assume that the receiver cosine has the form $\cos(\omega_c \pm \Delta\omega)t$, where $\Delta\omega$ is the error in frequency. Picking up the analysis at around Eq. (3.2.3), our $E(\omega)$ will now be

$$E(\omega) = \tfrac{1}{4}F(\omega \pm \Delta\omega) + \tfrac{1}{4}F(\omega \mp \Delta\omega) + \text{high-frequency terms} \qquad (3.2.11)$$

that is, the two terms around $\omega = 0$ will not add up to produce our $F(\omega)$. To get a better picture of what really happens, the reader should sketch the $E(\omega)$ above for both $\pm\Delta\omega$ (see Problem 3.3). In any event, this type of error (even if small) has a tendency to produce garbled speech, and as such it is much more serious than a phase error. For this reason, a small amount of the carrier is often sent together with $m(t)$, so that the receiver can detect it, amplify it, and use it to demodulate the incoming DSB wave. Actually, there need not be much effort in doing that, as practical multipliers are not ideal and usually do it on their own. In fact, the opposite may need doing (i.e., suppressing the carrier to a small amplitude), since (officially) this is a suppressed carrier system.

Now we come to an issue that will be of concern throughout this text, that of bandwidth. We wish to make the following interesting observation. If the original information-carrying signal has a certain bandwidth B (say $B = \omega_m$ in Fig. 3.2.2b), the signal that traverses the channel $m(t)$ has bandwidth $2B$, or in other words uses up $2B$ of the channel's bandwidth. So what, you say? Is that bad or good?

There is an anecdote about a clergyman who received a member of his parish one evening, named Jasper. Jasper started complaining about his business partner Cosmos, whom he described as a scheming, unreliable so and so, who could not be trusted. "Yeah, you are right, you are right," agreed the clergyman. When Jasper left, in comes Cosmos, who in turn describes Jasper in exactly the same colors, backing up his accusations with incidents. "Yeah, you are right, you are right," consented the clergyman. Upon his departure, the clergyman's wife becomes exasperated. "How can you agree with both of them?" she demanded. "Isn't that completely outrageous?" "You are right, you are right," answered the clergyman thoughtfully.

So it goes with questions about bandwidth, also. Everybody is right and everybody is wrong, depending on his point of view. For the time being, we will say that the doubling of the bandwidth is bad, only to reverse ourselves later when other consideration such as noise enter the picture. Still we have to explain why it is bad, from the present point of view of our analysis, of course.

The reason it is bad is that it takes up quite a bit of space in the spectral allowance of the channel. In a given area, there are many transmitters who may wish to use the same channel with spectral band $[\omega_1, \omega_2]$. If every signal needs a lot of room to get through, the channel is not going to be able to accommodate too many signals. This whole business of sending a lot of signals through the same channel without causing a stampede in the FT domain goes by the name of frequency multiplexing. We do not want to give everything away about this concept, since we will cover it all by itself in a later section of the chapter. So it is bad to create an $m(t)$ with twice the original bandwidth of the information signal. But can you do anything about it? Read on (the next section) and see. But before you read on, can you make any connection between what we just said and the results of Example 3.2.1?

One more thing. The DSB system is the first communication system that we have analyzed. The reader would do well to stop and ponder the overall block diagram structure of this system, and compare it with Fig. 1.2.1. Are the various blocks of Fig. 1.2.1 easily identifiable in the DSB system? If there is some ambiguity, could it be interpreted as a granting of freedom to the scientist to group things together in a manner that suits his or her analysis?

3.3 THE SINGLE-SIDEBAND SUPPRESSED CARRIER SYSTEM

In ancient Greece, just outside present-day Athens and at the entrance to the valley of Kiffissos, there was a narrow passage guarded by a wild bandit, Procroustis. Whenever someone tried to cross the passage, Procroustis would capture him and lay him on top of a bed, presumably of a certain size, some say the width of the passage. Any portion of the passerby that exceeded the size of this bed would be chopped off by Procroustis with an ax. A rather gruesome affair, but signals sometimes suffer the same fate if their bandwidth is too large.

The single-sideband (SSB) suppressed carrier system, the topic of this section, differs from the preceding system in only one aspect. Just before the modulated wave $m(t)$ is ready to head off for the channel, one of its sidebands is chopped off. And unlike the passersby of the Greek myth, who were probably unable to continue their voyage, the now single-sideband wave continues on unperturbed and reaches its destination in fine spirits, reproducing at the final output of the receiver the original $f(t)$.

A block diagram of this system is portrayed in Fig. 3.3.1. Let us quickly run through a pictorial analysis of this system and then see what can be done about a mathematical one to please the exacting reader.

Let us consider an $f(t)$ with the $F(\omega)$ shown in Fig. 3.3.2a. After the multiplication with the carrier, $M(\omega)$ will have the form of Fig. 3.3.2b, a result of the shift-and-half routine. Now we chop off one of the sidebands, say the

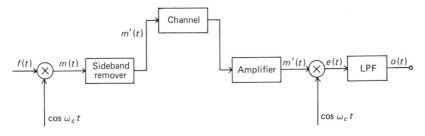

Figure 3.3.1 Block diagram of SSB communication system.

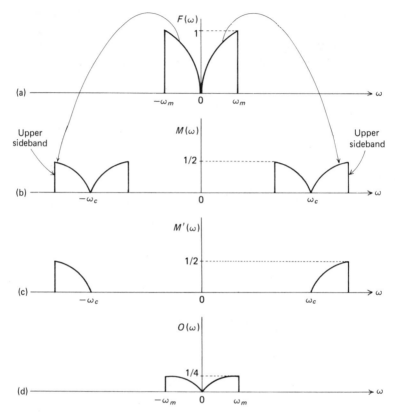

Figure 3.3.2   Pictorial analysis of SSB system (a) $F(\omega)$; (b) $M(\omega)$; (c) $M'(\omega)$; (d) $O(\omega)$.

lower. The result is $M'(\omega)$, shown in Fig. 3.3.2c. It is interesting to note at this point that the upper sideband actually *contains all the information about the original signal*, and that is the reason why this method works. The part of $M'(\omega)$ sitting on the $\omega > 0$ is actually a replica of the part of $F(\omega)$ at $\omega > 0$, and $M'(\omega)$ for $\omega < 0$, a replica of $F(\omega)$ for $\omega < 0$. And something similar (what exactly?) happens if we eliminate the upper sideband, a fact which points out that only one sideband is necessary; the other is redundant. All this, of course, we could have guessed from the beginning. We never wanted to end up with both sidebands; we were sort of stuck with them by multiplying with a cosine. It would be nice if we could just shift only the $F(\omega)$ part for $\omega > 0$ to ω_c [and the $F(\omega)$ for $\omega < 0$ to $-\omega_c$], but we do not know of any method to do so. Maybe the reader can find one. As it is, we shift the whole thing, end up with both sidebands, and chop one off (easy to do with a filter, for example).

   The rest of the pictorial analysis moves along easily. After the amplifier, the output of the multiplier has a FT whose part around $\omega = 0$ is what is shown in Fig. 3.3.2d as $O(\omega)$ and other parts (around ω_c and $-\omega_c$), which do not make it through the LPF. Thus

$$o(t) = \tfrac{1}{4}f(t) \tag{3.3.1}$$

In comparison with the DSB, and if all other factors remain the same, we have suffered loss of amplitude by one-half, since we chopped an equal but redundant part from the $m(t)$. Regardless, the system works, and it certainly demands only half the bandwidth for $m(t)$, in comparison with our DSB system.

Now we come to the problem of a rigorous mathematical analysis. The problem lies, of course, with expressing $m'(t)$ in terms of $f(t)$ [or $M'(\omega)$ in terms of $F(\omega)$]. The mathematics we are familiar with at this point do not help us in doing that; we need more—the theory of the **Hilbert transform**. To see exactly where the problem lies, let us go through the math for a *specific* $f(t)$, one that can be done with what we already know.

Let us assume that

$$f(t) = \cos \omega_m t \tag{3.3.2}$$

with ω_m a low frequency (say around 1 kHz). Then

$$F(\omega) = \pi[\delta(\omega - \omega_c) + \delta(\omega - \omega_c)] \tag{3.3.3}$$

Now $m(t)$ will be

$$m(t) = \cos \omega_m t \cos \omega_c t$$
$$= \tfrac{1}{2}[\cos (\omega_c - \omega_m)t + \cos (\omega_c + \omega_m)t] \tag{3.3.4}$$

with $M(\omega)$ as shown in Fig. 3.3.3. Let us next assume that we pass $m(t)$ through an ideal BPF which cuts off the upper sideband. Here there is no problem identifying the time form of this sideband; it is the second term in Eq. (3.3.4).

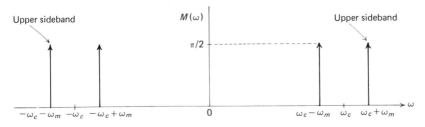

Figure 3.3.3   The $M(\omega)$ for $m(t) = A \cos \omega_m t \cos \omega_c t$.

The SSB wave now has the form

$$m'_-(t) = \tfrac{1}{2} \cos (\omega_c - \omega_m)t \tag{3.3.5a}$$

or

$$m'_-(t) = \tfrac{1}{2}(\cos \omega_m t \cos \omega_c t + \sin \omega_m t \sin \omega_c t) \tag{3.3.5b}$$

where the subscript $(-)$ symbolizes the *lower* sideband.

This last innocent-looking expression is, in fact, tremendously interesting, and we will have to come back to it as soon as possible. For the present, let us proceed with the analysis.

At the receiver, use of Eq. (3.3.5a) gives

$$e(t) = m'_-(t) \cos \omega_c t = \tfrac{1}{4}(\cos \omega_m t + \cos 2\omega_c t) \qquad (3.3.6)$$

and after the LPF,

$$o(t) = \tfrac{1}{4} \cos \omega_m t = \tfrac{1}{4} f(t) \qquad (3.3.7)$$

so all is fine. The key to the whole thing was our ability to express $m'_-(t)$.

Now let us go back to expression (3.3.5b), which we claimed to be of fundamental importance, since it must contain in it the key to the puzzle of how to write $m'_-(t)$ for a *general* $f(t)$.

First we note that $\sin \omega_m t = \cos(\omega_m t - 90°)$. If we denote this last expression by $\hat{f}(t)$, then we can write

$$m'_-(t) = \tfrac{1}{2}[f(t) \cos \omega_c t + \hat{f}(t) \sin \omega_c t] \qquad (3.3.8)$$

So here is why the expression is important. The thought that arises in everybody's mind is this. If you can write Eq. (3.3.8) for a cosine (and obviously a similar one for a sine as the reader should verify), why can't we extend it for all $f(t)$'s? And if it can be done (and we would not be bothering otherwise), what is this $\hat{f}(t)$, and how do we find it?

Well, we are going to give the answer to this question and then show that it is the correct one. Here it goes.

The $\hat{f}(t)$ we are after exists, and is, in fact, called the **Hilbert transform** (HT) of $f(t)$. The formula that provides it is

$$\hat{f}(t) = \frac{1}{\pi} \int_{-\infty}^{+\infty} \frac{f(\tau)}{t - \tau} \, d\tau \qquad (3.3.9)$$

which is nothing more than the convolution of $f(t)$ with the function $1/\pi t$. Actually, the term "transform" is an unfortunate one in this case, as $\hat{f}(t)$ and $f(t)$ are both defined in the same domain. The Hilbert *brother* would have been more appropriate.[3] In view of the convolution that we mentioned, the HT of $f(t)$ can be thought of as emerging at the output of a filter with impulse response $1/\pi t$ when the input is $f(t)$. The filter is called the **Hilbert filter** (no argument with this name).

Now we are about ready to prove Eq. (3.3.8) for a general $f(t)$. Actually, we will prove this in the frequency domain; that is, we will prove that $M'_-(\omega)$ obtained from Eq. (3.3.8) possesses only the lower sideband. To do so, we first note that from the time convolution property of the FT,

$$\hat{F}(\omega) = F(\omega)[-j \operatorname{sgn}(\omega)] \qquad (3.3.10)$$

since $-j \operatorname{sgn}(\omega)$ is the FT of $1/\pi t$. In case the reader does not remember, we

---

[3]Actually, it is also called the quadrature function of $f(t)$, which is fine with us.

remind him or her that

$$\text{sgn}\,(\omega) = \begin{cases} 1 & \omega > 0 \\ -1 & \omega < 0 \end{cases}$$

and therefore

$$\hat{F}(\omega) = \begin{cases} -jF(\omega) & \omega > 0 \\ jF(\omega) & \omega < 0 \end{cases} \tag{3.3.11}$$

All this is not really all that new. We already know the expressions above for sines and cosines. If Eq. (3.3.8) holds (and we will show that it does), $\cos \omega_c t$ and $\sin \omega_c t$ are obviously HT of each other [since Eq. (3.3.5b) certainly holds]. Well, it is not hard to show Eq. (3.3.11) for $f(t) = \cos \omega_c t$ and the reader should try to show it on the margin of this page (with a pencil). Note also that the HT is manifested as a $90°$ phase angle shift of the spectral components in the frequency domain, and this is the reason for its other name (quadrature function).

After the above, the rest of the proof that $M'_-(\omega)$ is the lower sideband of $M(\omega)$ is routine. Indeed, the FT of $m'_-(t)$ can be found from Eq. (3.3.8) and the usual FT properties as

$$M'_-(\omega) = \frac{1}{4}[F(\omega - \omega_c) + F(\omega + \omega_c)] + \frac{1}{4j}[\hat{F}(\omega - \omega_c) - F(\omega + \omega_c)] \tag{3.3.12}$$

Let us take the terms for $\omega > 0$. They are

$$\frac{1}{4}F(\omega - \omega_c) - \frac{j}{4}\hat{F}(\omega - \omega_c) \tag{3.3.13}$$

Now, in view of Eq. (3.3.11),

$$\hat{F}(\omega - \omega_c) = \begin{cases} -jF(\omega - \omega_c) & \omega > \omega_c \\ jF(\omega - \omega_c) & \omega < \omega_c \end{cases} \tag{3.3.14}$$

If we insert Eq. (3.3.14) into expression (3.3.13), the result is that $M'_-(\omega)$ for $\omega > 0$ is given by

$$\begin{cases} \frac{1}{2}F(\omega - \omega_c) & \text{for } \omega < \omega_c \\ 0 & \text{for } \omega > \omega_c \end{cases} \tag{3.3.15}$$

that is, that only the lower sideband has been kept for $\omega > 0$. The same reasoning leads to similar results for $\omega < 0$, completing the proof that Eq. (3.3.8) holds in general.

Incidentally, if we kept the upper sideband, expression (3.3.8) would have a negative sign on the right-hand side, so we can combine the two and write

$$m'_\pm(t) = \frac{1}{2}[f(t) \cos \omega_c t \mp \hat{f}(t) \sin \omega_c t] \tag{3.3.16}$$

Once this expression has been found, the rest of the mathematical analysis of the SSB systems is child's play. At the receiver, the only term that will eventually pass through the LPF is the one with $\cos \omega_c t$, as the other term contains a sine, and Example 3.2.1 has convinced us that its contribution will be zero. The final output will, of course, be as shown in Eq. (3.3.7).

The time has come for a couple of examples.

Example 3.3.1

Find out the effect of a frequency error in the demodulation of $m'_-(t)$.

Solution. Let us assume that the error is positive, that is, that the cosine at the receiver has the form $\cos (\omega_c + \Delta\omega)t$. If $F(\omega)$ is the one shown in Fig. 3.3.2a and the $m'(t)$ is of the $m'_-(t)$ variety, the final output $O(\omega)$ will have the form shown in Fig. 3.3.4a, as can easily be verified by the reader by the shift-and-half routine. A negative error will produce the $O(\omega)$ shown in Fig. 3.3.4b. If the incoming signal is $m'_+(t)$, the results are reversed. In any event the overall situation is not as bad as in the DSB case. When the error in Fig. 3.3.4a occurs (and if $\Delta\omega$ is not too big), a male base singer may sound like a female soprano.[4] The other error is, however, more serious and usually leads to unintelligible speech, unless $\Delta\omega$ is very small.

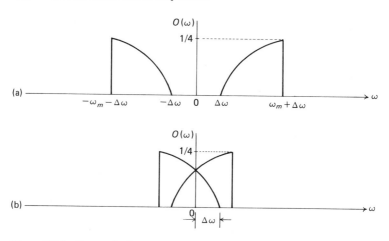

Figure 3.3.4 Errors in frequency: (a) positive error with incoming $m'_-(t)$; (b) negative errors with incoming m'_-t.

Example 3.3.2

What happens if the product detector makes a phase error?

Solution. Let us assume that

$$e(t) = m'_-(t) \cos (\omega_c t + \varphi) \qquad (3.3.17)$$

Use of Eq. (3.3.8) and lots of trigonometry (which we leave for the reader as an exercise) leads to

$$o(t) = \tfrac{1}{4}[f(t) \cos \varphi - \hat{f}(t) \sin \varphi] \qquad (3.3.18)$$

which is called *phase distortion*. Even though Eq. (3.3.18) looks formidable, the practical effect of a phase distortion on voice does not disturb the intelligibility of speech. On

---

[4]Actually, a Donald Duck effect is produced, which is exploited by cartoon film makers.

other types of data, however, this distortion can be quite serious, a fact that limits the overall popularity of the system.

Another weakness of the SSB system is the added complexity. Canceling a sideband completely with an ideal filter is easy, but no ideal filters exist in practice. Practical BPF filters do not have a sharp cutoff at ω_c. Now, if the informational signal is voice, this does not create serious problems because the signal itself has a FT which is tapered away from ω_c. For other types of signals, though, which have all the frequencies around $\omega = 0$ (and therefore around $\omega = \omega_c$ in their modulated versions), pratical BPF filters can cause problems. We will discuss this problem again in Section 3.6.4.

Actually, there are other ways of canceling a sideband, albeit not very simple. One of them, based on Eq. (3.3.16), is shown in Fig. 3.3.5. The blocks entitled "$-90°$ phase change" can be thought off as Hilbert filters, so the system is a realization of the $m'_{\pm}(t)$ of Eq. (3.3.16).

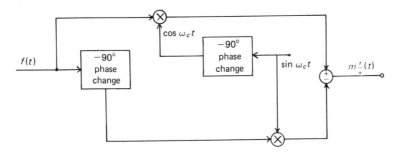

Figure 3.3.5 Sideband canceler for SSB.

In closing this section we should remark that reducing the bandwidth has, of course, its advantages, but it also carries its price. Aside from the observed loss of amplitude, complexity, and so on, the SSB system has a somewhat worse performance in the presence of additive noise, a characteristic attributed to the reduced bandwidth that appears here as an advantage. We will have more to say on this in a later chapter.

3.4 THE DOUBLE-SIDEBAND LARGE CARRIER SYSTEM

So far we have analyzed two amplitude modulation systems and we have never looked at a figure of the time shape of a typical $m(t)$. Our principal reason for this is that time sketches of $m(t)$ are difficult to draw, and we have therefore postponed them until now, when we have a bona fide use for them. The truth is that without looking at a sketch of $m(t)$, we will be unable to understand the present (AM) system.

It should be stated outright that AM is the most popular system, the one used by commercial (or goverment) radio stations, and the only one generally known to the lay person. It was also the first one to be used for mass communications back in the early part of the twentieth century. It is, therefore, the real McCoy, and this is why we refer to it as AM (amplitude modulation) and reserve its other name for the specialist.

The main idea behind this system is *money*. Its claim to fame comes from the use of a very cheap receiver, called a *peak* (or *envelope*) *detector*. This cheap receiver led to the mass production of (relatively) inexpensive home radios, which in turn established this system on a worldwide basis. If the reader wonders why this system is not the only one in use, considering its advantages, he or she must wait for the completion of the discussion.

As we mentioned earlier, we first have to look at $m(t)$. The reason is that it is the nature of $m(t)$ which holds the key to the use of the cheap receiver, and therefore the key to the whole system.

Let us assume that the information signal $f(t)$ is a *positive* pulse, as shown in Fig. 3.4.1a. Now, our $m_1(t)$ from the previous systems has been

$$m_1(t) = f_1(t) \cos \omega_c t \qquad (3.4.1)$$

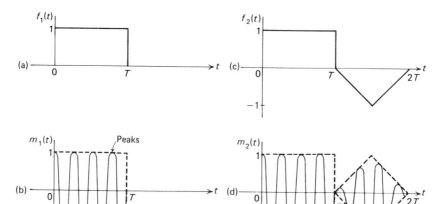

Figure 3.4.1 Looking at $m(t)$.

whose sketch is given in Fig. 3.4.1b, assuming, of course, that ω_c is large, somewhere in the vicinity of 500 kHz.

Let us try it once more. This time our $f_2(t)$ has the form of Fig. 3.4.1c; that is, *it is not everywhere positive* as $f_1(t)$ was. Now the sketch of

$$m_2(t) = f_2(t) \cos \omega_c t \qquad (3.4.2)$$

is the one shown in Fig. 3.4.1d. What is the difference between $m_1(t)$ and $m_2(t)$ as to how they carry the corresponding information signals, and why?

If we look at the positive peaks of $m_1(t)$, we note that a line through them reproduces the original $f_1(t)$. These peaks, in other words, carry approximately the information signal $f_1(t)$. This is not, however, the case with $m_2(t)$. The positive peaks here trace out $f_2(t)$, but *only wherever this signal is positive*. As soon as $f_2(t)$ becomes negative, the positive peaks of $m_2(t)$ no longer trace the original signal but rather, its positive mirror image with respect to the t axis. Here, then, is the crux of the matter. *If the information signal is always positive, the positive peaks of $m(t)$ trace out this signal*, and a device that can follow these peaks (the peak detector we mentioned above) can reproduce it easily at the receiver. If this device turns out to be easy to construct and quite inexpensive, so much the better.

But, the reader may object, how many information-carrying signals are always positive? The human voice certainly is not. No problem, no problem at all, we answer. Just add a constant, equal to the largest negative value of the signal (or bigger) and *make it* everywhere positive. Are there any objections? We will see.

A block diagram of the AM system is shown in Fig. 3.4.2. The first operation is the addition of a constant A. In view of what we remarked above, it is obvious that A is such that

$$f(t) + A \geq 0 \tag{3.4.3}$$

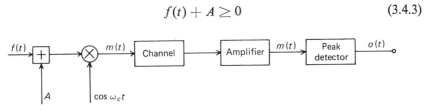

Figure 3.4.2   Block diagram of an AM system.

Next we multiply the output of the adder with the carrier $\cos \omega_c t$. The result is

$$m(t) = [A + f(t)] \cos \omega_c t = A \cos \omega_c t + f(t) \cos \omega_c t \tag{3.4.4}$$

whose FT is

$$M(\omega) = A\pi[\delta(\omega - \omega_c) + \delta(\omega + \omega_c)] + \tfrac{1}{2}[F(\omega - \omega_c) + F(\omega + \omega_c)] \tag{3.4.5}$$

Both Eqs. (3.4.4) and (3.4.5) state unequivocally that a portion of the carrier is now sent along to the channel and receiver. That fact, coupled with the two sidebands of the second bracketed term of Eq. (3.4.5), accounts for the complete name of the system.

The signal $m(t)$ traverses the channel and reaches the receiver. After some amplification that restores its original amplitude (or more), it enters the now famous detector (a diode–RC combination in its simplest form: How are they connected?). The peak detector traces out the positive peaks; that is, it extracts from $m(t)$ the term $A + f(t)$ (approximately, of course). Finally, then,

$$o(t) \approx A + f(t) \tag{3.4.6}$$

Needless to say, the original dc level (if any) of the $f(t)$ has been lost, without, however, any practical consequences (a simple capacitor gets rid of A).

We now become a little more precise about the terminology of this system, more specifically, about the term *positive peaks* that we have been using. We can define the **envelope** of signals like our

$$m(t) = f(t) \cos \omega_c t \qquad (3.4.7)$$

as the absolute value[5] of $f(t)$ [i.e., $|f(t)|$]. The inexpensive peak detectors can now be renamed as envelope detectors (or demodulators), as they receive $m(t)$ and spill out $|f(t)|$. It is quite obvious that if $m(t)$ has the form of Eq. (3.4.4), its envelope is $|A + f(t)|$, or $A + f(t)$, since this quantity is always positive. Therefore, the envelope detector will provide us with the $o(t)$ of (3.4.6), a fact that we knew but which we re-obtained without using the word "peak."

Up to now we have had nothing but praise about this system. The time has come to show its main and most serious flaw. Let us return to Eq. (3.4.4), which represents the signal emitted from the transmitter, and find its average power. We assume that $f(t)$ has zero average value (no dc term), for if it did, it could be included in the constant A. The average power of $m(t)$ is

$$\overline{m^2(t)} = \overline{A^2 \cos^2 \omega_c t} + \overline{f^2(t) \cos^2 \omega_c t} + \overline{2Af(t) \cos^2 \omega_c t} \qquad (3.4.8)$$

where the time averages have been indicated by bars.

Now the last term of Eq. (3.4.8) is pretty close to zero since it can be broken up into two terms (due to the $\cos^2 \omega_c t$), the first of which is definitely zero [no dc in $f(t)$] and the second can be argued to be nearly zero (argue it out). Therefore,

$$\overline{m^2(t)} = \frac{A^2}{2} + \frac{\overline{f^2(t)}}{2} + \frac{\overline{f^2(t) \cos 2\omega_c t}}{2} \qquad (3.4.9)$$

Again we can argue out the third term (do it), and finally,

$$\overline{m^2(t)} = \frac{A^2}{2} + \frac{\overline{f^2(t)}}{2} \qquad (3.4.10)$$

This last equation tells the story, a rather sad one for this system. The power that is available at the broadcasting station for $m(t)$ does not all go into the information signal $f(t)$. A part of it (and a great part, as we shall soon see), goes into the constant A, the carrier. Now, this certainly was not the case with our previous two systems. There, all of the available power went into the sidebands, which carry the useful information (no carrier).

To get a better feeling for how much power is *wasted* (not really wasted, though, since if we do not put it there, no cheap receiver, no AM system) in the carrier, we take a specific $f(t)$ and crank out some numbers. Let

$$f(t) = K \cos \omega_m t \qquad (3.4.11)$$

our usual friend, the low-frequency cosine. It is obvious that the system's

[5]If we want to get more precise yet, we can define the envelope as the magnitude of $z(t) = m(t) + j\hat{m}(t)$, where $\hat{m}(t)$ (see Problem 3.16) is the Hilbert transform of $m(t)$. But we do not need that here.

constant A must be

$$A \geq K$$

since K is the maximum negative value of $f(t)$. In this case $m(t)$ can be written as

$$m(t) = A\left(1 + \frac{K}{A} \cos \omega_m t\right) \cos \omega_c t \qquad (3.4.12)$$

We usually define

$$m = \frac{K}{A} \qquad (3.4.13)$$

as the **modulation index of the system.** Since its maximum value is unity and its minimum zero (why?), the m is often given in percent.

Now the total available average power in the transmitter can be easily shown as

$$\overline{m^2(t)} = \frac{A^2}{2} + \frac{A^2 m^2}{4} = \frac{A^2}{2}\left(1 + \frac{m^2}{2}\right) \qquad (3.4.14)$$

of which $A^2/2$ is in the carrier and the rest in the sidebands. The relative average power in the sidebands is therefore

$$\eta = \frac{m^2}{2 + m^2} \qquad (3.4.15)$$

where η can be viewed as the **efficiency** of the system. Now the best that we can do is when this η reaches its maximum. This will take place when $m = 1$ (or 100% modulation), that is, when we add just enough A, and no more. This fact, which the reader must verify, leads to

$$\eta = 0.33 \qquad (\text{or } 33\%) \qquad (3.4.16)$$

which means that 67% of the available average power goes to the carrier—an alarming quantity, but it is the best that you can do, in the case of a cosine. Usually we do not even know the maximum negative value of $f(t)$, so we add enough A to take care of things, with the result that we are operating at less than 33% efficiency.

The notion of a modulation index can also be extended to all signals, as long as they are normalized so that their maximum negative value is unity. Then we can write

$$m(t) = [1 + mf(t)] \cos \omega_c t \qquad (3.4.17)$$

where $m \leq 1$. The interpretation of m is the same. If $m = 1$ (or 100% modulation), we are adding enough constant to raise the maximum negative value of $f(t)$ to zero. Otherwise ($m < 1$), we are adding more than necessary, thus suffering loss of power.

As we mentioned above, this system is used for broadcasting by radio stations. Such stations are forced by international agreements to broadcast in the spectral region 535 to 1605 kHz. This, of course, means that the carriers used are with frequencies inside this region. Now every station is allowed a 10-kHz bandwidth, which means that the original voice signals must be limited by filtration to a maximum frequency of 5 kHz. Since the human voice (including music) reaches frequencies of 15 kHz (or even higher), it follows that AM stations do not have very high fidelity. We shall see later that FM stations are superior in this respect.

We close this section with an example.

Example 3.4.1

In an AM system the information signal is a pulse,

$$f(t) = u(t) - u(t - 1) \tag{3.4.18}$$

Find (a) the value of the constant that must be added for 100% modulation and (b) the sketch of $|M(\omega)|$, assuming that the carrier's frequency is ω_c.

Solution. (a) The pulse is never negative. Therefore, the constant $A = 0$ (the only chance that $m = 1$, mathematically).

(b) The FT of a pulse centered around $t = 0$ is given in Table 2.6.1. Using the time-shifting property of the FT we obtain

$$F(\omega) = \frac{\sin (\omega/2)}{\omega/2} e^{-j\omega/2} \tag{3.4.19}$$

Now $m(t)$ is of the DSB type,

$$m(t) = f(t) \cos \omega_c t \tag{3.4.20}$$

and therefore

$$M(\omega) = \frac{\sin \left(\dfrac{\omega - \omega_c}{2}\right)}{\omega - \omega_c} \exp -j\left(\frac{\omega - \omega_c}{2}\right) + \frac{\sin \left(\dfrac{\omega + \omega_c}{2}\right)}{\omega - \omega_c} \exp j\left(\frac{\omega + \omega_c}{2}\right)$$

It is a good thing that we are not asked to sketch $M(\omega)$, because it is complex and cannot be sketched. Luckily, the exponentials have unity magnitude, so the sketch of $|M(\omega)|$ is as shown in Fig. 3.4.3. It should be noted that there is no impulse at $\omega = \omega_c$. No constant added, no carrier, no impulse. Even so, this signal can be detected with a peak detector.

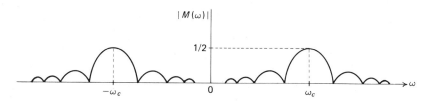

Figure 3.4.3   Sketch of $|M(\omega)|$ for Example 3.4.1.

3.5 FREQUENCY MULTIPLEXING
AND THE SUPERHETERODYNE RECEIVER

The communications field is one of the most policed businesses in the world. A person cannot just pick up some spare parts, make a transmitter, and start broadcasting. That would be international communications piracy, and the person could quickly end up in jail. A license is needed and those are issued on the basis of strict national regulations and international agreements. Most important of all is the region of the frequency spectrum in which you can broadcast (i.e., the value of the frequency of the carrier that you must use). The entire useful electromagnetic spectrum has been divided up into bands, and specific types of communications and radar systems are allowed to operate only in exact, predetermined allocations within these bands (see Appendix A to get a better overall picture).

All AM commercial (or goverment) stations are forced to broadcast in the region 535 to 1605 kHz, as we mentioned toward the end of the preceding section. Furthermore, each station is allowed a bandwidth [for its modulated carrier $m(t)$] of only 10 kHz, a condition that forces the station to restrict its information signal $f(t)$ to 5 kHz by filtration. To ensure that no interference occurs between stations (no overlapping of spectra), carriers are usually 30 kHz apart in value; that is, there is a gap of 20 kHz between spectra, called a **guard band**. This whole scheme of frequency slot allotment to each station, which makes things nice and clean, goes by the name **frequency-division multiplexing** (FDM). Needless to say, the name applies irrespective of whether the stations use DSB, SSB, or AM (or even FM and PM, as we shall see). The key idea behind it is that signals are neatly separated and kept distinct from each other in the frequency domain, even though they are on top of each other in the time domain. It is a nifty way of doing things.

We can even give a block diagram of this scheme, and we have done so in Fig. 3.5.1. In this figure, it is assumed that the modulation is DSB, and therefore the receivers, are product detectors. It is also assumed that there is no guard band between stations, obviously an ideal and not a practical condition.

Let us assume that the channel allocation is the $[\omega_1, \omega_2]$ range. Then if each station is allowed bandwidth B and the guard band between stations is W, the number of stations that can operate in a given locality will be

$$n = \frac{\omega_2 - \omega_1}{B + W} \tag{3.5.1}$$

So typically there can be

$$n = \frac{1605 - 535}{30} \approx 35$$

AM stations.

The principle of FDM can have other forms, depending on the application. Consider, for example, Fig. 3.5.2a, another possible manifestation of the same principle. The signals $f_1(t), f_2(t)$, and so on, have maximum frequencies equal to ω_m. The first one is left unperturbed. The second one is multiplied with $\cos 2\omega_m t$, the third with $\cos 4\omega_m t$, and so on. This premultiplication with cosines called **subcarriers** places the spectra of their $m_i(t)$, each in a specific allocation, at a low frequency, of course (see Fig. 3.5.2b). Next the $m_i(t)$ are added and the sum is multiplied by the channel carrier $\cos \omega_c t$, which shifts the entire sum spectrum to the proper channel range. A system such as this, which uses many low-

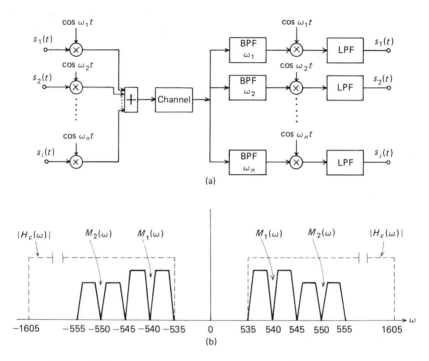

Figure 3.5.1 (a) FDM system; (b) spectra of the broadcast signals.

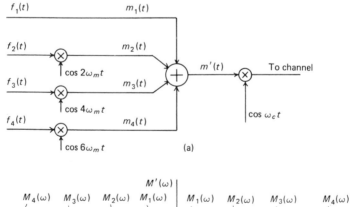

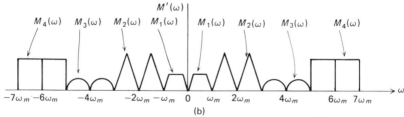

Figure 3.5.2 (a) FDM system with subcarriers; (b) spectrum of $m'(t)$.

frequency subcarriers and one main carrier, can be used in cases when one station wishes to send more than one signal, as in stereophonic broadcasts.

It should be reemphasized that FDM systems can be used by all the types of amplitude modulation systems. When the systems are large carrier AM, impulses will show up at the spectrum of $m(t)$. If the systems are SSB, the number of stations in a given locale can be doubled. Actually, they can be doubled with DSB also. (How?) Can they be doubled with AM?

At this point we would like to make an interesting observation. We stated in the introduction to this chapter that amplitude modulation was a way of matching the information signal to the space channel using a reasonable-size antenna. This is true, of course. But a little thinking will bring out a second important reason for shifting spectra, and that is the FDM system. For if we could indeed couple the voice signal *directly* to the space channel, and if the shift-and-half routine did not exist, the only thing that we could accomplish would be to send only one signal per locality. At least, this seems to be the answer at this stage of our understanding, which, of course, will be contradicted later when we come to the study of the time-division multiplexing principle. There are only a few absolute statements that you can make in this field.[6]

Now we come to the receiver part of an FDM system, and in the most common case of application, this brings us to the home-type AM radio. How do we make up a receiver capable of tuning in to any AM station we desire?

The answer to this question does not appear to be very difficult, and, in fact, it is not. After all, BPF filters exist and are being continuously designed to match the ideals more and more closely. So use a bunch of them, one for each station and you are done. Or if you do not like the idea of making a lot of them, make one that is *movable*; that is, its $H(\omega)$ can be moved along the AM broadcast range. Both ideas are feasible and have been used in the design of AM receivers of the home type. Their problem is amplification. If you have a lot of filters, you must also have a lot of amplifiers (stages) after each filter, and even though amplifiers can be the filters, that is still a lot of amplifiers. If you use one movable filter, you also need movable (or tunable) amplifiers. All that may be costly, or bulky, or difficult to do, depending on the state of the technology. In any event, the receiver that seems to combine ease of construction, inexpensiveness, and whatever else is taken into consideration is the one called **superheterodyne receiver**, and it is now going to be discussed below. This discussion will be given with reference to Fig. 3.5.3, which represents the key functions of this receiver in block diagram form.

Let us assume that there is an incoming modulated wave $m_1(t)$ whose $M_1(\omega)$ is centered around the frequency $\omega_c = 570$ kHz. It should be noted at

---

[6]"There are no absolutes in the world of science" declared the professor to the class. "In fact, anyone who insists that he is absolutely sure about anything is an idiot."
"Are you sure about that, sir?" asked a meek voice from the back of the room.
"Absolutely," asserted the professor.

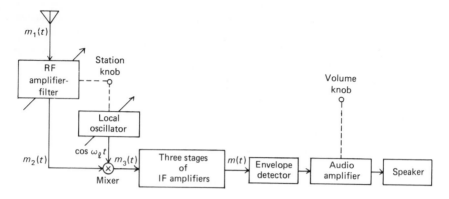

Figure 3.5.3 Superheterodyne receiver.

the start that after the block entitled "three stages of IF (intermediate frequency) amplification," the system is pretty much our familiar AM receiver, with an audio amplifier and a speaker. That much we know. The question is what the other blocks are doing in this receiver.

Well, let us start in the middle (the three IF stages), which represents the key to the understanding of this receiver. This amplifier (we can think of it as one) has an $H(\omega)$ that is *fixed* (in the frequency domain) in the range 450 to 460 kHz (i.e., with a center at 455 kHz). If it is fixed and cannot be moved, the incoming signals *must be moved to it*. That is exactly what takes place at the previous blocks. The incoming signal $m_1(t)$ goes through an amplifier first [the RF (radio-frequency) amplifier of Fig. 3.5.3], and is renamed $m_2(t)$. The next operation is a multiplication with a cosine generated inside the radio. This operation (called *mixing* or *heterodyning*) moves the frequency spectrum to the range 450 to 460 kHz, where the IF amplifiers are waiting. So, we repeat, instead of moving the amplifiers around, we keep them fixed and move the incoming signals around to their frequency range. That is the principle behind the super-heterodyne receiver.[7]

Now let us look at some numbers. If the incoming $m_1(t)$ is frequency centered at $\omega = 570$ kHz (and $\omega = -570$ kHz), how do we shift it to the range 450 to 460? Well, with a cosine of $\omega_\ell = 115$ kHz, of course. But is that the only one that can do the job? The answer is no! A cosine with $\omega_\ell = (570 + 455)$ $= 1025$ kHz (i.e., 455 kHz higher) will also do it, although in an unusual manner, by shifting the part sitting at $\omega = -570$ kHz to the right at the positive 450 to 460 kHz range, and the part sitting at $\omega = +570$ kHz to the negative 450 to 460 kHz range. Between the two, we pick the one with $\omega_\ell = 1025$ kHz, for

---

[7]This is one of many manifestations of the well-known religious aphorism, "If the mountain won't come to Mohammed, Mohammed will have to go the mountain."

practical design reasons. The local oscillator will have to have the range 0.1 to 1 MHz if we picked the first one, and this is presumably more difficult to construct than one in the range 1 to 2 MHz.

Still we are not out of the woods yet. The reader will have noticed that the first block of the receiver (the RF amplifier–filter) is a tunable one, tunable, in fact, by the same knob that controls the local cosine. Isn't this cheating? Aren't movable amplifiers–filters what this system is meant to avoid? Let us see.

The antenna and the first RF amplifier receive all stations in the AM range 535 to 1605 kHz. When all the incoming signals are multiplied by the local cosine of $\omega_\ell = 1025$, the station broadcasting at 570 kHz will not be the only one shifted to the range 450 to 460 kHz. The one sitting at 1480 kHz will also be shifted to this range. (Why and how?) In fact, a little figuring shows that for every station you are trying to shift to 455 kHz, you also get a second one sitting 910 kHz (twice the 455) higher. This second one, called the **image station**, must be filtered out, and this is the reason for the fact that the first block is a tunable amplifier filter. Its purpose is to reject the image station. However, its bandwidth is quite large and the tuning is rough. So it is only a little bit of cheating.

The superheterodyne receiver has, of course, some problems; everything does. One of them is the fact that it is forced to generate a local cosine which it (involuntarily) emits, acting as a transmitter. If another radio is in close proximity, and their local oscillators differ by 455 kHz, a difference beat frequency is created which passes through the IF amplifiers and reaches the speaker. The noise produced is not very pleasing to the human ear.

We close with a warning to the reader. The fact that he or she now understands (we hope) the operation of a home radio does not mean that he or she can fix it. In fact, looking at a radio's schematic diagram, the reader may not even be able to identify the blocks of Fig. 3.5.3.

3.6 OTHER AMPLITUDE MODULATION SYSTEMS: COMPARISONS

It seems to be part of the human condition never to be satisfied with the status quo and always to strive for improvement. So it is with amplitude modulation. After the three basic systems (DSB, SSB, AM) were designed and analyzed, ideas started pouring in about combinations, variations, and the like. After all, why not pick the advantages of each, and try to arrive to a variation that has no disadvantages? What about combinations of the variations, to create new variations?

This section is devoted to some such clever ideas, not all of which have yet proven practical. But who knows—the future is still wide open, and the final word is yet to come.

3.6.1 Compatible Single-Sideband System

The compatible single-sideband system is a noble attempt to combine the advantages of all the previous systems, with none of their major weaknesses.

What is the main advantage of DSB? That all the transmitted power is in the useful signal. What about SSB? Again, all its transmitted power is in the useful signal, and it requires half the bandwidth. Between the two, it appears that the SSB is preferable, since more signals can be packed in the same channel band. Now, think of the AM. What is its claim to fame? The cheap receiver. So, how about combining SSB with AM? Can we come up with a system that has the small bandwidth of SSB and the cheap receiver of the large carrier AM?

Such a system has actually been shown to exist, even though it cannot exactly be called an amplitude modulation system. It is called a compatible single-sideband (CSB) system, the name being due to the effort to make the SSB modulated signal compatible with an envelope detector.

The modulated signal of this system has the form

$$m(t) = \sqrt{f(t)} \cos(\omega_c t + k\hat{f}(t)) \qquad (3.6.1)$$

where k is a constant and $\hat{f}(t)$ the HT of $f(t)$. It should be noted that the information signal appears as amplitude modulation in the form of its square root, and that it also shows up as part of the phase of the carrier in the form of its HT. We will not discuss this system further, since it has not yet found much application, and the phase term brings it out of the scope of this chapter. The interested reader can study the CSB by looking up the papers by Logan and Schroeder (1962) and Voelcker (1966).

3.6.2 SSB with Carrier

So, the previous attempt has not quite been crowned with success. But what if we give in a little, and allow a carrier with the SSB signal? Could it then be demodulated with a peak detector? This way, at least, we would still be able to double the AM stations in the allowable band without rendering existing radios obsolete.

Let us not give away the answer to this possibility and try to work it out, in mathematical terms. The SSB form of the modulated wave is given by Eq. (3.3.16) as

$$m'_+(t) = \tfrac{1}{2}[f(t) \cos \omega_c t - \hat{f}(t) \sin \omega_c t] \qquad (3.6.2)$$

where the subscript $(+)$ designates the upper sideband and $\hat{f}(t)$ the Hilbert transform of $f(t)$.

If we throw in a carrier at the transmitter, the result will be $m(t)$,

$$m(t) = A \cos \omega_c t + m'_+(t) \qquad (3.6.3)$$

Now, as we know, the peak detector will extract the envelope of Eq. (3.6.2). Therefore, the final output of the detector will be

$$o(t) = \sqrt{\left[A + \frac{f(t)}{2} \right]^2 + \frac{[\hat{f}(t)]^2}{4}} = \sqrt{A^2 + Af(t) + \frac{f^2(t)}{4} + \frac{[\hat{f}(t)]^2}{4}} \qquad (3.6.4)$$

The only way you can make this expression resemble the original $f(t)$ is as follows. Assume first that A (the amplitude of the carrier) is so large that the last two terms of Eq. (3.6.4) can be ignored. This is bad business, of course, because it implies a lot of power in the carrier. Anyway, with this approximation,

$$o(t) \approx \sqrt{A^2 + Af(t)} \qquad (3.6.5)$$

Now use the binomial expansion and omit higher-order terms to arrive at

$$o(t) \approx A + \frac{f(t)}{2} \qquad (3.6.6)$$

which is acceptable, but with some distortion (remember the approximations). Even so, this system is not used in practice, because its efficiency is even less than that of the regular AM. But theoretically, at least, it does combine lowest bandwidth with inexpensive demodulation.

3.6.3 Carrier Reinsertion Systems

This is a kind of variation to the variation type of system that we mentioned at the start of this section. The key idea is this. If the main objection to the AM is the loss of power that goes to the carrier, why not transmit the signal as a DSB (or SSB) type and then *insert* the carrier at the receiving end, before the envelope detection? This way we would still use the inexpensive receiver, without the heavy loss of transmitted power that the AM system demands. It is actually a good idea, since the generation of a cosine term by a local oscillator is already in the superheterodyne receiver anyway, and this may not add up to a lot of new complexities. Let us take a quick look at the mathematics of this idea, to see if it works.

We start by noting that the received signal will have the form

$$m'(t) = f(t) \cos \omega_c t \qquad (3.6.7)$$

(i.e., it is the DSB type). At this point we add a carrier term, with enough amplitude so that the result is always positive and can be demodulated by an envelope detector. Denoting the new signal by $m(t)$, we have

$$m(t) = A \cos (\omega_c t + \theta) + f(t) \cos \omega_c t \qquad (3.6.8)$$

and the bugaboo is already apparent. It is easy to say *insert the carrier*, but this carrier will not be in phase with the received cosine, since it was not sent out at the same time at the transmitter. In fact, even the frequency of the new carrier may not be the same, but we will see about that later.

Equation (3.6.8) can be rewritten as (why?)

$$m(t) = A(t) \cos (\omega_c t + \varphi) \tag{3.6.9}$$

where

$$A(t) = \{[A + f(t)]^2 - 2Af(t)[1 - \cos \theta]\}^{1/2} \tag{3.6.10}$$

$$\varphi = \tan^{-1} \frac{A \sin \theta}{f(t) + A \cos \theta} \tag{3.6.11}$$

Now the envelope detector will extract the term $A(t)$. If $\theta = 0$ (i.e., no phase error), the result will be

$$A(t) = A + f(t) \tag{3.6.12}$$

which is fine. If, however, $\theta \neq 0$, things get messed up. Rearranging $A(t)$, we get

$$A(t) = A\left\{1 + \frac{2f(t)}{A} \cos \theta + \left[\frac{f(t)}{A}\right]^2\right\}^{1/2} \tag{3.6.13}$$

which for $|f(t)| \ll A$ reduces approximately to (why?)

$$A(t) \simeq A + f(t) \cos \theta \tag{3.6.14}$$

a result similar to Eq. (3.2.10). For large errors in θ (around 90°), we can end up with nothing.

Now, as we mentioned above, we could make an error in frequency, and the inserted carrier could have the form $A \cos (\omega_c + \Delta\omega)t$, or even the form $A \cos [(\omega_c \pm \Delta\omega)t + \theta]$ if both errors are made. We will not take up such an analysis here, but the reader can verify that the results are similar to those discovered in DSB synchronous detection.

So, you say, these could be serious problems. What can we do? Well, we might dream up a further variation of the system above as follows. Why not send just a small amount of the carrier at the transmitter, then pick it up, amplify it, and use it at the receiver? This way we will not have much loss of power, and at the same time no problems with phase or frequency errors.

This, too, is possible, and is actually done in practice, not only for this system, but also for the regular DSB to avoid the same errors that occur there. A system for extracting the small amount of carrier sent is covered in Problem 3.9. Actually, in practice, most modulators (even perfectly balanced ones) unwillingly produce a small amount of carrier, and steps must be taken to suppress it, if it is not wanted. That is the reason why the term *reinsertion* is often used in cases when the carrier is added at the receiver.

3.6.4 Vestigial Sideband Systems

We mentioned at the closing of the SSB analysis that a major weakness of the system is the added complexity of the sideband canceler. It would be nice if the job could be done with a simple filter, even though no one would expect it to be ideal.

Well, practical filters, as we already mentioned, do not have sharp cutoffs in their transfer functions $H(\omega)$. It can even be proven,[8] by the Paley–Wiener criterion, that such filters are not physically realizable. So we just have to accept that the filter will have a tapered cutoff and see what can be done. The main thing is to be able to recover the original $f(t)$ after the product demodulation.

When the filter is not ideal, one of the sidebands cannot be completely eliminated. A small portion of it, a trace (that is what "vestige" means) will remain. Even so, intelligent filter design can make the thing work, and the result is called a vestigial sideband system (VSB), as is obvious by now. Let us see what this clever design is all about.

Let us assume that the DSB modulated wave $m(t)$ has a FT $M(\omega)$ as shown in Fig. 3.6.1a. It should be noticed that this $m(t)$, unlike voice signals, is flat around ω_c. It is more like the video signal of commercial TV, which, incidentally, uses a form of modulation similar to VSB.

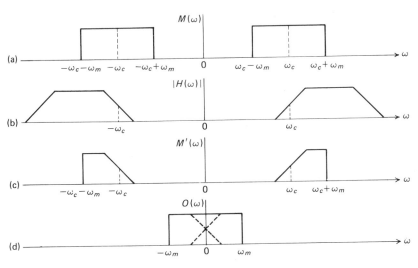

Figure 3.6.1   Vestigial sideband system.

Figure 3.6.1b shows a possible filter with tapered response around ω_c. After $M(\omega)$ traverses this filter, it becomes $M'(\omega)$, as shown in Fig. 3.6.1c. It should be noted that $M'(\omega)$ is not a clear upper single-sideband signal. It contains a vestige of the lower sideband. But (and this is important) the filter has *eaten up* part of the upper sideband as well. This is precisely what makes it work. For as the reader can ascertain by looking at Fig. 3.6.1d, this last effect eventually cancels out the vestige, and the final output $O(\omega)$ resembles (ideally, anyway) the original $F(\omega)$. The system works even with an envelope detector if a large carrier signal is sent along with $m'(t)$.

Some mathematics is now in order, to see if we can specify this filter a

[8] See Lathi (1968, p. 121).

little more analytically. With $H(\omega)$ the transfer function of the filter, we have

$$M'(\omega) = \tfrac{1}{2}[F(\omega - \omega_c) + F(\omega - \omega_c)]H(\omega) \qquad (3.6.15)$$

as the signal that enters the channel. At the synchronous detector, after the multiplication with $\cos \omega_c t$ we will have

$$E(\omega) = \tfrac{1}{4}F(\omega)H(\omega + \omega_c) + \tfrac{1}{4}F(\omega)H(\omega - \omega_c) + \text{high-frequency terms}$$
$$(3.6.16)$$

The final operation is passage through the LPF. All high-frequency terms will vanish, and the output will be

$$O(\omega) = \tfrac{1}{4}F(\omega)[H(\omega + \omega_c) + H(\omega - \omega_c)] \qquad (3.6.17)$$

We would very much like to have the $O(\omega)$ above equal to $F(\omega)$, within a constant, of course. This means that

$$H(\omega + \omega_c) + H(\omega - \omega_c) = C \qquad \text{(a constant)} \qquad (3.6.18)$$

Now let us recall that our original informational signal $f(t)$ was assumed such that $F(\omega) = 0$ for $|\omega| > \omega_m$. A little reflection will show that for this reason, Eq. (3.6.18) need only hold for $|\omega| < \omega_m$ [i.e., in the spectral region of $f(t)$]. Thus

$$H(\omega + \omega_c) + H(\omega - \omega_c) = C \qquad \text{for } |\omega| < \omega_m \qquad (3.6.19)$$

Equation (3.6.19) is the desired specification of the filter. It tells us that $H(\omega)$ should be such that when shifted right and left, the portions around the origin should add up to a flat transfer function. This can be accomplished if the $H(\omega)$ has odd symmetry around $\omega = \omega_c$, with the 50% response level at $\omega = \omega_c$. That is why we drew it as we did in Fig. 3.6.1c.

As we mentioned earlier, the VSB system has enjoyed some practical application in television and some data transmission systems. Actually, TV uses a slightly different form of VSB. The sideband elimination filter is not carefully designed as we described above, but other filters reshape the final output at the receiver. Stark and Tuteur (1979) have an excellent analysis of the U.S. television system (black and white and color), and the interested reader is referred to them for further study.

This ends our presentation of amplitude modulation communication systems. A note of cautious warning is now due. Comparisons of various systems (of any type) are not possible in absolute terms, or rather, are not meaningful. Only the specific application provides the basis for comparisons, and every new application has a surprisingly new set of constraints. For example, the fact that commercial stations use the AM system (for economic reasons) does not necessarily make it the best system for communications between two aircraft flying on a mission. Other constraints, such as close proximity of the aircraft, immunity to jamming (interference noise), or cryptographic capabilities, probably suggest other systems, much more suitable for the application. This is the reason why every system must be studied in detail, and both its strengths and weaknesses must be well understood. If the best system for every application were known in advance, the communications field would have been reduced to a cookbook of recipes; it would be dead. And it is not dead, by any means; it is not even ailing.

PROBLEMS, QUESTIONS, AND EXTENSIONS

3.1. Show that Eq. (3.2.2) can also be obtained by using the frequency convolution property of the Fourier transform.

3.2. Go through the entire mathematical analysis of the DSB system assuming that the carrier is a sine (not a cosine), and so is the signal used in the product detector. What is the implication of the result in frequency-division multiplexing of DSB systems?

3.3. Sketch $E(\omega)$ of Eq. (3.2.11) if $f(t)$ is a voice signal. Find and sketch $O(\omega)$ for the system of Example (3.2.2).

3.4. Frequency translation (and therefore amplitude modulation) of an information signal $f(t)$ with low-frequency content can also be achieved by multiplication of the $f(t)$ with a periodic signal and then proper filtration. Show that this is true by multiplying $f(t)$ (a voice signal) with:

(a) An infinite train of impulses with period equal to $T = \pi/\omega_m$ with ω_m as shown in Fig. 3.2.2b (ideal sampling).

(b) A train of periodic pulses of the same period (natural sampling).

Discuss the possibility of doing the same for product demodulation.

3.5 The idea suggested in Problem 3.4 gives rise to the **gating circuit**, whose simplest form is shown in Fig. P3.5. How often should the switch close, and for how long, if a voice signal's spectrum is to be translated to $\omega_c = 100$ kHz? Where should a BPF filter be located, and what bandwidth should it have for SSB modulation?

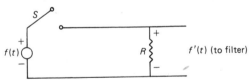

Figure P3.5   Gating circuit.

3.6. A nifty way to create the product of two signals $[f(t)$ and $\cos \omega_c t]$ is by squaring their product and then correctly filtering the result. Assume that

$$f(t) = 10 \frac{\sin 10^3 t}{10^3 t}$$

Find the FT of

$$f'(t) = (f(t) + \cos 10^6 t)^2$$

What should be the bandwidth of an ideal filter that could extract the product $f(t) \cos \omega_c t$, and where should it be located in ω?

3.7. Consider the circuit of Fig. P3.7, which can come in other forms and is called a **balanced modulator**. Assume that the diodes are identical, with a v–i characteristic

$$i = av + bv^2$$

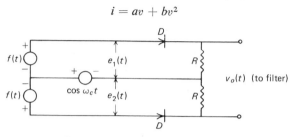

Figure P3.7 Balanced modulator.

Show that the output $v_o(t)$ is given by

$$v_o(t) = 2R[2bf(t)\cos\omega_c t + af(t)]$$

which means that proper BPF filtration can produce a modulated signal of the DSB variety.

3.8. A signal $f(t)$ is bandlimited up to ω_m [i.e., $F(\omega) = 0$ for $\omega < -\omega_m$ and $\omega > \omega_m$]. Now $f(t)$ is multiplied by $\cos 2\pi f_c t$, and the new signal is denoted by $m(t)$. Find f_c if the bandwidth of $m(t)$ is 5 % of f_c.

3.9. We noted that the DSB system requires a cosine of the same frequency and phase angle for product demodulation. Often, the transmitter sends out a small amount of the carrier (not enough for peak detection, however), so that the receiver can extract it and use it for its homodyne detector. Show that the system of Fig. P3.9 creates ω_c, assuming that $m(t) = f(t)\cos\omega_c t$. The symbol NBF designates an ideal narrowband filter. The 2:1 frequency converter is a device that halves the frequency of its input signal. The signal $f(t)$ is human speech.

Figure P3.9 System for extracting the carrier.

3.10. Another method for generating the cosine needed at the receiver was proposed by Costas (1956). A simplified analysis of the Costas loop can be found in Stark and Tuteur (1979) and a more detailed one in Stiffler (1971). Look up one of these references and prepare a lecture on this subsystem.

3.11. A student suggested the following system for reducing the bandwidth of an amplitude-modulated system. The information signal $f(t)$ should first pass a scale change and be turned into $g(t) = f(bt)$, with b such that $G(\omega)$ has a smaller bandwidth than $F(\omega)$. Then it should be sent over the channel. At the receiver, after demodulation, a reverse scale change would restore it to its original form. Is it possible? Comment on the realizability of such a system.

3.12. Assume that $f(t)$ is such that it can be written as

$$f(t) = \text{Re}\,[z(t)]$$

where $z(t)$ is a complex signal with one-sided FT [i.e., $Z(\omega) = 0$ for $\omega < 0$]. Show that

$$z(t) = f(t) + j\hat{f}(t)$$

where $\hat{f}(t)$ is the HT of $f(t)$. Think about what this implies about the HT.

3.13. For Example 3.3.1, work out the output using Hilbert transforms.

3.14. Prove Eq. (3.3.18) in the text.

3.15. Consider an $f(t)$ with $F(\omega)$ as shown in Fig. P3.15. This $f(t)$ is first multiplied by $\cos \omega_m t$ and passed through an ideal LPF with cutoff at $\omega = \omega_m$. Then it is multiplied by $\cos \omega_c t$, where $\omega_c \gg \omega_m$. The first operation is meant to "scramble" the signal. Find the FT at each stage of the transmitter. Design an unscrambler for the receiver.

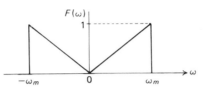

Figure P3.15 A scrambler.

3.16. Consider the framework of Problem 3.12. The $z(t)$ defined there is usually called the **complex envelope** of $f(t)$, and its magnitude the **envelope**. Show that the definition of the envelope below Eq. (3.4.7) agrees with this new definition.

3.17. Argue out the third term of Eq. (3.4.9).

3.18. The modulated AM signal

$$m(t) = [A + f(t)] \cos \omega_c t$$

enters a synchronous detector, where instead of a cosine, the multiplication is done with a periodic signal $p(t)$. Under what conditions [period, Fourier coefficients of $p(t)$] can the demodulation take place?

3.19. Consider the signal $f(t)$ with

$$F(\omega) = [u(\omega + 1) - u(\omega - 1)]e^{-j\omega}$$

Find the value of the constant A that is needed for 50% modulation if $f(t)$ is to be used in an AM system.

3.20. We have the signals $f(t), f^2(t), f^4(t), \ldots, f^{2n}(t)$ for n positive integers. If

$$f(t) = \left(\frac{\sin 10^3 t}{10^3 t}\right)^2$$

how many such signals can be frequency multiplexed in the AM spectral range? Assume a guard band of 2 kHz. [*Hint:* If $f(t)$ is bandlimited up to $\omega = \omega_m$, then $f^2(t)$ is bandlimited up to $\omega = $? See Problem 2.25.]

3.21. In a SSB system (upper sideband only), the information signal is a human voice with maximum frequency ω_m. The synchronous detector makes a frequency error of $\Delta\omega = +\omega_m$. What will be the effect on the output of the detector?

3.22. Consider the signal

$$f(t) = \frac{\sin 2\pi \times 10^3 t}{2\pi \times 10^3 t}$$

This signal is processed so that it becomes $f(3t - 5)$, and then it enters a DSB system. Find the FT at each point in the system, all the way to the output of the demodulator. Next assume that $f(3t - 5)$ is to enter an AM system. Find the constant A needed to produce 60% modulation.

=====4=====

AMPLITUDE MODULATION: RADAR SYSTEMS

4.1 INTRODUCTION

Creatures that lack a certain sense often have other senses highly developed. Blind persons usually reach out to recognize objects (or people) by touch, or listen with a highly developed sense of hearing. Radar is an artificial way of extending human vision. Like an arm that can extend millions of miles away, radar can "see" for us objects that we cannot observe with the naked eye. It is no surprise, then, that radar is often called a *sensing device*.

As we have already mentioned in the introductory chapter (read it again), radar performs its functions by transmitting a known signal in the form of electromagnetic energy, and then waiting for a return, an echo. The echo, which presumably comes back from some object (usually called a *target*), carries back information about this object in the form of changes in some of the transmitted signal's original parameters. Careful analysis of these changes in the receiver, coupled with other features of the entire system (antenna orientation, for example) can actually lead to a good overall picture of the target even in environments that prohibit observation by the naked eye (or a telescope) such as darkness, cloud covers, or large distances.

There is a plethora of radar systems, and it has become very difficult to classify them into various categories. As the name of this chapter implies, we are about to begin a study of AM radar systems. By this we mean that the signal that the radar emits to the environment will be of the AM type. This type of classification is not really all that well accepted, but we feel that it is appropriate for this book, where radar and communication systems are presented in a unified way.

The most fundamental job of a radar is to detect the target, that is, to state that *something is out there*, with certainty. After that has been decided come the chores of measuring its range, its complete location (azimuth and height), its velocity, and so on. Often there is more than one target, and the radar must detect them all (*resolve them* as it is called) and zero in to measure the parameters of all of them. Doing all these things is not only difficult for a radar, it may even be impossible. For this reason many radars are often employed in a given location, together with a computer that coordinates their activities. Of course, this cannot be done in all occasions. You can hardly place many radars in a single aircraft and expect it to do much more than carry them around. When the space is limited, a *specialist* radar is chosen suited to the purpose on hand, and one hopes for the best after that.

This chapter considers systems whose main functions are to detect and measure the range and velocity of a target. All these jobs are strongly related to the type of signal that the radar uses, and the type here is AM. The resolution problem is not covered here in detail, as we feel that it is better treated with the presence of noise—and noise we are presently ignoring for pedagogical reasons.

Before we begin the analyses of various radar systems, we shall have a few words to say about antenna characteristics and the problems associated with wave propagation. These things could have been, of course, discussed in Chapter 3, as they are important in understanding the overall behavior of communication systems as well. Nevertheless, we feel that they are more important for radar systems, *particularly in understanding their weaknesses*, and choose to cover them in this chapter. Our presentation will be very cursory, for such things are presumably covered in other courses, as we mentioned in the introduction to Chapter 3.

4.2 ANTENNA CHARACTERISTICS

Antennas play a more important role in radar than in communication systems. Some radars (conical scan, monopulse, etc.) even owe their existence to the type of antenna they use. It is not out of place, then, even in a systems book, to cover briefly their key characteristics, for they will come in quite handy later, particularly in understanding tracking radars and the principles of countermeasures.

The antenna is the last device in the transmitter of both radar and communication systems. It is a transducer that converts electrical power to electromagnetic power, enabling the transmitted signal to traverse the channel. It is also the first device at the receiver side, where, of course, it performs the opposite function. Either way, however, its characteristics are the same, a fact expressed by the **reciprocity theorem**, which the reader can find in Silver (1949). With this in mind we will discuss its characteristics as a transmitting device, and we will let the reader conceptualize the results for the case when the antenna is used for receiving.

A key antenna characteristic that concerns us here is its radiation intensity pattern, which describes how the antenna radiates power in various directions in space, and gives us an idea about the directivity of the radiated power.

Let us assume that we place an antenna A at the origin of a coordinate system as shown in Fig. 4.2.1a, and "face" it toward the direction R. A typical antenna will radiate power as shown in Fig. 4.2.1b. This sketch of power intensity $P(\theta, \varphi)$ in watts per solid angle versus θ and φ, the angles defined in Fig. 4.2.1a, is called the **radiation intensity pattern** of the antenna. As can be seen there, there is one large main lobe pointed toward R, with many other sidelobes pointed toward other directions. If the maximum value of $P(\theta, \varphi)$ at the center of the main lobe is normalized to unity, this pattern is called simply the **radiation pattern** of the antenna, and this is what we shall be assuming in our discussion from now on. In fact, we will simplify matters even further by ignoring the three-dimensional nature of the pattern and consider its projection on a plane passing through the origin and containing its maximum, as shown in Fig. 4.2.1c. In view of the overall symmetry of the pattern, this last picture appears to be satisfactory.

We mentioned earlier that the radiation pattern gives us an idea about the directivity of the transmitted power. To make this more precise, we can define the **directive gain** of an antenna as

$$G_D = \frac{\text{maximum radiation intensity}}{\text{average radiation intensity}} \qquad (4.2.1)$$

which, as the name implies, gives a measure of how directive the pattern is. In view of the fact that the average radiated power is the total radiated power divided by 4π, we can also write

$$G_D = 4\pi \frac{\text{maximum radiation intensity}}{\text{total radiation intensity}} \qquad (4.2.2)$$

Another useful quantity dealing with the directivity of the antenna is the **power gain**, which includes antenna losses as well, something omitted in the definition above. The power gain of an antenna can be expressed as

$$G = KG_D \qquad (4.2.3)$$

where K is called the **radiation efficiency factor** and takes values in $0 \le K \le 1$ ($K = 1$ designates the ideal case, no losses). Practical values for G and G_D are close to each other. Often, G is expressed in decibels by the well-known relationship

$$\text{dB} = 10 \log_{10} G \qquad (4.2.4)$$

The antenna gain G can also be expressed approximately as

$$G = \frac{4\pi A_e}{\lambda^2} \qquad (4.2.5)$$

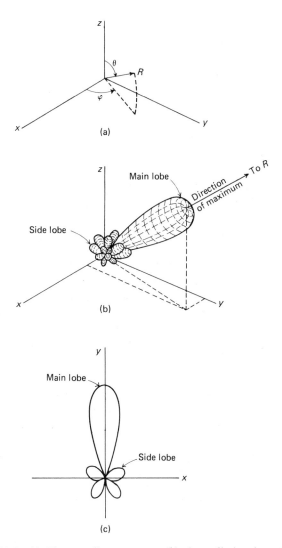

Figure 4.2.1   (a) The coordinate system; (b) the radiation intensity pattern; (c) the radiation intensity pattern on one plane.

where λ is the wavelength of the transmitted signal and A_e is the **effective aperture or effective area** of the antenna, related to the actual physical radiating area A through

$$A_e = \rho_\alpha A \tag{4.2.6}$$

where ρ_α is called the **aperture efficiency** and takes values in $0 \le \rho_\alpha \le 1$. Equations (4.2.5) and (4.2.6) give us an idea of how the gain G is related to the physical characteristics of the antenna and the signal frequency it is meant to transmit. Now we can see why low frequencies require antennas with gigantic dimensions as we remarked in the introduction to Chapter 3, and why modulation is so useful.

Actually, we will not have occasion to use these relationships and that is the reason we are giving them here without proofs or elaboration. We simply want to get a few basic things down about antennas, things that will make later material more comprehensible.

The most important idea here is the radiation pattern of the antenna. Even in the most *directive* antenna patterns the power is not all concentrated in one direction only. The sidelobes also emit (and receive) power in other directions, and this can cause problems, particularly in an electronic countermeasure regime. Thus it would seem worthwhile to design antennas with no sidelobes in the radiation pattern. However, this is not possible without suffering directivity. If we try to decrease the side lobe maxima, the main-lobe beam width increases so that the gain at the maximum is reduced—another one of those cases where you have to accept a compromise.

The radiation intensity pattern of an antenna gives us an idea of how it behaves as a spatial (three-dimensional) filter. But the antenna can also be thought of as a regular filter of a time signal, and as such it has a transfer function, frequency bandwidth, and the like, things which, as we remarked above, are related to its physical characteristics. The frequency bandwidth of an antenna is usually expressed as a percentage of its center frequency. Generally speaking, antennas with a $\pm 5\%$ bandwidth are considered quite broad and are hard to design, and this can be a problem in various applications where broad-frequency-bandwidth antennas are required (in reconnaisance, for example; see Chapter 15).

Our final remarks will deal with the **polarization** of the antenna's emitted wave, that is, with the direction of its electric field vector relative to the earth's surface. The simplest types of polarization are **vertical** or **horizontal**, where the directions of the electric field are obvious by their names. When the direction of the electric field vector is other than the above (as it is in most radar antennas), the polarization is called **linear**, and it can obviously be resolved into a vertical and a horizontal component. Some antenna systems are so designed that they can transmit two linearly polarized waves of the same frequency, traveling in the same direction, but not necessarily of the same amplitude and phase. This is called **elliptical** polarization and has not been used much in practice. A special case of this is **circular** polarization, when the amplitudes are equal and their phase difference is $90°$. This last type has seen a lot of application because it appears to reduce undesired effects (mainly echoes) from rain, snow, and other types of precipitation in the atmosphere.

With this we close our short discussion of antenna characteristics. There will be more material on this issue in sections where we specifically need it.

4.3 A WORD ABOUT WAVE PROPAGATION

The electromagnetic waves emitted from the radar's antenna start their journey through the channel, the environment around the radar's location. This channel, the earth's troposphere (up to 20 km from the surface of the earth), is a far cry from an ideal one. All sorts of things take place there, which make our mathematical analyses of the performance of a radar rather weak in predictive capabilities. Some of these things will be discussed in this section, mainly the important ones. The subject of wave propagation through an inhomogeneous medium with boundaries (as the case is here) is too broad and complicated to be done any justice here. In fact, serious research is still going on to extend our rather meager knowledge of this area.

We shall distinguish three basic phenomena that effect propagation and therefore the radar's performance. These are **reflection** of the waves on the earth's surface, **refraction** of the waves in the troposphere, and **attenuation** of the waves due to the various gases that make up the atmosphere. Our treatment is meant only as a brief oversimplification of these phenomena. Our goal is that the reader develop an appreciation of the fact that a radar's performance is extremely difficult to predict to any generality, as it is closely tied to the particular environment that it is installed to operate, and this environment is dynamically changing.

4.3.1 Reflection

Let us explain this phenomenon with the aid of Fig. 4.3.1. In this figure the emitted wave starts at point A and the target is located at point B. The antenna is assumed omnidirectional (it transmits equally to all directions) and the earth's surface a plane, not a curve. With these assumptions, it is seen that the target is hit not only by a direct wave emanating from the antenna, but also by a delayed one emanating from the same source and reflected from the earth. The target, then, receives the sum of these two waves (after a while, anyway) and this sum may or may not be in phase. Depending on the phase, the energy that arrives at the target can be from zero to twice the amount that would reach there without the earth's reflection. This amount actually depends not only on

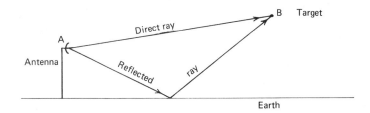

Figure 4.3.1 Illustration of reflection.

the phase but on other factors as well (polarization, coefficient of reflection which varies with the locale, height and angle of the antenna, etc.) and the results are messy, asymmetrical, and difficult to predict. Add to that that the same factors are involved on the return echo, that the earth is not flat (mountains, trees, etc.), that there may be other reflecting surfaces besides the earth, and so on, and the matter becomes rather depressive. Nevertheless, one has to live with it.

This whole business of the reflection phenomenon can be thought of as a change in the radiation pattern of the antenna. To comprehend this point of view, we refer the reader to Fig. 4.3.2, where an example of this change is illustrated. The antenna coverage is drastically changed if reflections are taken into account, showing clearly areas where the coverage is strong and weak (zero to double the amount). Of course, this is only an example, and the radiation pattern is difficult to sketch, even for a given locale, since conditions there can be dynamically changing (if it is a sea environment, the sea waves can change the reflecting surface). This is, of course, bad for the radar, but also bad for an aircraft that may try to take advantage of the weak points in the coverage. But this is another topic, left for the final chapter of this book.

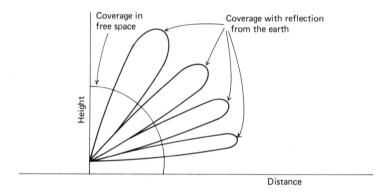

Figure 4.3.2   Antenna radiation pattern with reflection.

4.3.2 Refraction

Waves do not travel in a straight line when they traverse the atmosphere, as they do in free space. To begin with, their velocity is C/n, where C is their velocity in a vacuum and n the **index of refraction**. Now n changes value, since the atmosphere is inhomogeneous, and it is smaller at higher altitudes. The end result of this is that the upper portions of a wavefront travel faster than the lower ones, and the whole front is bent downward, as if it is following the curvature of the earth.

The upshot of the above is that the radar wave can curve and reach points that are not on the line of sight of the antenna, as illustrated in Fig. 4.3.3. This means that the range capability of the radar is extended (good). However, it

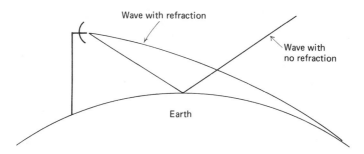

Figure 4.3.3 Illustration of refraction.

also means that the measurement of the *height* of a given target cannot be measured with simple trigonometry. In practice, height deviations caused by refraction are compensated by considering the earth's radius to be $\frac{4}{3}$ of its actual value, and it seems to work. Also, studies have shown that better accuracy can be had in the basis of the **exponential reference atmosphere** proposed by Bean and Thayer (1959).

It should be remarked here that refraction is not the only cause of the curving of the wave paths. There are others as well (warm air ducts, for example) which will not be discussed in this brief presentation.

4.3.3 Attenuation

Attenuation is, of course, the decrease in the signal's amplitude (or energy or power) with distance. Most of it is caused by the well-known $1/R^2$ law, particularly for low-frequency waves. When the frequency content of the waves is placed at higher and higher frequencies, the major cause of attenuation becomes the molecular absorption from water vapors and oxygen molecules. This increase in attenuation is even irregular. As the wavelength decreases, the attenuation increases until the wavelength reaches the value of 1.35 cm. After that, the attenuation drops a little, to increase dramatically again (18 dB/km) when the wavelength reaches 0.165 cm. A similar thing happens when the attenuation is caused by oxygen molecules.

Aside from the reasons above, attenuation depends heavily on weather conditions (rain, fog, snow, etc.), particularly for frequencies over 10 GHz, and it is caused by wave scattering. Even mild rain can cause an attenuation of 1 dB/km if the wavelength is 1 cm or smaller.

4.4 THE RADAR EQUATION AND ITS IMPORTANCE

There is one more item of importance that we must consider before we start our analyses of various radar systems. This item is the **radar equation**, an equation that relates the power of the echo to the other parameters of the entire system (radar, channel, target).

Let us assume that the radar emits a signal with initial (before the antenna) average power P_T watts. If the signal traverses a homogeneous channel medium in a straight line (no reflections, refractions, etc.) and meets an object at a distance R from the radar, the return power of the echo from the object, P_R, is given by

$$P_R = K\frac{P_T}{R^4} \tag{4.4.1}$$

where K is a constant depending on the units and (more important) on the various other parameters of the entire system, such as antenna gain, target cross section, and so on.

Now, in radar, when we utter the phrase the "radar equation," we do not mean a specific form, and this causes confusion. So let us expand a little on this issue, come up with a couple of other forms, and understand the whole thing a little better.

Again we assume that P_T represents the power of the transmitted signal. Let us assume that the antenna radiates equally in all directions (omnidirectional). Then, from field theory, we know that the power at a distance R from the transmitter will be $P_T/4\pi R^2$ watts per unit area. This already accounts for part of the R^4 in Eq. (4.4.1), as well as part of the constant K.

At the distance R, a target is met, and a reflection is developed for the return trip to the radar. Now the amount of power that is reflected back depends on the shape of the object, its size, its reflectivity, and so on. We lump all of that crudely into a constant σ which we call the **radar cross section of the target**. Furthermore, the reflected signal, the echo, can be thought of as emanating from the target (i.e., the target acts as a radiator). This being the case, the power received back at the radar, a distance R again from the new source will be

$$P_R = \frac{P_T\sigma}{(4\pi R^2)^2} \tag{4.4.2}$$

accounting for the R^4 in the simplified version (4.4.1). We can make it even more specific yet. Most antennas are not omnidirectional and radiate toward a given direction with a gain G_0. This will lead to

$$P_R = \frac{P_T\sigma G_0}{(4\pi R^2)^2} \tag{4.4.3}$$

Furthermore, the received signal goes through an **effective area** of the antenna (A_e) (assuming that it is the same antenna, as is usually the case), which is given by

$$A_e = \frac{G_0\lambda^2}{4\pi} \tag{4.4.4}$$

where λ is the **wavelength** of the center frequency of the signal's spectrum. This leads to the form

$$P_R = \frac{P_T G_0^2 \lambda^2 \sigma}{(4\pi)^3 R^4} \tag{4.4.5}$$

and there is no end to this story. The interested reader can pursue the matter further in the bibliography on radar systems [e.g., Skolnik (1962, 1970), Nathanson (1969)]. There are enough forms to issue a whole book on it, although a doubtful best seller. And many forms are not just rehashes of the fundamental form of Eq. (4.4.1) as we may have unwittingly implied. There exist forms taking into account noise (in the channel or devices), jamming effects, and so on. Some of these forms we will see later, after Chapter 13. For the time being, Eq. (4.4.1) will serve us just fine—with only one variation, which will enable us to appreciate the importance of the radar equation.

We already know that the most fundamental act of radar is to detect the target. There is no meaning to any measurement of the characteristics of the target (distance, velocity, etc.) unless you can first decide that something is there. To do so, you must be able to recognize an echo when it has arrived, or in other words, the power of the echo must be sufficient to make it recognizable. There are many effects that contribute to the difficulties of recognizing the return signal, most of which are caused by noise, which we are presently ignoring. In any event, if all deleterious effects are taken into account, we can arrive at a minimum value that P_T must have in order for the echo to be recognizable in the receiver. If the power of this **minimum detectable signal** is denoted by S_{min}, the radar equation, when solved for R, will provide us with the maximum range of the radar. Indeed, denoting this range by R_{max}, we have

$$R_{max} = \left[\frac{KP_T}{S_{min}} \right]^{1/4} \tag{4.4.6}$$

another form of the radar equation, and a useful one, at least theoretically. It tells us how far the radar can detect targets, a fundamental parameter in the design of any radar. Now some things become obvious. To increase the maximum range of the radar, we might think of increasing P_T. Since P_T is average power, this can be done by shortening the duration of the pulse. Unfortunately, there is a limit to this, since most high-powered radar devices are peak-power limited. So we abandon that and start tinkering with K. The K is made up of all sorts of constants, not all of which can be tinkered with, as for example σ, the radar cross section of a target. For this and other reasons that deal with wave transmission through an inhomogeneous medium (practical channels always are), the practical usefulness of (4.4.6) is very limited. Experiments under various atmospheric conditions and target configurations are often necessary to determine R_{max}, and even that is sort of an approximation, an average.

It should be mentioned that R_{max}, even though crudely estimated, plays a role in radar system design and analysis, a role that can be fundamental. Most radars emit some sort of periodic signal, the value of the period being quite important in the measurement of the range and velocity of the target. Well, this period has a lot to do with R_{max} also. But we are getting somewhat ahead of the game. We will take all that up soon enough, in the next section.

4.5 THE PULSE RADAR SYSTEM

The **pulse radar** (PR) system is the first radar system that we will discuss, so we must take it slow and easy; we must use velvet gloves in handling it. Some of the ideas behind the analysis will prove to be of immense importance in comprehending other radar systems, as well as the entire field. It might take us a little longer, but the time (and pages) will not be wasted.

The PR system's principal purposes are to search, detect, and measure the range of targets. However, these are only its primary functions. The system is actually a jack-of-all-trades, a general-purpose system. Modifications in the parameters of its signal and antenna, or additions of special subsystems, enable it to do other functions quite adequately. In this section we take up only its primary function—measuring range. Detection will be discussed in Chapter 14 and some of its variations will come under different names in later sections or chapters.

Let us first take an overall look at the system, with the aid of Fig. 4.5.1, and then we will get down to details concerning its main function and how it is accomplished. The PR system, as the name implies, sends out a pulse (actually, a group of pulses, as we will see soon enough) or rather, wants to send out a pulse. We know that such a thing cannot be done with a reasonable-size antenna.

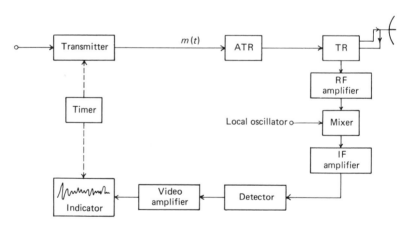

Figure 4.5.1 General pulse radar.

So this radar system (like the communication systems we covered) is forced to use a carrier, with frequency in the band agreed upon (see Appendix A). The pulse is latched on the amplitude of the carrier (AM)—otherwise, we would not be discussing it in this chapter. Nevertheless, it is worth remembering that the main signal of interest is the pulse (or pulses), and it is primarily on this signal that we will base our discussion of the system's subtleties.

The block called "transmitter" is made up of a pulse generator that modulates a high-powered tube (usually, a magnetron with peak power 1 to 10 MW) to produce the $m(t)$ suitable for transmission. The next two blocks, ATR and TR (together they are called a duplexer), are switches, whose purpose is to isolate the transmitter from the receiver, since only one antenna is used for both. When the transmitter emits a pulse, the TR (transmit–receive) switch breaks the connection with the receiver; otherwise, the receiver devices will be damaged from the high power of the transmitted signal. Needless to say, when the modulated pulse leaves, the TR reconnects the receiver, whereas at the same time ATR (anti-transmit–receive) is activated, disconnecting the transmitter from incoming echoes. Let us not forget that echoes coming from far away have miniscule powers, and we would not want to lose part of them to the transmitter.

Now let us assume that a single modulated pulse was emitted from the transmitter, passed the switches, traveled the wires (usually waveguides) to the antenna, left it, met an object, and an echo returned back to the radar. This echo is first RF amplified, and then it enters a mixer (multiplication by a cosine and filtering), which brings down its spectrum to the IF range, where more (and easier) amplification occurs. All this we can understand from a study of the superheterodyne receiver (Section 3.5). Then comes the detector. What could that be? Well, the echo is a small power version of $m(t)$, which was really a DSB wave. So it can be a product detector. Actually, however, since $m(t)$ is an amplitude-modulated wave by an always positive pulse, it can also be considered as an AM wave and can be envelope detected very nicely. This is what usually happens, and the original pulse is reconstituted, although in a much smaller magnitude. The result after amplification goes to a visual indicator, of which there are many kinds. The most common ones are the A-scope, which shows the received signal's amplitude (vertical axis) versus range (horizontal axis), and the PPI (plan position indicator), which shows the target as a blip in angle and range, on a polar map display (see Fig. 4.5.2).

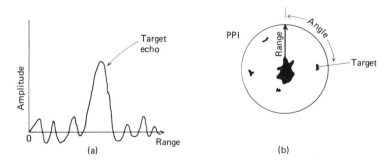

Figure 4.5.2   (a) A-scope; (b) PPI.

The timer synchronizes the start of the transmitted pulse with the visual indicator, so that the range information can be obtained. Antenna positions are also taken into account for angle location in the PPI display. All these matters will be better understood as the analysis of various radar systems evolves.

As we mentioned above, the same antenna is used by both receiver and transmitter with the aid of the duplexer (TR and ATR). It is obvious that since no reception takes place during transmission, no echo can be received; that is, objects lying in certain distances (called *blind ranges*) cannot be detected. This can be solved with two antennas, but then they have to be shielded. The entire system is then bulkier, certainly not a suitable choise for many applications (aircraft radars, for example).

Antennas come in various shapes and forms, as we mentioned in an earlier section. For small radar systems operating at a wavelength of 10 cm or shorter, the parabolic dish type is the most popular, and it can produce a "searchlight" beam or a fan-shaped beam, depending on the design. In most radars, the antenna is rotated mechanically so that it makes a complete 360° revolution every few seconds. A search radar with a vertically fan shaped beam can cover the space around it in fractions of a second. Its instantaneous location is also fed to the radar screen (if the screen is a PPI), so that the blip (of the echo) can be shown in the right azimuth angle (no height).

We mentioned that the A-scope and the PPI are the most popular visual indicators. This is true, but they are not the only ones. There is the type B scope, for example, which shows the target location in azimuth–range coordinates; type C, which shows it in azimuth–elevation cordinates; and many others. The interested reader is referred to Skolnik (1970, p. 6-3) for further reading.

Radar-transmitted peak powers are usually in excess of 100 kW, even for small radars. Since pulses typically last for a few milliseconds, average powers are around 100 W or more. The carrier frequency can vary from 200 MHz to 30 GHz or even higher. This range, of course, is split up into bands with World War II secrecy code names of X, K, and so on, as can be seen in Appendix A.

The discussion above covers in general the overall picture of the operation of a pulse radar. Not all pulse radars are, of course, exactly alike, even in the sense of the general block diagram of Fig. 4.5.1. Some do not have the RF amplifier, for example, and use IF stages only after the mixer. Others may have additional devices, such as automatic frequency control (AFC) or automatic gain control (AGC), which automatically adjust the receiver if the transmitter's frequency or gain change. But in any event, what should be striking is the similarity of this system to the AM communication systems. We are basically discussing a DSB communication system with an envelope detector, made possible by the fact that the informational signal (pulses) is always positive. But we must not let the similarities lull us into a sense of complacency. It is true that the radar sends out (and receives) pulses, in the same way as an AM communication system does. But the information of interest is not the pulses themselves here, but the changes they suffer in their parameters caused by their bounce from the target, the antenna position at reception, and so on. So the *approach* may be the same, but the *pursuits* are quite different.

Now it is time to get a little more specific about this system. We recall that radars are meant to measure target parameters, and this, the pulse radar, is designed primarily to measure range. How does it accomplish this goal?

Well, we said that the radar transmits an RF pulse and then waits until it receives an echo. Let us assume that such an echo is received, Δt seconds after transmittal. Since electromagnetic energy travels with the speed of light c, the distance to the target will be

$$R = \frac{c\,\Delta t}{2} \qquad (4.5.1)$$

where the factor 2 is meant to account for the round trip to the target and back, and $c = 3 \times 10^8$ m/s. So assuming that the echo can be detected and the delay Δt from transmission to reception can be measured, the range measurement is a trivial calculation.

Now the PR does not, of course, send out a pulse and then wait forever for an echo. That is not only boring, it is useless. After a little while, the pulse has gone so far away that even if it met a target, the echo from it would be undetectable. So the radar sends out a pulse, waits until this pulse can travel the distance and back of its maximal range $R_{\max}$ (i.e., waits time T dictated by $R_{\max}$), and then lets go of a new pulse. (That is where the various radar equations come in handy, in determining the waiting period.) So the radar actually emits a *periodic signal* (a periodic RF pulse), and its period, called the **interpulse period** (IPP), is determined by its maximal range, either theoretically (approximate) or by experimentation in a given terrain. The inverse of this period is called the PRF (the **pulse repetition frequency**) of the radar.

This parameter, the PRF, is a key one. It can save you or kill you. But we will wait a while before we discuss this, just for the suspense. We are still discussing the measurement of the target's range.

To measure the delay between transmission and reception, you must have a point of reference on the signal shape. The signal you sent out must somewhere have a distinct point, say its start or its finish. That is why we sent out pulses. Pulses have (theoretically at least) a sharp start (or finish), so we can measure the delay from start to start. Still, in practice, sharp edges get rounded off (they require a very large bandwidth in the devices to stay sharp), or they get obscured by noise. What can we do to ensure a good estimate of the range (delay in echo return) in the face of such uncertainties?

Well, that is where the idea of time correlation functions (or time ambiguity functions) enters the picture (reread Section 2.10). We can take the time cross-correlation function of the incoming echo signal with the original pulse, which results in a triangle with a sharp maximum, a point much easier to detect in the presence of noise. That is exactly what is done in practice, an operation that is not shown in the block diagram of Fig. 4.5.1 (see, however, Fig. 4.5.3b, where there is a triangle). Actually, since the echo is presumed similar to the pulse originally transmitted, this cross-correlation will be similar *in shape* (not in time location) to the autocorrelation of the original pulse. So this tells us something else, which is quite important. We must choose a signal for transmission whose time ambiguity function exhibits a sharp maximum at the origin (it is always there, although it can be in other places as well). For that is the point on which we will base our range reference during the cross-correlation operation, and it should be made as immune from uncertainties as possible.

The pulse has, of course, this characteristic, even though it was not originally chosen on the basis of the argument above. And the narrower the pulse is (an impulse?), the better, for measuring the target range. Still, we start to wonder, subconsciously anyway, if there could not be a better shape. We will see about that later. In any event, the upshot of this discussion is that narrow pulses are quite adequate for measuring range, and that this range measurement can be made more effective with a cross-correlation device after the envelope detector. Now the original signal and the echo are alike and in view of Eq. (2.10.18), this is equivalent to filtering the echo with a filter whose impulse response is

$$h(\tau) = s(-t) \tag{4.5.2}$$

where $s(t)$ is the original pulse. This type of filter is a shifted version of a *matched* filter, as we have already mentioned in Section 2.10. So in practice a matched filter will provide us with the sharp maximum we desire, and all will go well (don't bet on it). Usually, the amplifiers themselves serve the role of the matched filter, so that no extra step is necessary.

The time has come to discuss the importance of the PRF. Let us say that you set it at a value Δ, which corresponds to a given R_{max}. Now suppose that a target is located farther than R_{max}, but for various reasons (high σ, good weather conditions, etc.) it produces an echo. This echo will show up at the radar receiver, after the second pulse has been emitted, even though it was caused by the first pulse. In fact, it can even show up after the third pulse has been emitted, if it is far enough and can still produce a detectable echo. The radar is in no position to know which one of its emitted pulses caused this echo. In Fig. 4.5.3, for example, the triangle received could have come from the first pulse, in which case the range would be $R_1 + R_2$, or from the second pulse, in which case the range is R_2. This phenomenon is called **range ambiguity**. In actuality, you are never absolutely sure whether a certain range is R or $R + nR_{max}$, assuming, of course, that the PRF corresponds to the R_{max}.

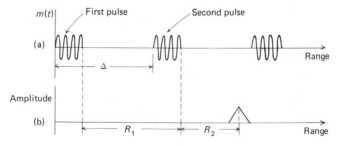

Figure 4.5.3 Ambiguity in range: (a) emitted signal; (b) received signal (A-scope).

What can we do about this ambiguity problem? Well, we can try setting the PRF at a very large value, so as to take care of very large distances, and have no possible detectable echo from a target farther than that. If the pulse radar is meant only to detect and measure range in one direction, we can certainly try this approach. This not being the usual application, however, other parameters (antenna rotation speed, for example) limit the PRF value. And even if they did not, one can never be absolutely sure that some huge object could not return a detectable echo from even farther anyway. So we can reduce the range ambiguity problem to something we can live with, but not eradicate it altogether. Furthermore, pulse radars (or modified forms of them) perform other parameter measurements as well. And as we will see in another section, decreasing the PRF may save you from some range ambiguities, but it can kill you in measuring velocity. So much for that solution.

Is there anything else we can do? Well, we could, if we suspect an ambiguity, change the PRF for a short while. Could that clear it up? The reader should try to come up with the answer.

Now we come to another interesting problem, the last one we discuss in this section. This problem is called **range resolution**, and has to do with the ability of the radar to distinguish between two targets that are closely positioned to each other.

It is quite obvious that in order for the pulse radar to be able to distinguish them (to *resolve* them, as it is called), the return signals must not overlap to such a degree that they cannot be recognized as two separate echoes. This can be quite a problem, because there is also noise in the system which obscures the signals received. That is where the time ambiguity function comes in handy again, and helps us understand this problem a little better. The output of the correlator (matched filter) of a single echo must be a high sharp peak, *and only a single peak*, for if there are more than one, we might take them as other echoes. Luckily again, very narrow pulses have such time ambiguity functions, even though they were not originally picked with regard to their time ambiguity functions. But it is not the first time that practical engineering came through with the correct answer, and then some theoretician came along to prove it.

There has been no mathematics in this section up to now, and this may leave the reader with an uneasy feeling about the conclusions of our discussion. It is time to correct all that and put a bit more rigor in the analysis. In what follows we will try to come up with mathematical expressions for the emitted signal and its time ambiguity function, to see whether the statements we made above actually hold water. We will do so in a series of steps, so that the final expressions are well founded and understood by the reader.

As we stated earlier, a first crack at an understanding of the PR system can be had by assuming that it transmits a single ideal pulse, which we can denote by $s(t)$. If this pulse sits symmetrically around the origin and has amplitude A and duration T_P, it can be written as

$$s(t) = A\left[u\left(t + \frac{T_P}{2}\right) - u\left(t - \frac{T_P}{2}\right)\right] = \begin{cases} A & \text{for} \quad -\frac{T_P}{2} \leq t \leq \frac{T_P}{2} \\ 0 & \text{elsewhere} \end{cases}$$

Another way to write it is by using Woodward's (1953) compact notation. This is

$$s(t) = A \, \text{rect}\left(\frac{t}{T_P}\right) \tag{4.5.3}$$

which is meant to symbolize the more detailed form above.

We also know from Table 2.6.1 the nature of the FT $S(\omega)$ of such a pulse, so

$$A \, \text{rect}\left(\frac{t}{T_P}\right) \longleftrightarrow AT_P \, \text{Sa}\left(\frac{\omega T_P}{2}\right) \tag{4.5.4}$$

are a transform pair, where we recall that

$$\text{Sa}(\omega) = \frac{\sin \omega}{\omega} \tag{4.5.5}$$

If the pulse starts at $t = 0$, it can be written as

$$\text{rect}\left(\frac{t - T_P/2}{T_P}\right) \tag{4.5.6}$$

and has a FT of

$$AT \, \text{Sa}\left(\frac{\omega T_P}{2}\right) e^{-j\omega T_P/2} \tag{4.5.7}$$

as we already know.

What about the time ambiguity $R_s(\tau)$ of such a pulse? Well, we know quite well that it is a triangle, don't we? And if the pulse gets very narrow, so narrow that it becomes an impulse, so does the $R_s(\tau)$. (Why?) So even this simple first mathematical formulation tells us that a pulse is an adequate signal for measuring range and resolving two or more targets. In fact, it looks as though in the limit it is ideal, since it (and its correlation function) becomes an impulse and all can be done in a perfect manner.

Now the radar does not really send out a single pulse, but something more like a periodic pulse train. A periodic pulse train with period T that lasts from $-\infty$ to $+\infty$ can be written as

$$\sum_{n=-\infty}^{+\infty} s(t - nT) \tag{4.5.8}$$

where $s(t)$ is the fundamental pulse of (4.5.3), which sits around $t = 0$. Or we can write it as

$$\sum_{n=-\infty}^{+\infty} s(t - nT) = s(t) * \sum_{n=-\infty}^{+\infty} \delta(t - nT) = s(t) * \text{comb}_T \tag{4.5.9}$$

that is, as a convolution with a comb in time (an infinite periodic impulse train with period T).

A compact notation for the above is

$$[\text{rep}_T\, s(t)] = \sum_{n=-\infty}^{+\infty} s(t - nT) \qquad (4.5.10)$$

also suggested by Woodward. It is actually an operational type of notation, which means that $s(t)$ is infinitely repeated, at intervals of T. The brackets are necessary to avoid confusion. For example,

$$[\text{rep}_T\, s_1(t)s_2(t)] \neq [\text{rep}_T\, s_1(t)]s_2(t) \qquad (4.5.11)$$

since the left-hand side is the function $s_1(t)s_2(t)$ repeated at intervals of time, whereas the right-hand side is the product of $s_2(t)$ with the repeated-every-T-seconds $s_1(t)$. The reader should be sure to understand the difference.

What about the FT of the signal of Eq. (4.5.10)? In view of Eq. (4.5.9) and the convolution property of the FT, we have that

$$[\text{rep}_T\, s(t)] \longleftrightarrow 2\pi S(\omega) \sum_{n=-\infty}^{+\infty} \delta(\omega - n\omega_0)$$

where $\omega_0 = 2\pi/T$.

If we now use our familiar delta function property, Eq. (2.3.6), the above can be written as

$$[\text{rep}\, t_T s(t)] \longleftrightarrow 2\pi \sum_{n=-\infty}^{+\infty} S(n\omega_0)\, \delta(\omega - n\omega_0) \qquad (4.5.12)$$

where $S(\omega)$ is, of course, the FT of the single pulse given by (4.5.4).

If we ignore the 2π factor, the result is similar to the Fourier series coefficients of the periodic pulse discussed in Section 2.5, the only difference being that delta functions are now attached to each "coefficient." Therefore, sketches of the FT of (4.5.12) can be found in Figs. 2.5.4 and 2.5.5, for specific values of period T and duration T_P of the impulse train. The reader should ensure that he or she understands how the density and magnitude of the spectrum change with changes in T and T_P by restudying Example 2.5.5.

So expression (4.5.10) resembles more closely the signal emitted from a PR radar signal. What about its time ambiguity function $R(\tau)$? Well, that should be an infinite series of triangles, spaced T seconds apart. Now if the pulses are made as narrow as impulses, $R(\tau)$ becomes an infinite series of impulses. (How?) This is very interesting. This $R(\tau)$ does not have a single maximum at $\tau = 0$, but many of them, spaced T seconds apart. So we should have ambiguities about the range every T seconds, as we mentioned earlier. The mathematics shows it quite well. There is no way you can get away from them, even if your pulses are turned into impulses, something that cannot be done in practice anyway. Nevertheless, the narrower the pulses, the better the resolution of two or more closely situated targets, as the reader can easily ascertain.

We are not out of the woods yet. The signal of Eq. (4.5.10) is not exactly our final expression. As the antenna rotates, its beam *illuminates* a certain area for a little while. So if our interest is in what is emitted to a given small area where a target might lie, then we do not have a single pulse, or an infinite number of pulses. In a pulse radar, antenna beam width and rotation and T and T_p values are such that the signal that is actually emitted toward a specific point is a finite number of pulses N.[1] How do we represent them, and what is their FT or amplitude spectrum? We will do that next under the assumption that they all have the same amplitude (i.e., antenna gain effects during rotation will be ignored). It should be noted that if there is an object in the area of illumination, the return signal will also be a finite number of pulses, although delayed (and partly deformed, which we ignore). Let us consider a finite number of pulses, as shown in Fig. 4.5.4a, where again we have ignored the carrier. For simplicity we have assumed them to be symmetrical around $t = 0$. We can conceive of them as an infinite number of pulses multiplied by a rectangular pulse of the form rect (t/T_i), where T_i is approximately the illumination time of the target. This being the case, we can denote them as $s_N(t)$ (N is their number) and write

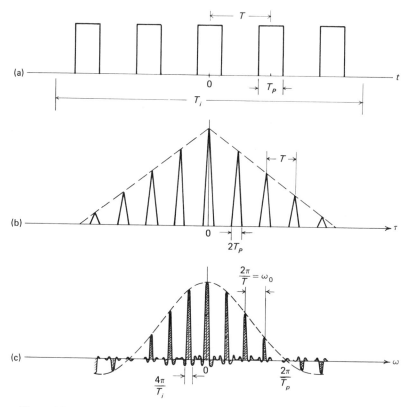

Figure 4.5.4 (a) Finite number of pulses; (b) their time ambiguity function: (c) their FT.

[1] If, for example, the horizontal beam width is $\theta = 1.5°$, PRF = 350 pulses/s and $\omega = 6$ revolutions/s, typical values for such radars, then $N = 15$ (figure it out).

$$s_N(t) = \text{rect}\left(\frac{t}{T_i}\right)\left[\text{rep}_T \text{ rect}\left(\frac{t}{T_P}\right)\right] \qquad (4.5.13)$$

with

$$\left[\text{rep}_T \text{ rect}\left(\frac{t}{T_P}\right)\right] = [\text{rep}_T s(t)] \qquad (4.5.14)$$

where $s(t)$, the single pulse, is written in the rect notation. Expression (4.5.13) is the most realistic one in representing the radar signal emitted toward a target, or received by a pulse radar, except, of course, for a possible time shift and multiplication with a carrier. It is rather messy, but there is not much we can do about it.

Let us look now at the time ambiguity function of the signal of Eq. (4.5.13) (shown in Fig. 4.5.4b), because the story is the most interesting one so far. There is the main maximum at $\tau = 0$, on which we hope to base our range measurement. But there are the other secondary maxima, which define the ambiguity ranges at T seconds (in range) apart, as was the case with the infinite pulse train we considered earlier. The interesting part (not shown anywhere up to now) is that these maxima decrease in value, telling us, in essence, that the ambiguity about the echo having originated at a distant targe. decreases as the distance increases. This, of course, makes sense, since the echo from such targets decreases in magnitude due to the increased distance. Anyway, we should now feel quite confident that this last mathematical expression is a fairly close approximation to the actual practical happenings of a PR system.

What about the FT of the signal $s_N(t)$? We already know the FT of the bracketed term; it is given in expression (4.5.12). Multiplying it by rect (t/T_i) produces a convolution in the frequency domain divided by 2π. Thus

$$S_N(\omega) = T_i \,\text{Sa}\left(\frac{\omega T_i}{2}\right) * \sum_{n=-\infty}^{+\infty} S(n\omega_0)\,\delta\left(\omega - \frac{n\pi}{T}\right) \qquad (4.5.15)$$

which has been sketched in Fig. 4.5.4c. Equation (4.5.15) has an easy interpretation (relatively easy anyway). First we note that it is a convolution of two terms. The term on the right is exactly the same as Eq. (4.5.12), and therefore it represents the FT of an infinite series of pulses. The other term [Sa $(\omega T_i/2)$] is nothing but the FT of the *wide* pulse (with width T_i), and thus it is a rather *narrow* continuous signal, as we can ascertain from the scaling property of the FT. Convolving now this last continuous signal with the infinite series of delta functions has the effect of reproducing the signal Sa $(\omega T_i/2)$ around each delta function, as shown in the figure. So, starting with a pulse (continuous spectrum), we went to an infinite series of pulses (discrete spectrum) and then to a finite number of pulses (continuous spectrum). The last step can be thought of as taking each discrete point of the discrete spectrum and stretching it out into a narrow Sa $(\omega T_i/2)$ shape.

This last FT is, naturally, quite important, and the reader should study it closely, noticing its differences from the others, its maxima and where they occur, its zeros, and so on. We will see it again soon when we take up target velocity measurements with the PR system in a later section.

One final remark. The FT of all our signals have been found and shown, without the carrier in them (i.e., before the modulation). If they are multiplied by a $\cos \omega_c t$, as is always done in pulse radar, their shapes are naturally shifted to $+\omega_c$ and $-\omega_c$.

4.6 THE CONTINUOUS-WAVE RADAR SYSTEM

A narrow rectangular signal (or a group of such signals) may be a good choice for measuring target range (shift in time), but is it a good choice for measuring its radial velocity?

To answer this question, we must first learn how velocity is measured, and this brings us to the **Doppler effect**. Let us see what this is all about.

We already know that if the radar sends out a signal $s_e(t)$, the return from a stationary target will be

$$s_r(t) = As_e(t - \tau_0) \tag{4.6.1}$$

where A takes care of losses and the delay τ_0 is related to the target's distance R by

$$\tau_0 = \frac{2R_0}{c} \tag{4.6.2}$$

as we learned in the preceding section [see Eq. (4.5.1)]. Note, however, that we are now stressing the fact that (4.6.2) holds for stationary targets, a detail that we chose to ignore in Section 4.5.

Let us now assume that the target is moving with radial velocity v_r (with respect to the radar). If this is the case, the distance R_0 is changing, even during the short illumination time that the radar signal is bouncing off the target. Let us assume that v_r is uniform (no acceleration). Under these conditions, the target distance is not a constant R_0, but some $R(t)$, where

$$R(t) = R_0 \pm v_r(t - t_0) \tag{4.6.3}$$

where t_0 is the time at which the distance was R_0 (i.e., when $t = t_0$). This being the case, the delay τ_0 is not a constant either. In view of Eq. (4.6.3), it is

$$\tau = \frac{2}{c} R(t) = \frac{2R_0}{c} \pm \frac{2v_r}{c}(t - t_0) = \tau_0 \pm \frac{2v_r}{c}t \mp \frac{2v_r}{2}t_0 \tag{4.6.4}$$

Thus the received echo will now have the form

$$s_r(t) = As\left(t - \tau_0 \mp \frac{2v_r}{c}t \pm \frac{2v_r}{c}t_0\right)$$

$$= As\left[\left(1 \mp \frac{2v_r}{c}\right)t - \tau_0 \pm \frac{2v_r}{c}t_0\right] \tag{4.6.5}$$

This equation pretty much tells the story of the Doppler effect. The main difference between Eqs. (4.6.5) and (4.6.1) is a time-scaling factor

$$\alpha = 1 \mp \frac{2v_r}{c} \tag{4.6.6}$$

which compresses the signal in time if the target is approaching ($+$ sign), much as a ball is squashed if it is bounced off an approaching object. And the opposite happens (time expansion) if the target is moving away from the radar, just as a ball is squashed less by an object that is moving away than by a stationary object. This time compression (expansion) is small since $2v_r \ll c$, unless the targets are moving with very high velocity.

Besides this scale change, the Doppler effect also causes a time delay on the return signal by an amount $\pm 2v_r t_0/c$, which we will do well to remember, even though it does not show up on the amplitude spectrum of the return signal. Anyway, let us ignore this for the time being and concentrate on the effects of the change in time scale.

The question that comes to mind is: How do we measure this α? For if, indeed, we could measure it, we could, effectively, have v_r, the relationship between the two being an easy one, given by Eq. (4.6.6). Of course, this whole development is based on the idea of uniform velocity, but similar things happen in the nonuniform case, except that the α is more complicated.

Here is another instant where understanding the FT properties comes to play its part. A scale change in time is, of course, reflected as an inversely proportional scale change in frequency. This causes not only expansion or compression, but could also cause a shift.[2] Let us see about that below.

Let us assume that the emitted signal is of the AM variety, that is,

$$s_e(t) = s(t) \cos \omega_c t \tag{4.6.7}$$

where $s(t)$ is a constant, a pulse, a group of pulses, or what have you. The FT of $s_e(t)$ will be

$$S_e(\omega) = \tfrac{1}{2}[S(\omega - \omega_c) + S(\omega + \omega_c)]$$

If the target is moving with radial velocity v_r (uniform), the return signal will be [in view of Eq. (4.6.5)]

$$s_r(t) = s\left(\alpha t - \tau_0 \pm \frac{2v_r}{c} t_0\right) \cos\left(\alpha \omega_c t - \tau_0 \pm \frac{2v_r}{c} t_0\right) \tag{4.6.8}$$

and this equation illustrates quite well what happens. The original signal suffered not only distortion in time (and therefore in frequency), but also suffered a shift in frequency. The cosine that it is multiplied by no longer has a frequency of ω_c, but of $\alpha \omega_c$, which means that the (distorted) FT of $s(t)$ has been shifted to a new value, $\alpha \omega_c$.

---

[2]We have seen this before, haven't we? Where?

To see this more clearly, let us ignore the $(2v_r/c)t_0$ and τ_0 (they add only a phase angle to the FT) and find the new $S_r(\omega)$. Recalling the time scale property, we have that

$$S_r(\omega) = \frac{1}{2\alpha}\left[S\left(\frac{\omega}{\alpha} - \omega_c\right) + S\left(\frac{\omega}{\alpha} + \omega_c\right)\right] \tag{4.6.9}$$

which means that the distorted form $[S(\omega/\alpha)]$ is now sitting at $\omega = \alpha\omega_c$, not at $\omega = \omega_c$ as was the case with a stationary target. So, indeed, $S_r(\omega)$ has experienced a shift ω_d, given by

$$\omega_d = \omega_c - \alpha\omega_c = \omega_c - \left(1 \mp \frac{2v_r}{c}\right)\omega_c$$

or

$$\omega_d = \pm\frac{2v_r\omega_c}{c} = \pm\frac{4\pi v_r}{\lambda} \tag{4.6.10}$$

which is usually called the **Doppler shift**. The sign of the shift depends on the direction of the movement of the target, and it is positive if the target is approaching the radar.

The upshot of this discussion is that the radial velocity of a moving target can be measured by measuring the shift that it causes in the spectrum of the echo. Of course, it causes a distortion as well, but it is usually slight and very difficult to measure, anyway. We simply ignore it, and zero in on the Doppler shift ω_d, the value of which easily provides us with v_r by Eq. (4.6.10).

But we have already talked about the key ideas behind measuring shifts, in the preceding section. Sure, it was a shift in time and we called it a delay. So what? A shift is a shift is a shift,[3] whether it is in time, in frequency, in space, or whatever. So we forget about the fact that the variable is now ω (actually we should not forget it, but simply ignore it) and proceed to apply the ideas of Section 4.5 to this case as well.

To be able to measure a shift, the spectrum of the emitted signal must have a sharp point of reference. Better yet, its autocorrelation function must have a sharp maximum and preferably, only one. Now we are talking, of course, about the autocorrelation function *of the FT* of the emitted signal $m(t)$ (with or without the carrier), not of $m(t)$ itself. This function is called the *frequency autocorrelation* (or *ambiguity*) *function* and is given by

$$\mathcal{H}_m(v) = \int_{-\infty}^{+\infty} M(\omega)M(v - \omega)\,d\omega \tag{4.6.11}$$

assuming that $M(\omega)$ is like an energy signal in ω, or the other usual expressions otherwise (see Section 2.10).

So a good signal for measuring v_r would be one whose $\mathcal{H}(v)$ has a sharp maximum, as sharp as possible. What, then, would be an excellent candidate in

[3] May Gertrude Stein forgive us for the horrible misquotation.

this case? Why, a complex exponential, of course. If

$$m(t) = e^{j\omega_c t} \qquad (4.6.12)$$

then $\mathcal{H}_m(\nu)$ is a delta function at the origin—an ideal choice, but an impractical one, since complex exponentials are only a figment of our imagination and cannot be produced with a device. So we try the next best thing, the cosine. Thus

$$m(t) = \cos \omega_c t \qquad (4.6.13)$$

should make a fairly good second choice. Its FT

$$M(\omega) = \pi[\delta(\omega - \omega_c) + \delta(\omega + \omega_c)] \qquad (4.6.14)$$

certainly has a sharp point of reference at $\omega = \omega_c$ (if we measure the shift for positive frequencies that exist in practice), and its $\mathcal{H}(\nu)$ (find it) has a very sharp maximum at $\nu = 0$, on which you can base your measurement if you desire.

That is the basic idea behind continuous-wave (CW) radar, even though frequency ambiguity functions did not figure in its original conception. The signal it uses is simply a carrier, with frequency in the radar band, of course. It can still be considered an AM signal, a trivial case with constant-amplitude modulation, and that is why we are discussing it in this chapter. In fact, if you think of it that way, the original signal (before modulation) is a constant whose FT is $2\pi\delta(\omega)$, a nice sharp signal as demanded in our shift measurements. So is its $\mathcal{H}(\nu)$; and this is actually a better way to think of it, since the comparison with the pulse radar analysis is then more direct. The ideal there was also an impulse, but in time, of course.

We should also remark that even though theoretically the original $s(t)$ is an everlasting constant, [or $m(t)$ an everlasting sinusoid], in practice, the illumination time of a given area is finite. Therefore, the emitted signal that can bounce off a target is more closely portrayed by

$$m(t) = \cos \omega_c t \, \text{rect}\left(\frac{t}{T_i}\right) \qquad (4.6.15)$$

and we are back to a pulse, multiplied by a cosine, whose FT has the Sa (ω) form, shifted to $+\omega_c$ (and $-\omega_c$). Still, there is a lot of difference between Eqs. (4.6.15) and (4.5.13) (aside from the fact that the carrier is missing there)— which is that now we have a constant, lasting T_i seconds, not N pulses. The difference is, of course, in the FT, where it counts. The FT of (4.6.15) has a narrow continuous Sa (ω) form, shifted to ω_c (and $-\omega_c$), whereas the FT $S_N(\omega)$ of Eq. (4.5.15) shown in Fig. 4.5.4c is not very suitable for measuring spectral shifts. But we will see some more on this toward the end of the section when we try to answer the question posed at the start of our present discussion. In any event, the actual signal used in a CW radar does have the proper FT and $\mathcal{H}(\nu)$ characteristics, although in a rather approximate way. Nothing in practice is as in theory, anyway.

It is time to start looking at the actual CW radar system. A simplified version of such a system is shown schematically in Fig. 4.6.1. It should be noted that the system is shown with two antennas, since the emitted signal is never interrupted and this is considered one of its disadvantages. Actually, CW radars with one antenna can also be constructed, based on the fact that the return echo is of different frequency and thus the receiver can, if it wishes, filter out the emitted signal.

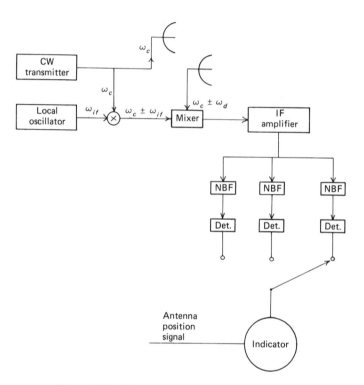

Figure 4.6.1   Block diagram of a CW radar system.

In the system shown in Fig. 4.6.1, the transmitter (magnetron, klystron, etc.) emits a signal, which for simplicity we assume to be a pure sine wave $\cos \omega_c t$ (the figure shows the frequencies at various points in the system). The return echo from a moving target with radial velocity v_r is a signal $\cos (\omega_c \pm \omega_d)t$ (we ignore its phase). This signal is mixed (multiplication and filtering) with a $\cos (\omega_c \pm \omega_{if})t$, produced by multiplication of the transmitted signal and a $\cos \omega_{if}t$, generated locally. The result is a cosine in the IF region of the form $\cos (\omega_{if} \pm \omega_d)t$, which is amplified and heads for a group of NBF filters. One of these filters extracts out the ω_d component. Its output is turned into a constant (dc) by an envelope detector of the AM type, and via an electronic sampling switch and the proper adjustments, the final output is a velocity value at the correct direction (the antenna position is also fed to the indicator). So if a target is there, and is moving with some v_r, the direction (usually only an azimuth angle) and the value of the velocity will show up on the screen of the indicator. Objects that do not move (mountains, trees, etc.), or targets with no v_r, show up with zero velocity. Negative or positive v_r's are measured by the output of the filter as well, since the middle NBF filter measures zero velocity and the others are designed to its left and right. Sometimes, the radar is designed so that the sign of the ω_d cannot be measured, and then other subcircuits can do this particular job (see Problem 4.17).

The NBFs must be made as narrow in bandwidth as the desired accuracy of the v_r measurement. Of course, this is a function of the state of the art of filter design as well. To get an idea of the bandwidth required, we take a look at the Doppler effect equation. It can easily be calculated that if $f_c = 10$ MHz and $v_r = 2000$ ft/s, then $f_d = \pm40$ kHz.

So the analysis of a CW radar system was fairly straightforward, what with all this background that we have developed in studying communication systems. The shift-and-half routine is quite high on the list of tools, as is filtering, envelope detecting, and so on. The same story will reappear in the analysis of other AM systems, as we will see in the following sections.

Now we must return to the fundamental question posed at the beginning of this section and add a few others as well. We hope that the reader finds such issues stimulating. They lie at the heart of the matter and are the basis of further research and improvement. To do these issues justice, we should have noise as well in the analysis of the system. But, no matter. Adding noise makes the thing mathematically complicated and obscures the ideas. It is pedagogically better, we think, to be exposed to these ideas now so that their reexamination later in the presence of noise will be less painful.

Let us therefore start by simplifying our radar signals so that the entire discussion is stripped of all its confusing practical matters. Let us consider that our signals are looked at before the modulation and without regard for the target illumination time T_i. For CW radar, this means, then, that we will stick to

$$s_c(t) = 1 \quad \text{(a constant)} \tag{4.6.16}$$

as adequately representing the emitted signal. For the PR system we will even make a further simplification and assume that only one pulse is emitted, that is,

$$s_p(t) = \text{rect}\left(\frac{t}{T_p}\right) \tag{4.6.17}$$

which means that we also ignore the fact that maximum range dictates a PRF. Our conclusions will be approximate this way, but valid in general, anyway.

Let us start with the CW system. We are by now convinced that the signal it emits is excellent for measuring spectral shifts. We cannot even think of a better one, since its $\mathcal{K}_c(v)$ is an impulse at $v = 0$ [see Eq. (2.3.10b)], and getting sharper than that is impossible. But how about measuring shifts in time (i.e., the ranges of targets)? Pretty poor, we are afraid. Its time ambiguity function is also a constant, with no beginning or end for a reference point. So do not expect much from CW radar as far as ranges go; it just cannot do it. It can be modified to do it (CW with FM, for example), but then it is not CW radar, it is something else.

Now, let us go on to pulse radar. To make it measure the range very accurately, you try to make the pulse as narrow as possible. In fact, you would make the pulse into an impulse if you could—the ideal situation—since its time ambiguity function $R_p(\tau)$ would also be an impulse at $\tau = 0$. But what about its ability to measure spectral shifts? Zilch again. An impulse has a FT which is a constant. Its $\mathcal{K}_p(v)$ would also be a constant, which is the bottom in spectral shift measurements. Very interesting.

These two systems then seem to be exactly opposite (at least theoretically) in their ideal forms. The argument above is more solid for the CW system since Eq. (4.6.16) is close to the actual signal. For the PR signal it holds, but to a lesser degree. The fact that we send a series of pulses creates a spectrum with peaks. We can try to zero in on one of its peaks and measure its shift, for example. But if the IPP is very long, the peaks get too close together. Even if the IPP is short, the $\mathcal{K}_p(v)$ has many maxima, which create ambiguities. We will see more on this later when we discuss the pulse-Doppler radar system.

The upshot of the above is that $R(\tau)$ and $\mathcal{K}(v)$ can serve us well. We notice that they seem to go in opposite ways. If $R(\tau)$ is sharp, $\mathcal{K}(v)$ is smooth, and vice versa. Wouldn't it be nice to find a signal that has both of them as sharp as possible? Does such a signal exist? To study this matter, Woodward (1953) combined the two ambiguity functions into one containing both variables (time and frequency shifts). The combined function, called the **time frequency ambiguity function** (or **signal ambiguity function**) is such that it reduces to $R(\tau)$ for $v = 0$, and to $\mathcal{K}(v)$ for $\tau = 0$. It is well that Woodward tried this. His idea gave the optimum radar signal design field a status it deserved. Up to then it was only empirical—hit and miss. We mentioned this earlier (see the end of Section 2.10) and we will see it again later when we reexamine the same issues in the presence of noise.

We close this section with a few remarks on applications. CW radar systems are old; in fact, some of the first experimental radars were CW. They are used primarily to detect targets in clutter (noise from immovable objects). The *radio proximity* (VT) *fuse* used to fuse artillery projectiles was this type of radar. CW radars are often used for the detection of tornadoes, as an aircraft navigation aid, as a rate-of-climb meter for vertical-takeoff aircraft, and for other purposes. Radar speed traps along the highways are probably the closest that most people come to a CW radar system.

4.7 THE PULSE-DOPPLER RADAR SYSTEM

If pulse radar can measure range well, and CW can measure velocity well, why not combine the two and come up with a radar system that can do both? That is the key idea behind pulse-Doppler radar. Before we get too enthusiastic about the idea, however, we should recall the classical maxim that "you never get anything for nothing." A price will have to be paid. The new system might well do both simultaneously, but it will end up being "a jack-of-both-trades and master of neither."

Let us start by looking at the *basic* signals that PR and CW use, and see what we mean by *combining the two*. By "basic" we mean the signal before the modulation. We will also ignore the illumination time T_i and bring it in later to verify our conclusions. Under these conditions, the CW signal is

$$s_{cw}(t) = K \tag{4.7.1}$$

and the PR signal is

$$s_{pr}(t) = [\text{rep}_T \, s(t)] \tag{4.7.2}$$

where

$$s(t) = A \, \text{rect}\left(\frac{t}{T_p}\right) \tag{4.7.3}$$

and T_p and T are pulse width and IPP, respectively.

So, one signal is a constant, and the other is a pulse train with period T, or a *constant with gaps in it*. How do we combine the two, or, more accurately, reach a compromise between them? Well, by decreasing the IPP of the PR signal, obviously. For if we keep decreasing it, we will eventually turn the PR signal into a CW signal. Naturally, our aim is to decrease it enough so that the PR signal gains some of the capability of measuring target velocity, while keeping a good amount of its capability to measure target range. That is the whole idea behind pulse-Doppler radar—a fairly simple one, the reader will agree. And adding the illumination time T_i takes nothing away from the argument above, since even within this time, the only way that you can compromise between the two signals is by decreasing the IPP.

Now let us go to the frequency domain and see what a decrease in the IPP does to the spectrum of the PR signal. Ignoring illumination time, the FT of a periodic pulse leads to the line spectrum studied in Example 2.5.5 and discussed again in our study of the PR system. The lines have amplitudes dictated by the values of a $\sin(a\omega/a\omega)$ function, with maximum at $\omega = 0$. We learned in Section 2.5 that the effect of decreasing the period of the pulse train is to increase the distance between these lines. If the distance between these lines gets big enough, we can zero in on one of these lines and measure *its shift* caused by the Doppler phenomenon, and therefore the velocity of the target. That is exactly how the pulse-Doppler radar measures target velocity without losing its ability to measure range, since it is still a pulse radar.

If we include the illumination time, things change a bit, but the central idea behind pulse-Doppler radar remains the same. The spectrum in question is now the one shown in Fig. 4.5.4c (reproduced later as Fig. 4.7.2), which is not, of course, of the discrete type, but still has distinct rounded peaks. Increasing the PRF brings these peaks farther apart. The radar can then measure the Doppler shift of one of them (usually the middle one, with the highest amplitude) and this measurement can easily be converted to target velocity with the use of Eq. (4.6.10).

Naturally, pulse radar's ability to measure range suffers a decline due to the increase in the PRF. That is the price we must pay for the compromise. Increasing the PRF causes a corresponding increase in the *blind* ranges, since the *transmit* time is increased and the *receive* time decreased. And that is not all. The decrease in the interpulse period decreases the distance between the peaks of the time ambiguity function (Fig. 4.5.4b), a result that increases the ambiguities in range measurements. That is why we remarked earlier that tampering with the PRF must be done carefully.

The time has come to stop the discussion and take a look at a block diagram of the pulse-Doppler radar system, as shown simplified in Fig. 4.7.1. This diagram shows that this system is a combination of pulse radar and CW radar. It has one antenna and ATR and TR switches just like pulse radar, but the

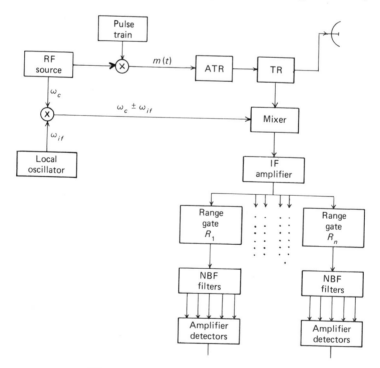

Figure 4.7.1 Pulse-Doppler radar system.

return signal is processed as in CW radar. The return signal is not, of course, a shifted (by ω_d, the Doppler frequency) sinusoid, but a series of pulses whose entire spectrum has been shifted by ω_d. The mixer shifts the spectrum of the pulses to the IF region where it is amplified. At this point, the signal will enter one of n range gates, depending on its time of arrival (i.e., the target distance). These range gates are filters that open and close at intervals of time corresponding to target distances, and in actuality they measure target range (see also Section 10.4).

Once target range can be ascertained by the above-mentioned system (or some other), the signal leaves the gate and enters a system of NBF filters centered around the middle of the spectrum of the pulses (the carrier frequency shifted to the IF region). Each filter is located at some anticipated shift (right or left) of the middle peak of the spectrum of the signal. Thus the system measures the Doppler shift, and therefore the target radial velocity, by identifying the filter with an output, much like the CW system does, as we saw in the preceding section. The rest of the system (local oscillators, etc.) is also typical of the CW system and its purpose was explained in Section 4.6.

Let us now take a look at how well this system can measure Doppler shifts, (i.e., identify its weak points on this issue), much as we did above for range measurements. One cannot expect a lot, of course. We know that it is a compromise, and therefore that it cannot measure both with extreme accuracy.

Theoretically speaking, even the frequency ambiguity function of the signal would tell us that this system will have problems with measuring frequency shifts. Of course, we do not have it. We have, however, the FT of the signal (Fig. 4.5.4c) and we could get it, if we wanted (the reader should try it). But we could guess what it would look like without actually finding it. It would look similar to the spectrum, in terms of peaks, anyway, and too many peaks, of course, indicate ambiguities.

Let us return to Fig. 4.5.4c, which we have reproduced in Fig. 4.7.2 for easy reference. In the system's receiver this FT is shifted to an IF frequency but this does not affect our argument here.

To measure the Doppler shift the system *concentrates* on the middle peak

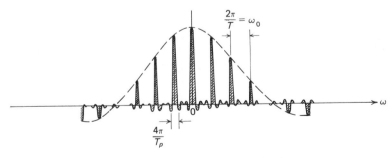

Figure 4.7.2 The FT of a group of pulses.

and follows its shift to the right or left with the NBF filters. However, the system cannot really tell whether what comes out of a NBF is the *shifted* middle peak or a shifted *other* peak. For example, there may be a target out there with such velocity that the entire spectrum is shifted by ω_0. This would cause the NBF filter located at $\omega = 0$ (actually, $\omega = \omega_{if}$) to give an output resulting in a measurement of zero radial velocity. These ambiguities are much like range ambiguities, if the reader can think of the emitted signal in its FT form. There, too, we look to measure the shift of a time pulse (a frequency pulselike signal, here), but we were not sure if we were measuring it or the shift of the pulse before it. To correct that, one can increase the PRF so that the peaks of the FT sit so far apart that no object can have such a speed as to create ambiguities. But that causes more *blind* ranges and more ambiguities in range. There just is no way that you can get this system to be very good in both measurements simultaneously, as we warned at the start of the section.

The pulse-Doppler system has other weaknesses, which we will not take up in this book. Let us just mention one more, though, which is important and easy to understand. The velocity measurement is based on a small part of the return echo (the middle peak of the FT), and therefore on only part of the entire return signal energy. That is not very efficient, particularly in the presence of noise, and more errors must be anticipated than in the case of a CW radar with the same return energy.

The central issue of the pulse-Doppler system is, of course, the PRF. Generally, this radar is designed so that the velocity measurement ambiguities are minimal. One can start by deciding the maximum radial velocity that he or she wishes to measure in a particular application without ambiguity. This sets the spacing between the FT peaks and therefore the PRF of the radar.

It can be shown by the use of the Doppler shift equation (4.6.10) that the minimum value of the PRF, f_{min}, is given by (show it as an exercise)

$$f_{min} = \frac{4v_{max}}{\lambda} \tag{4.7.4}$$

where v_{max} is the maximum anticipated closing velocity and λ is the wavelength of the carrier frequency.

We close this section with an example that shows some "ballpark" figures for the various quantities, for better familiarity with the system.

Example 4.7.1

Find the PRF necessary to measure velocities up to 4000 ft/s if the carrier frequency is 10,000 MHz.

Solution. After fighting some with the units, one obtains [using (4.7.4)]

$$f_{min} = 80 \text{ kHz}$$

4.8 THE PULSE RADAR SYSTEM WITH
MOVING TARGET INDICATION

Let us return to the regular pulse radar system and look at it with a distant philosophical eye, void of small details that cloud the issues. The emitted PR signal is an AM signal, where its modulating part is a pulse or a group of pulses. Since pulses are positive, we know that the emitted signal behaves like an *AM with carrier* type of signal (no added constant needed to make it positive, i.e., 100% modulation), and thus it can be demodulated with an inexpensive peak detector. All this was said before during our discussion of the PR system. Why bring it up again? Because although what we said above is true, all right, it is also true that the emitted PR signal is actually a double-sideband suppressed carrier signal, and as such it can also be demodulated with a synchronous-type detector. But why bother, you say? Why should we try to kill a fly with a machine gun when there is a fly swatter available? Well, what if synchronous-type detection helped in the problem of measuring the target velocity? Would it not, then, be worthwhile?

Synchronously detecting the return echo is the basic idea behind the pulse radar system with MTI (moving target indication). As the name implies, this type of radar attempts to *distinguish moving targets from stationary ones*. In an advanced form, it can also measure the actual velocity of moving targets.

To explain the mathematical idea behind this type of radar, we would first like to recall Example 3.2.2, which considers the case of a DSB system when the transmitter and receiver cosines have the *same phase* but differ slightly in carrier frequency ω_c. We found there that the FT of the final output of the receiver was

$$O(\omega) = \tfrac{1}{4}F(\omega \pm \Delta\omega) + \tfrac{1}{4}F(\omega \pm \Delta\omega) \qquad (4.8.1)$$

where $f(t)$ is the original informational signal and $\Delta\omega$ is the difference in frequency between the carrier and the locally generated cosine. In the time domain,

$$o(t) = \tfrac{1}{2}f(t) \cos \Delta\omega\, t \qquad (4.8.2)$$

rather than

$$o(t) = \tfrac{1}{2}f(t) \qquad (4.8.3)$$

as is the case without frequency error. Equation (4.8.2) represents the principal mathematical idea behind pulse radar with MTI. Let us see how.

We already know that the fundamental difference between the echoes received from stationary targets and those received from moving targets is that the latter add a frequency shift ω_d to the carrier of the radar signal. Thus the return echo from a moving target is (more or less)

$$m_r(t) = p(t) \cos(\omega_c \pm \omega_d)t \qquad (4.8.4)$$

where $p(t)$ is a finite series of pulses dictated by the illumination time. If we were now to synchronously detect the signal of Eq. (4.8.4) using a $\cos \omega_c t$ in the product detector, the result would be

$$o(t) = \tfrac{1}{2} p(t) \cos \omega_d t \qquad (4.8.5)$$

much as in Example 3.2.2, bringing the Doppler frequency ω_d out where we can deal with it, as we will see in a minute. When the target is stationary, the synchronous detector output will be

$$o(t) = \tfrac{1}{2} p(t) \qquad (4.8.6)$$

much like Eq. (4.8.3) in the DSB system case.

Equations (4.8.5) and (4.8.6) pretty much tell the story of the PR system with MTI. Their difference is the $\cos \omega_d t$ term. If the target is stationary, the output comprises pulses of the same amplitude. If the target is moving (i.e., it has a radial velocity component), the amplitude of the pulses varies. Conversely, if we can detect no variation in amplitude from pulse to pulse, the target is stationary. If there is variation, the target is moving. This gives us the ability to *distinguish* between stationary and moving targets, which is the main function of the system. Of course, the variation in amplitude is related to $\cos \omega_d t$ and further refined circuitry can also measure ω_d, although with difficulty and limitations, but these issues will not be discussed in this book.

In most MTIs the output of the synchronous detector enters another subsystem which makes the operation of the radar even simpler. A typical such subsystem is a delay line canceler, shown in Fig. 4.8.1. The modulated (by $\cos \omega_d t$) series of pulses is delayed by one pulse period, T_p, and subtracted from itself. If the target is stationary, we have the situation in Eq. (4.8.6), where the amplitude of the pulses is the same, and $x(t)$ will be zero. If the target is moving, $x(t)$ will have an output. Even a monkey can be trained to differentiate moving targets from stationary ones.

Figure 4.8.1 Delay line canceler.

The presentation above is, of course, a simplification of the MTI system, as all of our systems are in this book. The *practical* application of the foregoing idea is quite difficult and took many years to develop. As we remarked earlier, the echo and the locally generated cosine must be in phase, something not easily accomplished, owing to the nature of the devices that are used in the radar frequency and power ranges. Then there is the Doppler phase-angle shift, which cannot be omitted altogether as we usually do in analyses, not to mention higher-order effects of the Doppler phenomenon if the velocity is not uniform.

MTI subsystems (or circuits, as they are often called) are of many types and can be found in the radar literature. If the radar is sitting on a moving platform (ship, aircraft, etc.), then even stationary targets are in relative motion, and this must be taken into account, usually by synchronizing with a return from such targets and not with the carrier's frequency. The latter circuits are called **asynchronous MTIs** to distinguish them from the regular ones (on stationary platforms), which are called **synchronous**. There are even some types called **area MTIs**, which do not use Doppler information directly but rather, the actual movement of the target from scan to scan.

Even if all phase problems are ignored, MTI radar still has some problems inherent in its nature. The most important one is *blind speeds*, target velocity values that cannot be detected by the radar. The reasoning behind this problem is quite simple. It is possible to have zero output from the delay line canceler, even though there is a $\cos \omega_d t$ in the input. The signals $p(t)$ and $\cos \omega_d t$ are both periodic (within the illumination period) and their product (under certain conditions) can also be periodic with the same period, giving zero out of the subtractor. The conditions have to do with T_p and ω_d. Usually, T_p is fixed, so they have to do with ω_d (see Problem 4.22). In other words, certain ω_d's (i.e., certain velocities) cannot be detected, and these are called blind speeds. That is why we often refer to MTI radar as having blind speeds, in contrast with pulse-Doppler radar, which has *blind ranges*, as we mentioned earlier.

With this, we close our first chapter on radar systems, the ones whose emitted signal is of the AM type. We should like to emphasize that the AM-ness of these systems is in the signal emitted from the radar. The notion of AM is used in other types of radar systems (tracking types, for example), but in those the AM modulation of interest is caused by the nature of the antenna and target location, not by the inherent nature of the transmitter, as in the systems discussed in this chapter.

PROBLEMS, QUESTIONS, AND EXTENSIONS

4.1. Consider a PR system that can detect signals with $S_{\min} = 10^{-11}$ mW. It uses one antenna for transmitting and receiving, with $G_0 = 35$ dB. It broadcasts a signal with $\lambda = 3$ cm and with $P_T = 1$ MW.
   (a) How far can it detect a small plane with $\sigma = 1$ m^2?
   (b) What should be the IPP of this pulse radar so that there are no range ambiguities up to the distance obtained in part (a)?

4.2. How do we justify that the time ambiguity function of an impulse is itself an impulse?

4.3. Let us consider Eq. (4.5.12), which gives us the FT of a periodic pulse $s(t)$ with period $T = 2\pi/\omega_0$.
   (a) Assume that the fundamental pulse $s(t)$ is

$$s(t) = 10 \text{ rect} \left(\frac{t}{10}\right)$$

with time in microseconds, and that $T = 1$ ms. Find and sketch the FT.

(b) Repeat part (a) with $T = \frac{1}{2}$ ms and note the difference in the spacing of the lines.

(c) Sketch the time ambiguity function in parts (a) and (b).

4.4. Repeat Problem 4.3 if the fundamental signal $s(t)$ is not a pulse but a triangle. Chose your own parameters for part (a).

4.5. Equation (4.5.15) gives us the FT of N pulses, where T_i is the illumination time which determines their number. Using the parameters of Problem 4.3 for the pulse duration and IPP:

(a) Sketch this FT (or amplitude spectrum) if $T_i = 200 \ \mu s$.

(b) Repeat with $T_i = 100 \ \mu s$ and note the difference.

4.6. Find the FT of N triangle signals. Each triangle lasts $10 \ \mu s$ with maximum amplitude 5 V. The IPP [actually, ITP (intertriangle period)] is $500 \ \mu s$.

4.7. Repeat Problem 4.6 if the number of triangles is $N/2$ (i.e., illumination time halved), but the rest of the parameters remain the same.

4.8. Repeat Problems 4.6 and 4.7 with the IPP changed to 1 ms. What do you observe?

4.9. Find the time ambiguity function for all the signals defined in Problems 4.6 through 4.8.

4.10. Consider the signal

$$f_1(t) = \left[\text{rep}_T \text{ rect} \left(\frac{t}{T_p}\right) \cos \omega_c t \right]$$

How does it differ from the signal

$$f_2(t) = \left[\text{rep}_T \text{ rect} \left(\frac{t}{T_p}\right) \right] \cos \omega_c t$$

in the time domain? Find the FT of both and compare them. The differences should be studied well. Often, in radar, we want to emit $f_1(t)$ and the devices produce $f_2(t)$, and vice versa.

4.11. To refresh your memory on time scaling and phase shifts, try the following problems.

(a) If

$$s(t) = \text{rect} \left(\frac{t-10}{5}\right)$$

find its FT and the FT of $s(3t - 20)$.

(b) Repeat for

$$s(t) = \text{rect} \left(\frac{t-10}{5}\right) \cos 10^6 t$$

(c) Repeat parts (a) and (b) if

$$s(t) = \begin{cases} 1 - \dfrac{|t|}{5} & \text{for } |t| \leq 5 \\ 0 & \text{elsewhere} \end{cases}$$

4.12. In Section 4.6, immediately above Eq. (4.6.12), it is stated that a complex exponential is an excellent choice for a signal with a sharp maximum in its $\mathcal{H}(v)$. But what about a constant? How is a constant related to a complex exponential? Can't we really consider that Eq. (4.6.13) is an AM signal with amplitude a constant?

4.13. How do we justify that the frequency ambiguity function of a constant signal is an impulse?

4.14. Find the frequency ambiguity function of $s(t) = 10(\sin 10^3 t / 10^3 t)$ and of $s(10t - 20)$.

4.15. Find the frequency ambiguity function of $s(t) = 100(\sin 10^4 t / 10^4 t)^2$.

4.16. Consider the signal

$$s(t) = \cos \omega_c t \operatorname{rect}\left(\frac{t}{T_i}\right)$$

which is more representative of the return signal in CW radar. T_i is, of course, the target illumination time. Find its FT $S(\omega)$. Find $\mathcal{H}_s(v)$. Discuss the radar's capabilities to measure target velocity in view of this signal.

4.17. Sometimes a CW radar has narrowband filters on only one side of the carrier's frequency. It still works (why?), but one cannot tell whether a target is approaching or receding. To correct this, the system of Fig. P4.17 is often used (borrowed

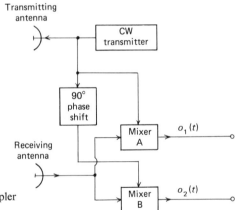

Figure P4.17 Measurement of Doppler direction.

from SSB communication systems, as can be seen by comparing it with Fig. 3.3.5). Assuming that the emitted signal is

$$s_e(t) = A_0 \cos \omega_c t$$

and the received one is

$$s_r(t) = KA_0 \cos ((\omega_c \pm \omega_d)t + \varphi)$$

show that the direction of the target can be determined by checking the phase-shift difference in the output $o_1(t)$ and $o_2(t)$. The two mixers filter out high-frequency components.

4.18. Prove Eq. (4.7.4).

4.19. Find and sketch the frequency ambiguity function of the signal in Eq. (4.7.2). Use your own signal parameters (pulse duration and IPP). Play around with the values of the parameters and note the changes.

4.20. What is the range ambiguity (in feet) resulting from using the PRF of Example 4.7.1?

4.21. One type of synchronous MTI system used is given in Fig. P4.21. Try to analyze its operation. You can get help in Barton (1976, p. 192) or Skolnik (1962, p. 118). COHO (coherent oscillator) is a stable oscillator whose frequency is the same as the IF used. STALO (stable local oscillator) is meant to be what its name implies.

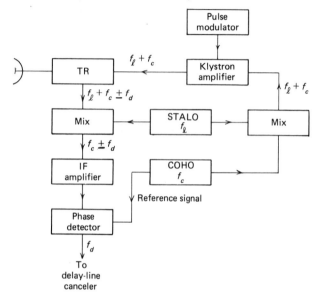

Figure P4.21   MTI block diagram.

4.22. Consider a synchronous MTI radar system with a given T_p. What are the conditions on ω_d that create blind speeds?

=5=

ANGLE MODULATION: COMMUNICATION SYSTEMS

5.1 INTRODUCTION

We have already noted that low-frequency-content analog signals (baseband signals, as they are often called) cannot travel on their own through the space channel. They have got to get on a carrier, sit back and relax, and let the carrier do the actual traveling. Amplitude modulation was a mechanism for such travel, and it led to *simple, efficient,* and *highly economical* receivers, as we stressed in Chapters 3 and 4.

Epithets such as *simple, economical,* and the like, are reminiscent of *tourist class* travel. And tourist class is not the only way to go, particularly if you are well-to-do and you like creature comforts such as more room to spread out, less noise from misbehaving children, or less jarring due to weather conditions. Well, for the discriminating traveler, we also have *first-class* travel, and first class in signal traveling is angle modulation, which includes both the *phase* and *frequency* types.

We mentioned in the introduction to Chapter 3 that a sinusoidal carrier $A \cos(\omega_c t + \theta_0)$ has three places in which to put the information signal: A, ω_c, and θ_0. If the carrier puts it in A, it is AM, and we are through discussing that. If the carrier puts it in ω_c, it is **frequency** (radian) **modulation**. If the carrier puts it in θ_0, it is **phase modulation**. Since the carrier can be written as

$$A \cos(\omega_c t + \theta_0) = A \cos \theta(t) \tag{5.1.1}$$

where

$$\theta(t) = \omega_c t + \theta_0 \tag{5.1.2}$$

141

is often called the *angle of the sinusoid*, and since both ω_c and θ_0 are in this angle, both frequency and phase modulations are often called **angle modulations**. All this is *loosely speaking*, of course. Precise mathematical definitions will be given in the next section.

Angle modulation may be first class for the traveler, but it is a *miserable* class for the analyst. In fact, except for a finite number of cases, angle modulation systems cannot be analyzed mathematically. Even these few cases, which, by necessity, we will have to present, are quite complex.

But that is part of the price you pay for first class. Still, we cannot help wishing that the price were paid entirely elsewhere. Teaching is difficult enough as it is and does not need complex mathematical analyses to confuse the basic issues. The reader must anticipate, then, a lot of approximations and hard-to-swallow generalizations. If we were lecturing on the topic, we would, of course, shout very loudly during all the weak points in the analysis, to show conviction, as clergymen are presumably taught to do during the weak points of their sermons (not really true, just part of an old joke).

5.2 PRELIMINARY NOTIONS OF FM AND PM

Let us start with a section that defines terms, discusses ideas, and points out some of the difficulties in angle modulation. Sometimes, appreciating the difficulties, knowing the *root of the problem*, so to speak, appears to solve it.

Consider first the carrier $c(t)$,

$$c(t) = A \cos (\omega_c t + \theta) \tag{5.2.1}$$

For a reason that will become obvious later, let us write this as

$$c(t) = A \cos \theta_i(t) \tag{5.2.2}$$

where we will call $\theta_i(t)$ the **instantaneous phase** (or angle) of the carrier.

Now let us differentiate this $\theta_i(t)$ with respect to t, denote the result $\omega_i(t)$, and call it the **instantaneous frequency** (radian) of the carrier. Thus

$$\omega_i(t) = \frac{d\theta_i(t)}{dt} = \dot{\theta}(t) \tag{5.2.3}$$

When the carrier is unmodulated, then

$$\theta_i(t) = \omega_c t + \theta_0 \tag{5.2.4}$$

and

$$\omega_i(t) = \omega_c \quad \text{(a constant)} \tag{5.2.5}$$

which simply means that the instantaneous frequency of such a carrier is a constant.

Note also that if you have the $\omega_i(t)$, you can get $\theta_i(t)$ by integration:

$$\theta_i(t) = \int \omega_i(t) \, dt = \int_0^t \omega_i(t) \, dt + \theta_i(0) \tag{5.2.6}$$

which comes from calculus. Indeed, for our carrier, starting with Eq. (5.2.5) and using Eq. (5.2.6), we obtain

$$\theta_i(t) = \omega_c t + \theta_0 \qquad (5.2.7)$$

and everything is fine and dandy.

The key idea in this discussion is that we have mathematically formalized the phase and frequency ideas in terms of each other, with Eqs. (5.2.6) and (5.2.3), respectively. After all, phase and frequency modulations are mechanisms whereby the information signal $s(t)$ will be inserted in these two things. Now we are ready to see how.

Let us take phase modulation first.

1. *Phase modulation.* Here the information signal $s(t)$ is placed as a *linear* term in the instantaneous phase of the carrier. In other words, in this case we have a box (phase modulator) with two inputs, the carrier and $s(t)$ (see Fig. 5.2.1a), and out of the box comes the original carrier with $\theta_i(t)$ modified to

$$\theta_i(t) = \omega_c t + \theta_0 + k_p s(t) \qquad (5.2.8)$$

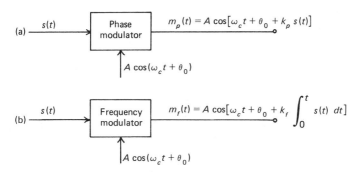

Figure 5.2.1   (a) Phase modulation; (b) frequency modulation.

where k_p is a constant of the modulating device.

So the expression for the output of the modulator, the modulated wave $m_p(t)$, will be

$$m_p(t) = A \cos(\omega_c t + \theta_0 + k_p s(t)) \qquad (5.2.9)$$

So far, so good. Now we look at the frequency modulation case.

2. *Frequency modulation.* Here the information signal gets inserted as a *linear* term into the $\omega_i(t)$ of the carrier. Out of this new box (Fig. 5.2.1b) comes the original cosine with $\omega_i(t)$ no longer ω_c but

$$\omega_i(t) = \omega_c + k_f s(t) \qquad (5.2.10)$$

where k_f is a constant due to the modulator. To write an expression for the output of the modulator, we must find the new $\theta_i(t)$. That is where Eq.

(5.2.6) comes in handy. Using it, we obtain

$$\theta_i(t) = \omega_c t + \theta_0 + k_f \int_0^t s(t)\, dt \qquad (5.2.11)$$

and with this,

$$m_f(t) = A \cos\left(\omega_c t + \theta_0 + k_f \int_0^t s(t)\, dt\right) \qquad (5.2.12)$$

somewhat messy, as we already warned the reader.

Now let us discuss these things simultaneously and get a better insight into their similarities and differences. To simplify matters, let us assume that $\theta_0 = 0$. With this,

$$m_p(t) = A \cos(\omega_c t + k_p s(t)) \qquad (5.2.13)$$

and

$$m_f(t) = A \cos(\omega_c t + k_f g(t)) \qquad (5.2.14)$$

where

$$g(t) = \int_0^t s(t)\, dt \qquad (5.2.15)$$

so we do not carry the integral around.

Let us take the $m_p(t)$ and find its $\omega_i(t)$. Equation (5.2.3) gives us that

$$\omega_i(t) = \omega_c + k_p \frac{ds(t)}{dt} \qquad (5.2.16)$$

This is very interesting. It tells us that in the PM case, the $\omega_i(t)$ has a linear term proportional to the derivative of $s(t)$. So we can say that in the PM case, *we have FM, but with the derivative of $s(t)$ and not $s(t)$ itself.* In other words, if we have a device that can produce FM, we can make it give us PM by putting in $\dot{s}(t)$ rather than $s(t)$, that is, by differentiating $s(t)$ before we insert it. That is nice.

Now look at the FM case. We have already found $\theta_i(t)$ in this case [Eq. (5.2.11)], and if we look at it closely, we note that it has a linear term proportional to the integral of $s(t)$. So we can say that *FM is really PM, with the modulating signal being $g(t)$ rather than $s(t)$.* In other words, if we have a device that produces PM, we can make it produce FM by inserting into it the integral of $s(t)$ rather than $s(t)$ itself. That, too, is nice.

The upshot of the two observations above is that we need not discuss both FM and PM, but only one of them. The one we will choose is FM, since this is the one that is mainly used for transmitting baseband analog signals such as speech or music. PM has some problems (we will see one of them later) and has found application primarily in transmitting digital signals (phase-shift keying), which we will cover in a later chapter. So FM it will be, and now and then we will refer to PM to see if our conclusions hold, and how.

Next, let us take a look at the typical FM communication system and see where the difficulties in the analysis lie. Such a system is given in block diagram form in Fig. 5.2.2. The first block is the FM modulator we mentioned earlier,

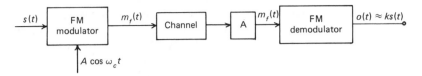

Figure 5.2.2   FM communication systems.

the nature of which the reader is assumed to be familiar with from other studies (we will have a little to say about it later). Out of this block comes $m_f(t)$ of the form (5.2.14), which gets amplified and pushed to the antenna (not shown), and traverses the channel. In the receiver, after filtering and amplification (block denoted as A) it goes through the demodulator, another device (also assumed known to the reader) that extracts the $s(t)$ from inside the angle of $m_f(t)$ and produces a reasonable facsimile of the original $s(t)$. So where is all this trouble in the analysis that we talked about earlier?

Well, it lies in that innocent phrase *traverses the channel*, which we sneaked in there when the reader was not watching. How do we know that $m_f(t)$ can traverse the channel? After all, not everything you sent out there can go through [$s(t)$ cannot]. In order to be able to do it, its frequency content must be around ω_c, something we do not really know about $m_f(t)$, and will not know unless we can produce $M_f(\omega)$, the FT of $m_f(t)$. That is where the bugaboo lies. Finding $M_f(\omega)$ is a very difficult task; in fact, it cannot be done in general for any $s(t)$. It *can* be done for a few $s(t)$'s, and after that, zilch! About the only thing we can do is try to generalize (by waving our arms violently) and hope that the reader believes us. But where exactly is the trouble? Let us see.

We are trying to find the FT of $m_f(t)$ of Eq. (5.2.14). This $m_f(t)$ can be written as

$$m_f(t) = A \cos \omega_c t \cos (k_f g(t)) - A \sin \omega_c t \sin (k_f g(t)) \qquad (5.2.17)$$

by breaking up the angle sum using a known trig relation. Now if we could find the FT of the two terms $\cos (k_f g(t))$ and $\sin (k_f g(t))$, the rest would be easy, since products with sines and cosines follow the half-and-shift routine.

It is exactly these two terms that lead us to a brick wall. The FT of $\cos (k_f g(t))$ or $\sin (k_f g(t))$ cannot be found in general for any $g(t)$ [or any $s(t)$ since $g(t)$ is its integral] and have, in fact, been found for only a few $s(t)$'s. Even for those that they have been found, the result is usually messy, an infinite series. There will be approximations galore, as we will see in the sections that follow.

To get a deeper insight into the trouble, consider the term $\cos (k_f g(t))$, and write it out in Taylor series form with respect to its argument. This gives

$$\cos (k_f g(t)) = 1 - \frac{k_f^2 g^2(t)}{2!} + \frac{k_f^4 g^4(t)}{4!} - \frac{k_f^6 g^6(t)}{6!} + \cdots \qquad (5.2.18)$$

which clearly points out the trouble we are in. If $s(t)$ is known [and therefore $S(\omega)$], $G(\omega)$ can be found by the integral property of the FT. But the term above has $g^2(t)$ in it, whose FT is the convolution of $G(\omega)$ with $G(\omega)$ (why?), and $g^4(t)$, whose FT is the convolution of the convolutions, and so on. So we can see that what we are after is a sum of such convolutions ad infinitum, or at least of a group of terms, if the rest can be ignored. Recall now that every time you convolve, you increase the bandwidth by about a factor of 2 (where did we first discover this?), and the result is that the $m_f(t)$ above is quite spread out in frequency; that is, it has a bandwidth which is a multiple of the bandwidth of $g(t)$ depending on the significant terms.

The discussion above not only points out the difficulties; it also gives insight. For one thing, the FT of $m_f(t)$ is around ω_c, as shown by Eq. (5.2.17), and therefore it can go through the channel, at least part of it. For another, this FT will have a large bandwidth [not at most twice the bandwidth of $s(t)$ as in the AM case], depending on the significant terms of Eq. (5.2.18), that is, on the value of k_f and the maximum value of $g(t)$. If we assume that $g(t)$ is normalized to a maximum value of unity, then the bandwidth of $m_f(t)$ clearly depends on k_f. Both of these are significant conclusions and will be proven correct in the sections that follow. Of course, $m_f(t)$ has another term just as difficult to deal with, the $\sin(k_f g(t))$ term, but a similar approach leads to the same conclusions—just more terms around ω_c.

Not all $s(t)$'s are impossible to do, or lead to infinite series. Some can be done, by using our knowledge of signals and a little ingenuity. Consider, for example, the following case.

Example 5.2.1

Find the FT of $m_f(t)$ if $A = 1$, and

$$s(t) = B \operatorname{rect}(t - \tfrac{1}{2}) \tag{5.2.19}$$

Solution. Let us first take a look at $\omega_i(t)$.

$$\omega_i(t) = \omega_c + k_f s(t) = \begin{cases} \omega_c + k_f B & \text{for } 0 \le t \le 1 \\ \omega_c & \text{elsewhere} \end{cases} \tag{5.2.20}$$

But this last equation tells us that $m_f(t)$ is a $\cos \omega_c t$ in the $(-\infty, 0)$ and $(1, +\infty)$ interval, and a $\cos(\omega_c + k_f B)t$ in the $(0, 1)$ interval. Now ingenuity comes into play. We can write $m_f(t)$ as

$$m_f(t) = \cos \omega_c t[u(-t)] + \cos(\omega_c + k_f B)t[u(t) - u(t-1)]$$
$$+ \cos \omega_c t[u(t-1)] \tag{5.2.21}$$

that is, as products of terms whose FTs are known. Now each term can be dealt with by using the FT properties and Tables 2.6.1 and 2.8.1. Thus we have that

$$u(-t) \longleftrightarrow \pi\delta(\omega) - \frac{1}{j\omega} \tag{5.2.22}$$

$$u(t-1) \longleftrightarrow \left[\pi\delta(\omega) + \frac{1}{j\omega}\right]e^{-j\omega} \tag{5.2.23}$$

and

$$u(t) - u(t-1) \longleftrightarrow \frac{\sin(\omega/2)}{\omega/2}e^{-j(\omega/2)} \tag{5.2.24}$$

Multiplying the first and last terms by $\cos \omega_c t$ shifts the spectra to $\pm \omega_c$. The middle term has its spectrum shifted to $\pm(\omega_c + k_f B)$. We will not bother to sketch $|M(\omega)|$, but it can be sketched. Note that the end result can be quite spread out, depending on k_f and B. Our initial guesses are borne out. Sketch it for various values of k_f and B if you do not believe it. This method can be used whenever $s(t)$ is one or at most n pulses, although it gets pretty messy. If the pulses are infinite in number (periodic or not), forget it.

We have said enough about the FT of $m_f(t)$, at least for the time being. It is time to say a few words about $m_f(t)$ itself, because it, too, is difficult to visualize or sketch.

Well, we all know what a regular cosine is and how it is sketched. It has a $\omega_i(t) = \omega_c$ (a constant), and this means that it is periodic, it crosses the zeros at regular intervals, and its maxima and minima are equidistant. But what if $\omega_i(t)$ is a function of time, as it is with the cosine of $m_f(t)$ [see Eq. (5.2.10)]? Obviously, all the above do not hold. The zero crossings (as well as maxima and minima) vary in distance with time in a manner proportional to $s(t)$. If $s(t)$ increases in value [$\omega_i(t)$ increases], the zero crossings get closer together, and vice versa. Let us look at a couple of examples.

Example 5.2.2

Sketch $m_f(t)$ if $A = 1$, $k_f = 10$, and

$$s(t) = \text{rect}\left(\frac{t}{2}\right) \tag{5.2.25}$$

Solution.   We go directly to $\omega_i(t)$ and ignore $m_f(t)$ altogether.

$$\omega_i = \omega_c + 10s(t) = \begin{cases} \omega_c + 10 & \text{for } |t| \le 1 \\ \omega_c & \text{elsewhere} \end{cases} \tag{5.2.26}$$

So $m_f(t)$ is basically a cosine with frequency ω_c from $(-\infty, -1)$ and $(1, +\infty)$ and a cosine with frequency $\omega_c + 10$ in the $|t| \le 1$ interval. Not too hard to sketch it, although we must watch out for a possible discontinuity at $t = \pm 1$. Try it!

Example 5.2.3

Do the same but with $s(t)$, as shown in Fig. 5.2.3a.

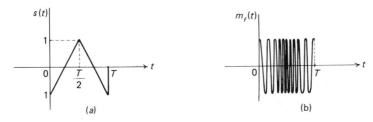

Figure 5.2.3   (a) $s(t)$; (b) $m_f(t)$ in the region $[0, T]$.

Solution.   Again we hit $\omega_i(t)$. Now it is

$$\omega_i(t) = \begin{cases} \omega_c & \text{for } t > T \text{ and } t < 0 \\ \omega_c + 10s(t) & \text{for } 0 < t < T \end{cases} \tag{5.2.27}$$

There is no problem sketching it when its frequency is ω_c. In between, it is practically impossible unless we do a lot of calculations. Anyway, Fig. 5.2.3b shows this difficult portion. The reader should gaze at this figure and verify the statements made earlier (increase of distance between zero crossing, etc.).

5.3 NARROWBAND FM AND PM

We now start a concentrated effort to find the FT of the FM modulated wave $m_f(t)$, which we rewrite here for easy reference:

$$m_f(t) = A \cos(\omega_c t + k_f g(t))$$
$$= A \cos \omega_c t \cos(k_f g(t)) - A \sin \omega_c t \sin(k_f g(t)) \tag{5.3.1}$$

recalling that

$$g(t) = \int_0^t s(t)\, dt \tag{5.3.2}$$

We already know that the naughty terms are $\cos(k_f g(t))$ and $\sin(k_f g(t))$, whose FT cannot be found for any $g(t)$. What can we do then? Well, one thing we can do is to see what happens when the naughty terms can be approximated by only their first term in their power series. The cases we are about to discuss go by the names **narrowband FM** and **narrowband PM**, for reasons that will soon become obvious. Narrowband FM has even found applications in ham and mobile communications, so what follows is not only of academic interest.

Let us assume that

$$|k_f g(t)| \ll \frac{\pi}{6} \tag{5.3.3}$$

In this case, we know from high school trigonometry that

$$\cos(k_f g(t)) \approx 1 \tag{5.3.4a}$$
$$\sin(k_f g(t)) \approx k_f g(t) \tag{5.3.4b}$$

and therefore,

$$m_f(t) = A \cos \omega_c t - A k_f g(t) \sin \omega_c t \tag{5.3.5}$$

whose FT is fairly easy to obtain. Indeed, using known facts,

$$M_f(\omega) = \pi A[\delta(\omega - \omega_c) + \delta(\omega + \omega_c)] - \frac{k_f A}{2j}[G(\omega - \omega_c) - G(\omega + \omega_c)] \tag{5.3.6}$$

If we now assume that

$$g(t) \longleftrightarrow \frac{S(\omega)}{j\omega}$$

which holds if $S(0)$ is zero (Table 2.7.1, Property 9), then

$$M_f(\omega) = \pi A[\delta(\omega - \omega_c) + \delta(\omega + \omega_c)] + \frac{k_f A}{2}\left[\frac{S(\omega - \omega_c)}{\omega - \omega_c} - \frac{S(\omega + \omega_c)}{\omega + \omega_c}\right]$$

$$(5.3.7)$$

which is a nice expression valid for all $s(t)$'s whose FT is known.

If the system is PM, the assumption $|k_f s(t)| \ll \pi/6$ leads to

$$m_p(t) = A \cos \omega_c t - A k_p s(t) \sin \omega_c t \qquad (5.3.8)$$

whose FT is

$$M_p(\omega) = \pi A[\delta(\omega - \omega_c) + \delta(\omega + \omega_c)] + j\frac{A k_p}{2}[F(\omega - \omega_c) - F(\omega + \omega_c)]$$

$$(5.3.9)$$

Now the reader can see why these cases go by the name *narrowband*. First look at Eqs. (5.3.8) and (5.3.9) and note that both the modulated wave and its FT are very similar to large-carrier AM systems. The only difference is in the phase shift caused by the $\sin \omega_c t$. Nevertheless, the bandwidth of this system is equal to the bandwidth of the AM systems (large carrier or DSB) and this is *narrow* as general PM systems go. The FM case, Eqs. (5.3.5) and (5.3.7), have the added difference that they deal with $g(t)$ rather than $s(t)$, but this hardly changes the bandwidth. In fact, it could reduce it some, depending on the definition of bandwidth one uses, since the shifted $S(\omega)$ terms are divided by $\omega \pm \omega_c$.

Expressions (5.3.5) and (5.3.8) even provide us with simple methods of producing these types of FM and PM, methods that have nothing to do with FM modulators. Can you draw block diagrams that use adders and multipliers and will produce FM and PM of the narrowband type? (See Problem 5.5.)

Example 5.3.1

The signal

$$s(t) = 0.01 \cos 10t \qquad (5.3.10)$$

enters an FM modulator with $k_p = 0.1$. Find the FT of the modulated wave if the carrier is $10 \cos 10^6 t$.

Solution.   We note that for this example,

$$k_f g(t) = 0.001 \sin 10t$$

and that $|k_f g(t)| \ll \pi/6$ for all t. This makes it narrowband FM. Thus

$$m_p(t) = 10 \cos 10^6 t - 0.001 \sin 10t \cos 10^6 t$$

whose FT is easy to obtain. There are a group of impulses at $\omega = \pm 10^6$, $\omega = \pm 10^6 - 10$, and $\omega = \pm 10^6 + 10$. Find them and sketch the amplitude spectrum.

We have postponed the regular FM case as long as we can. It is time to plunge into it with vigor.

5.4 WIDEBAND FM

There are two general methods that science uses to prove things. One, deduction, tries to hit the general case, and if that is proved, specific examples become subcases, easily deduced from the general proof. The other, induction, tries the problem out for a specific case, and if it works, the generalization can follow. Induction is the general idea behind what we are going to do in this section, although it is a bit weak in mathematical rigor. (Show it for $n = 1$, assume for n, and prove it for $n + 1$ will not be exactly our way.)

The specific case we shall take up will be when the modulating signal is

$$s(t) = a \cos \omega_m t \qquad (5.4.1)$$

which is not exactly Beethoven's Ninth Symphony, just a monotonous tone. Still, it is a sort of $n = 1$ case in the inductive method since most signals can be thought of as being made up of a countable (Fourier series), or uncountable (Fourier transform) sum of such cosines. So our arm-waving generalizations at the end may convince a lot of people of their validity, particularly since practice seems to bear them out.

With the information signal of Eq. (5.4.1), the $\omega_i(t)$ of the carrier turns into

$$\omega_i(t) = \omega_c + k_f a \cos \omega_m t \qquad (5.4.2)$$

where ω_c is high (f_c in megahertz) and ω_m low (f_m a few kilohertz).

Integrating $\omega_i(t)$, we obtain

$$\theta_i(t) = \omega_c t + \frac{k_f a}{\omega_m} \sin \omega_m t \qquad (5.4.3)$$

and therefore the FM modulated wave will be

$$m_f(t) = A \cos \left(\omega_c t + \frac{k_f a}{\omega_m} \sin \omega_m t \right) \qquad (5.4.4)$$

At this point we denote

$$\beta = \frac{k_f a}{\omega_m} \quad \text{(in radians)} \qquad (5.4.5)$$

which is called the **modulation index**, and with this new constant we have that

$$m_f(t) = A \cos \omega_c t \overbrace{\cos (\beta \sin \omega_m t)}^{\text{term A}} - A \sin \omega_c t \overbrace{\sin (\beta \sin \omega_m t)}^{\text{term B}} \qquad (5.4.6)$$

As we have already mentioned, terms A and B are the abominable ones. These two terms are actually the real and imaginary part of

$$e^{j\beta \sin \omega_m t} \qquad (5.4.7)$$

which is a periodic function with period $2\pi/\omega_m$, as the reader should try to prove as an exercise (Problem 5.7). This being the case, the function of (5.4.7) can be expanded on a Fourier series, and once this is done, the real part of this series will be our term A, and the rest our term B. So we seek constants C_n so that

$$e^{j\beta \sin \omega_m t} = \sum_{n=-\infty}^{+\infty} C_n e^{jn\omega_m t} \tag{5.4.8}$$

We know from Chapter 2 that

$$C_n = \frac{\omega_m}{2\pi} \int_{-\pi/\omega_m}^{+\pi/\omega_m} e^{j\beta \sin \omega_m t} e^{-jn\omega_m t} \, dt \tag{5.4.9}$$

Let us try a change of variable, namely $\omega_m t = x$. With this, we get

$$C_n = \frac{1}{2\pi} \int_{-\pi}^{+\pi} e^{j(\beta \sin x - nx)} \, dx \tag{5.4.10}$$

which is not exactly a bed of roses; in fact, it cannot be evaluated in closed form. Lucky for us, however, this integral comes up so often in engineering problems that it has been extensively calculated and tabulated for various β's and most n's of interest [see Jahnke and Emde (1945), for example]. It even has a name, **Bessel functions of the first kind**. Thus, if β is known (and it is in our problems), then each coefficient C_n is the value of the nth-order Bessel function $J_n(\beta)$ at the specific value of β that we have. Therefore, we can write

$$C_n = J_n(\beta) \tag{5.4.11}$$

and we can return to Eq. (5.4.8) and rewrite it as

$$e^{j\beta \sin \omega_m t} = \sum_{n=-\infty}^{+\infty} J_n(\beta) e^{jn\omega_m t} \tag{5.4.12}$$

often called the **Bessel–Jacobi equation**.

The next steps are as follows. We must take the real and imaginary parts of Eq. (5.4.12), which are our terms A and B, respectively (easy to do by splitting up the exponential into $\cos n\omega_m t + j \sin n\omega_m t$) and put them into Eq. (5.4.6). After that the multiplications by $\sin \omega_c t$ and $\cos \omega_c t$ can be taken care of by the known FT properties. We will not bother with all these steps (see also Problem 5.8 for an easier way to arrive at the final expression), but shall go directly to the final answer, which is

$$m_f(t) = A \sum_{n=-\infty}^{+\infty} J_n(\beta) \cos (\omega_c - n\omega_m)t \tag{5.4.13}$$

So we ended up with an infinite series, as we warned the reader earlier that we would. But there is no need for despair. The infinite series is actually finite, as we are about to show.

First, let us say a few things about the $J_n(\beta)$'s, which will make our work easier. For each value of n, these are tabulated (see Table 5.4.1) or plotted as shown in Fig. 5.4.1. So after β has been found, the value of each can be obtained from this plot, Table 5.4.1, or some more extensive table in the literature. It should be noted that the plots have the $J_n(\beta)$'s for positive values of n only,

TABLE 5.4.1   **Values of Some Bessel Functions**

n	1	2	3	4	5	6	7	8
					β			
0	0.7652	0.2239	−0.2601	−0.3971	−0.1776	0.1506	0.3001	0.1717
1	0.4401	0.5767	0.3391	−0.06604	−0.3276	−0.2767	−0.004683	0.2346
2	0.1149	0.3528	0.4861	0.3641	0.04657	−0.2429	−0.3014	−0.1130
3	0.01956	0.1289	0.3091	0.4302	0.3648	0.1148	−0.1676	−0.2911
4	0.002477	0.03400	0.1320	0.2811	0.3912	0.3576	0.1578	−0.1054
5		0.007040	0.04303	0.1321	0.2611	0.3621	0.3479	0.1858
6		0.001202	0.01139	0.04909	0.1310	0.2458	0.3392	0.3376
7			0.002547	0.01518	0.05338	0.1296	0.2336	0.3206
8				0.004029	0.01841	0.05653	0.1280	0.2235
9					0.005520	0.02117	0.05892	0.1263
10					0.001468	0.006964	0.02354	0.06077
11						0.002048	0.008335	0.02560
12							0.002656	0.009624
13								0.003275
14								0.001019
15								
16								

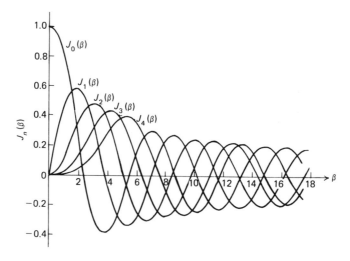

Figure 5.4.1   Plots of some Bessel functions.

whereas we need those for negative n's as well. This is because

$$J_n(\beta) = \begin{cases} J_{-n}(\beta) & \text{if } n \text{ is even} & (5.4.14a) \\ -J_{-n}(\beta) & \text{if } n \text{ is odd} & (5.4.14b) \end{cases}$$

another lucky break in the problem. And this is not all. A close look at the plots of Fig. 5.4.1 will reveal another pleasant surprise. The higher the β, the later the Bessel functions start, or to be more specific, if $\beta = 10$, then only the first 11 coefficients are nonzero—the rest are negligible. This is what makes the series finite. No matter how big β is, the significant $J_n(\beta)$'s will be only the first $n = \beta + 1$, approximately, of course.

Example 5.4.1

Assume that $\beta = 0.2$. Find the FT of $m_f(t)$ if the carrier's $A = 1$ and $\omega_c = 2\pi \times 10^6$, $\omega_m = 2\pi \times 10^3$.

Solution.   From the plots it is seen that only J_0 and J_1 are significant, and that

$$J_0(0.2) \approx 1$$

$$J_1(0.2) \approx 0.1 \qquad [\text{and } J_{-1}(0.2) = -0.1]$$

[incidentally, for small values of β, $J_1(\beta) \approx \beta/2$], so

$$m_f(t) = \cos 2\pi 10^6 t + 0.1 \cos 2\pi(10^6 + 10^3)t - 0.1 \cos 2\pi(10^6 - 10^3)t$$

whose FT can be found quite easily. Actually, we knew this result, since $\beta = 0.2$ is pretty close to the narrowband FM case, and we could have found it using the method of the preceding section. Still, it is nice that things check out.

Example 5.4.2

Now let us try $a = 100$, $k_f = 2$, and $\omega_m = 100$ rad/s. Sketch the amplitude spectrum of $m_f(t)$ if $A = 1$ and $\omega_c = 10^6$.

Solution.   The values given lead to $\beta = 2$. Table 5.4.1 gives us

$$J_0(2) \approx 0.22, \quad J_1(2) \approx 0.58, \quad J_2(2) \approx 0.35, \quad J_3(2) \approx 0.13$$

and the rest we ignore as being too small.

Taking into account the properties for obtaining the J_{-n}'s, we arrive at the final amplitude spectrum shown in Fig. 5.4.2.

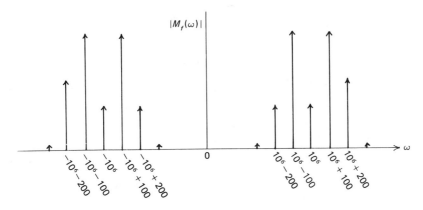

Figure 5.4.2   Amplitude spectrum of Example 5.4.2.

Having now gone through the pure cosine case, we might like to see what happens if $s(t)$ is

$$s(t) = a_1 \cos \omega_1 t + a_2 \cos \omega_2 t \qquad (5.4.15)$$

that is, a sum of two cosines not necessary harmonically related (after all, even in inductive proofs we sometimes try $n = 2$ to get a better flavor of the problem). Well, omitting the details (try them as Problem 5.9), we arrive at

$$m_f(t) = A \sum_{n=-\infty}^{+\infty} \sum_{k=-\infty}^{+\infty} J_n(\beta_1) J_k(\beta_2) \cos (\omega_c + n\omega_1 + k\omega_2)t \qquad (5.4.16)$$

where

$$\beta_1 = \frac{a_1 k_f}{\omega_1}, \qquad \beta_2 = \frac{a_2 k_f}{\omega_2} \qquad (5.4.17)$$

a real mess. Now we have impulses (in the spectrum) at ω_c, $\omega_c \pm n\omega_1$, $\omega_c \pm k\omega_2$ as well as $\omega_c \pm n\omega_1 \pm k\omega_2$ (cross-terms). A grand generalization this is turning out to be!

One can actually pursue this game even further, by assuming $s(t)$ to be periodic, that is, an infinite sum of sines and cosines [see Peebles (1976, p. 232), for example]. The result will hinge on your ability to get the Fourier series expansion of the bothersome terms A and B. But we will not cover this here, as its usefulness is minimal in practical problems, since periodic signals are not often sent. We will, though, see such a case in Chapter 6.

Now we turn our attention to the estimation of bandwidth of the modulated wave $m_f(t)$. This is of outmost importance, since FM stations must be frequency multiplexed in the same way that AM stations are, but in the range 88 to 108 MHz. After all, who cares about the actual shape of the spectrum as long as we know how wide it is for transmission purposes. We already know that it is sitting around ω_c, so it is suitable for transmission through the space channel.

We will start with our simple cosine case and then try to generalize. Let us take Example 5.4.2 as a start. In this example, $\beta = 2$ and the only significant terms are J_0, J_1, J_2, J_3, up to $n = 3 = \beta + 1$, as we remarked earlier. Since n also takes negative values resulting in impulses at $\omega_c - n\omega_m$, the bandwidth in this case is

$$B \approx 2 \times 3\omega_m = 6\omega_m$$

or

$$B \approx 2\omega_m(\beta + 1) \qquad (5.4.18)$$

Closer scrutiny and a little thought should convince the reader that Eq. (5.4.18) holds (approximately) for all single-tone modulation cases. Now recall Eq. (5.4.5) and use it in Eq. (5.4.18). The result is

$$B \approx 2(ak_f + \omega_m) \qquad (5.4.19)$$

which is another way to find B, if you do not want to bother finding β. The term ak_f is rather interesting. If we look at Eq. (5.4.2), we note that ak_f represents the **maximum frequency deviation** from ω_c (the $\cos \omega_m t$ cannot exceed the values

± 1), the deviation caused by the FM modulation. If we denote this by

$$\Delta\omega \triangleq ak_f = \max_t |\omega_i(t) - k_f s(t)| \qquad (5.4.20)$$

then

$$B \approx 2(\Delta\omega + \omega_m) \qquad (5.4.21)$$

often called *Carson's rule*, after John Carson, its originator.

Now comes the clincher—the grand generalization. We make the claim that Eq. (5.4.20) *holds for all modulating signals* s(t), with ω_m representing the *highest frequency present in* s(t) (and $\omega_m \ll \omega_c$). This claim (borne out in practice) can be justified with pseudorigorous arguments. They are okay, but we will not bother with them here (see Problem 5.10). It suffices to say that this rule has not yet been disproved by any practical or theoretically worked out example, so it is accepted. Maybe someday it will not be.

Doing some examples is probably worthwhile.

Example 5.4.3

Find the bandwidth of an FM wave if $k_f = 25$, $s(t) = 100 \sin 10^3 t$.

Solution.   We have that

$$\omega_i = \omega_c + 2500 \sin 10^3 t$$

We do not even have to know ω_c. We note that

$$\Delta\omega = 2500$$

and since $\omega_m = 10^3$, the result is

$$B \approx 2(10^3 + 2500) = 7.000 \text{ krad/s}$$

Example 5.4.4

Try it again with $k_f = 20$ and

$$s(t) = 10^3 \left[\frac{\sin 10^3(t - 10^6)}{10^3(t - 10^6)} \right]^2$$

Solution.   Here s(t) is not sinusoidal, but Carson's rule still holds. We need to find the maximum value of $|k_f s(t)|$, which will be the $\Delta\omega$. Now the maximum value of the term in parentheses is unity (at $t = 10^6$), so

$$\Delta\omega = 10^3$$

Next we need ω_m, the maximum frequency of s(t). The FT of s(t) is a triangle (see Table 2.6.1) with $\omega_m = 2 \times 10^3$ (we ignore the phase angle due to the shift), so

$$B = 2(10^3 + 2 \times 10^3) = 6.000 \text{ krad/s}$$

Now let us think about the implications of Eq. (5.4.20). Since $\Delta\omega$ depends on the constant of the modulating device, one can increase the bandwidth of the FM wave at will. You can start with human voice whose bandwidth is say $B = 15$ kHz and after modulation end up with whatever you like, not just at most 30 kHz as would be the case with AM. Typical FM stations use a bandwidth of around 180 kHz, as they are allowed $\Delta\omega = 75$ kHz, and presumably $\omega_m \approx 15$

kHz for good fidelity. Add another 10 kHz as a guard band, and stations use carriers 200 kHz apart. Actually, Carson's rule is approximate, so each station is forced to carry a NBF filter before transmission so that the prescribed bandwidth is not exceeded. Commercial television also uses FM for sound transmission, with each station allowed only 80-kHz bandwidth in this case.

So FM has a large bandwidth. What good is that? Here we went to great pains to come up with the SSB AM system to reduce the bandwidth, and now large bandwidth is acceptable? You bet it is; not only acceptable, but first-class travel for an $s(t)$, as we mentioned in the introduction. The main reason is that it behaves much better in the presence of additive noise, as we will see later. In fact, bandwidth and noise performance can be traded off, and this can be done very nicely with FM, whose bandwidth is at our will. With AM we are stuck with only once or twice the original bandwidth of $s(t)$. With FM we can easily change it by changing k_f. Of course, all these conclusions are to be taken with a grain of salt. FM is not always a panacea. In certain cases it behaves worse than AM (below a certain threshold in the signal-to-noise ratio). In other cases, we do not need the best system, either because it costs too much, or because we do not want an adversary to pick it up more easily at a far-off distance (electronic warfare). But we will see about such cases later.

We close this section with a word about PM, as we promised in Section 5.2.

Carson's rule holds for PM as well. However, note that in the single cosine case,

$$m_p(t) = A \cos(\omega_c t + k_p a \cos \omega_m t) \tag{5.4.22}$$

and

$$\omega_i(t) = \omega_c - k_p a \omega_m \sin \omega_m t \tag{5.4.23}$$

This means that

$$\Delta\omega = a k_p \omega_m \tag{5.4.24}$$

that is, here $\Delta\omega$ depends on ω_m (considered a disadvantage in commercial FM, where $\Delta\omega$ is fixed). Similar things happen for a general $s(t)$. Carson's rule holds approximately, but to find $\Delta\omega$, you must first differentiate the term $k_f s(t)$.

So it took some effort, but we managed to get some insight (and some approximate formulas) for FM and PM transmitted waves, which makes the overall system (Fig. 5.2.2) more understandable.

We will complete this chapter with a brief discussion on the nature of the devices that produce FM modulation and demodulation. This discussion will be quite succinct, as it is assumed that such knowledge is picked up elsewhere.

5.5 FM MODULATING AND DEMODULATING DEVICES

The state of the art of these devices is advancing by leaps and bounds, so there is no way to do this issue any real justice. New circuits are constantly coming out which can do the job better than before, for every application.

The only thing to do is to get to the heart of the matter: some fundamental notions about the nature of such devices.

5.5.1 Modulation

We have already noted that we can produce narrowband FM or PM by realizing Eq. (5.3.5) with AM-type subcircuits (see also Problem 5.5). Once the narrowband case is produced, it can be turned into a wideband by a frequency multiplier, which is nothing but a nth-law device [if $s(t)$ goes in, $s^n(t)$ comes out] followed by an appropriate filter. To see this more clearly, assume that a narrowband $m_f(t)$ enters a squarer, as shown in the first part of Fig. 5.5.1. The output of this squarer will be

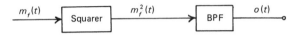

Figure 5.5.1   Frequency multiplier.

$$m_f^2(t) = A^2 \cos^2(\omega_c t + k_f g(t)) = \frac{A^2}{2}[1 + \cos(2\omega_c t + 2k_f g(t))] \qquad (5.5.1)$$

If this signal is now passed through a BPF filter that rejects the dc components, the result will be an FM signal with twice the amount of the original modulation. This routine can be repeated until the desirable amount of modulation is achieved. Of course, the frequency multiplier increases the value of the carrier as well, so either we start with a low ω_c or we must shift the final result to the proper carrier frequency by mixing.

Needless to say, if your starting system is narrowband PM, the signal $s(t)$ must be integrated first to produce PM, a case called **indirect FM**. This was the case with the system of E. H. Armstrong, who was the first to demonstrate its feasibility. For more information on this system, see Schwartz (1959).

Besides the AM-type narrowband systems, one can use any device that has the property of producing an output whose frequency varies linearly with the magnitude of the applied information signal. A voltage-controlled oscillator (VCO) is such a device, but we will not discuss it here. We will consider instead a simple oscillator, one of whose reactance elements is varied linearly with $s(t)$ (by using a varactor, say). The normal frequency of the oscillator is given by

$$\omega = \frac{1}{\sqrt{LC}} \qquad (5.5.2)$$

If one of the elements (say C) is varied linearly with $s(t)$, then

$$C = C_0 + as(t) = C_0\left(1 + \frac{a}{C_0}s(t)\right) \qquad (5.5.3)$$

where C_0 is its value when $s(t) = 0$, and a/C_0 is the maximum deviation from C_0.

With all this, the frequency of the oscillator will be

$$\omega_i(t) = \frac{1}{\sqrt{LC_0}[1 + (a/C_0)s(t)]^{1/2}} \tag{5.5.4}$$

If now a/C_0 is very small (much smaller than unity), the binomial theorem gives

$$\omega_i(t) \simeq \frac{1}{\sqrt{LC_0}}\left[1 - \frac{a}{2C_0}s(t)\right]$$
$$= \omega_c - k_f s(t) \tag{5.5.5}$$

which is the $\omega_i(t)$ of a narrowband $m_f(t)$. After this, frequency multipliers can take over the job again.

5.5.2 Demodulation

Here the situation is more complex, as there exist a plethora of FM demodulators (or detectors), some of which are even based on different fundamental notions. They go by such names as (1) slope detectors, (2) phase-shift discriminators, (3) phase-locked loops, (4) ratio detectors, (5) quadrature detectors, and (6) zero-crossing detectors. Most of them are based on the notion of differentiation, and this is the notion we discuss below.

Everybody knows at this stage that if you differentiate a sinusoid, you get the frequency up front in the amplitude [if $s(t) = a \cos \omega_c t$, $\dot{s}(t) = -a\omega_c \sin \omega_c t$] of a new sinusoid with the same frequency. So if we differentiate our $m_f(t)$, the result will be

$$\dot{m}_f(t) = -A[\omega_c + k_f s(t)] \sin (\omega_c t + k_f g(t)) \tag{5.5.6}$$

This new waveform is actually both an AM *and* an FM type, but the part that interests us is the AM. Indeed, our information signal $s(t)$ is part of the amplitude now, and an envelope detector can easily pick it up, just as in the large-carrier AM case (see also Problem 5.16). So a differentiator and an envelope detector can serve us well here, at least theoretically.

Now a differentiator has a system function $H(\omega) = j\omega$ with $|H(\omega)|$, as shown in Fig. 5.5.2 (i.e., a sloping line, a linear characteristic). And we want this type of line to be around ω_c, which is pretty high in value. So even though practically it is difficult to create such a differentiator, it is fairly easy to construct a circuit that will be like it (a sloping straight line) in a region around ω_c. Even a regular BPF located off center around ω_c does the job (see Fig. 5.5.3a). In fact, one can use two such BPFs, with their $|H(\omega)|$ located on either side of ω_c, and

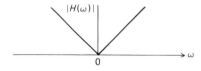

Figure 5.5.2 The $|H(\omega)|$ of a differentiator.

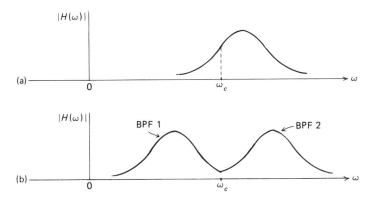

Figure 5.5.3   (a) A BPF as a differentiator; (b) two BPF's.

end up with a better approximation to the characteristic of Fig. 5.5.2, around ω_c, of course, as shown in Fig. 5.5.3b.

Variations of this theme result in many kinds of slope detectors, ratio detectors, Foster–Seeley discriminators, and so on. Even phase-shift detectors (or discriminators) are based on the idea of the differentiator, as can be seen in Fig. 5.5.4. For delaying $m_f(t)$ (by a small amount of time), and subtracting $m_f(t)$ from $m_f(t - \Delta t)$, is an approximation to differentiation.

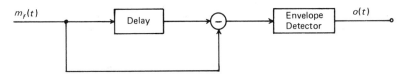

Figure 5.5.4   Block diagram of a phase-shift discriminator.

Not all FM detectors are based on the notion of differentiation. Zero-crossing detectors owe their existence to the fact that the multitude of zero crossings of $m_f(t)$ vary linearly with $s(t)$. Phase-locked loops (PLLs) are feedback devices that lock on the phase of $m_f(t)$. Details on these and many others can be found in electronics texts. A thorough treatment of PLLs can be found in the book by Viterbi (1966).

PROBLEMS, QUESTIONS, AND EXTENSIONS

5.1. Consider the signal

$$m(t) = 10 \cos (2\pi \times 10^6 t + 2\pi t^2)$$

(a) Is it FM or PM, and with what modulating signals?
(b) Find the instantaneous frequency of $m(t)$.
(c) Can you sketch $m(t)$?

5.2. The information signal $s(t)$ has the form

$$s(t) = 5 \sin 10^3 t$$

Find the output of an FM modulator if $k_f = 100$ and the carrier is $20 \cos 10^7 t$.

5.3. Consider Example 5.2.1 and assume that the modulating signal is two pulses (not one), that is,

$$s(t) = \text{rect}\left(\frac{t-1}{2}\right) + \text{rect}\left(\frac{t-5}{2}\right)$$

Find the FT of $m_f(t)$.

5.4. Find the FT and plot the amplitude spectrum of

$$m_p(t) = \cos(\omega_c t + 0.01 s(t))$$

for the following $s(t)$'s.
(a) $s(t) = \sin 10^3 t$
(b) $s(t) = \dfrac{\sin 10^3 t}{10^3 t}$
(c) $s(t) = e^{-|t|}$
(d) $s(t) = \text{rect}\left(\dfrac{t-3}{2}\right)$
(e) $s(t) = \left[\text{rect}\left(\dfrac{t-3}{2}\right)\right] * \left[\text{rect}\left(\dfrac{t-3}{2}\right)\right]$

5.5. Draw block diagram systems for producing narrowband PM and FM, using only adders, multipliers, integrators, and 90° phase shifters.

5.6. Try Problem 5.4(d) again with the assumption that the modulated signal is FM that is,

$$m_f(t) = A \cos\left(\omega_c t + k_f \int_{-\infty}^{t} s(t)\, dt\right)$$

Be careful with the integral property of the FT. [Is $S(0) = 0$?]

5.7. Prove that the function of expression (5.4.7) is periodic with period $2\pi/\omega_m$.

5.8. There is an easier way to arrive at Eq. (5.4.13). Start with the Bessel–Jacobi equation, multiply both sides with $e^{j\omega_c t}$, and then take the real part of both sides. What do you get? Using the same idea, arrive at the FT of

$$m_f(t) = A \sin(\omega_c t + \beta \sin \omega_m t)$$

that is, for the case when the carrier is a sine.

5.9. Prove Eq. (5.4.16).

5.10. Search the literature for arguments to support Carson's rule in the general case.

5.11. Consider wideband FM with $s(t)$ the periodic wave shown in Fig. P5.11. Show that the modulated wave can be written as

$$m_f(t) = A \sum_{n=-\infty}^{+\infty} C_n \cos(\omega_c + n\omega_m)t$$

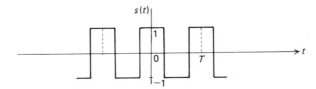

Figure P5.11 $s(t)$ for Problem 5.11.

where $\omega_m = 2\pi/T$ and

$$C_n = \frac{1}{2}\left\{\mathrm{Sa}\left[\frac{\pi}{2}(\beta - n)\right] + (-1)^n\,\mathrm{Sa}\left[\frac{\pi}{2}(\beta + n)\right]\right\}$$

[If you cannot do it, try the literature, for example, Lathi (1968, p. 225).]

5.12. Consider the modulated wave

$$m_f(t) = A\cos\left(\omega_c t + 100\int s(t)\,dt\right)$$

Find its approximate bandwidth if $s(t)$ takes the various forms of Problem 5.4.

5.13. (a) A 15-kHz single frequency modulates a 200-kHz carrier with maximum frequency deviation of 2 MHz. Find the approximate B of the modulated wave. Now halve the amplitude of the modulating signal and do it again. Then halve the frequency of the modulating signal and do it again. What are your observations?

(b) Assume that the same signal (with unit amplitude) phase-modulates the same carrier with $k_p = 50$. Try the same stuff as in part (a).

5.14. We have not said much about power in FM waves, so we must correct this omission here. We assume single-tone wideband FM, for which the pertinent expression is (5.4.13). Next, we make use of another nice property of Bessel functions; namely,

$$\sum_{n=-\infty}^{+\infty} J_n^2(\beta) = 1$$

With this, show that the total power in the FM wave is $A^2/2$; that is, it is independent of the modulating signal amplitude. Why?

5.15. Consider a wideband FM single-tone modulation with $\beta = 4$. Sketch the spectrum of the FM wave and then find the percent power in the carrier and in the sidebands (the rest of the impulses around the carrier).

5.16. Find the Hilbert transform and then the envelope of the signal in Eq. (5.5.6).

5.17. Search the literature for a description of the operation of the PLL and prepare a short lecture on it.

5.18. This problem deals with the transmitter and receiver of a typical stereophonic FM system (SFM). We feel that the reader should be able to analyze it alone, as it is a simple combination of AM, FM, and the ideas of frequency multiplexing.

(a) The transmitter of an SFM system is shown in Fig. P5.18a, in block diagram form. The signals $L(t)$ and $R(t)$ represent the sounds produced by the left and right sides of an orchestra or band. The LPFs limit $L(t)$ and $R(t)$ to 15 kHz.

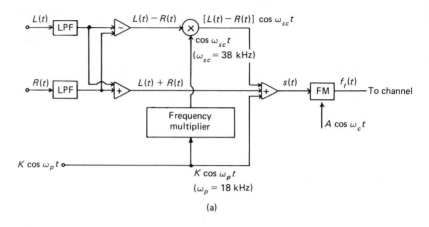

(a)

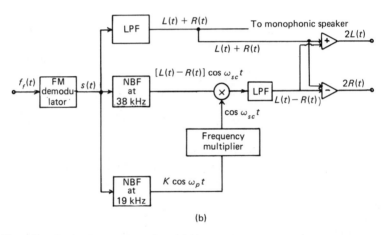

(b)

Figure P5.18   (a) Transmitter of a SFM system; (b) receiver of a SFM system.

Assuming that both $L(t)$ and $R(t)$ have $L(\omega)$ and $R(\omega)$ those of speech, find and sketch the FT of $s(t)$, that is, of the signal that enters the FM modulator. Incidentally, the bandwidth of $f_f(t)$ is still around 180 kHz.

(b) The receiver of a typical SFM has the form shown in Fig. P5.18b. Follow the signals through the blocks and verify that the outputs are as specified. Note that there is a monophonic version [the $L(t) + R(t)$ signal] which can be picked up by existing monophonic receivers, so that they do not become worthless (it was an FCC requirement).

(c) Can you design a quadraphonic FM system which includes a monophonic version for people who cannot buy a new radio?

=====6=====

ANGLE MODULATION: RADAR SYSTEMS

6.1 INTRODUCTION

Now that FM and PM modulation have become familiar to us, the field is open for studying the use of this type of modulation in radar systems. We have already seen that FM and PM have wide spectra, and this presumably improves a communication system's performance in the presence of additive noise—an advantage often worth paying for. But do these modulations improve any part of the performance of *radar* systems? If yes, what part, and at what cost? These are the types of questions that must be answered in a convincing manner, as convincing as the mathematical analysis will allow.

As we mentioned in the introduction to Chapter 4, a radar's first duty is to detect, and then to measure range, velocity, and complete location of a target. Detection is largely a problem better studied in the presence of noise and will be ignored here. The reader may still think of it in terms of the radar equation's implications of S_{min}, that is, in terms of losses due either to the wave traveling or to the parameters of the system. Complete location of a target (aside from its range) involves the measurement of azimuth and height, and this, too, will be ignored until we take up some of the tracking radars in Chapter 10. So range and velocity measurements will be our main issues here, as they were in Chapter 4. The problem of *resolving* targets will also be discussed in passing, but without fancy mathematical substantiation of the various assertions made.

FM is not used in radar only as a vehicle to transmit a signal from one place to another in a *first-class* type of accommodation as was the case in communication system. The signal itself does not carry the pertinent information —it is known to the receiver a priori, as we have often stressed. The information is carried by the changes that the target produces on this signal (delay, spectral shift, etc.). So lounging in first-class travel is not enough here. The signal must be sensitive to delay and spectral shift changes; otherwise, it is no good to us. Its time and frequency ambiguity functions had better have the right properties or else we do not use it. In other words, we do not use any signal for the FM vehicle here; we discriminate and pick the one we like.

We are going to discuss two types of radar here, both of which use FM (or PM) modulation in their emitted signals. Actually, both of them are already familiar to us, as they are basically the CW and the PR systems, now with FM modulations in their carriers. But FM (or PM) modulations do all kinds of crazy things to the time and frequency representations of the signals, things that are difficult to analyze mathematically, as we already know from Chapter 5. So expect a lot of arm-waving approximations and generalizations here as well. The road is rocky, sprinkled with integrals that cannot be solved in closed form.

6.2 FREQUENCY-MODULATED CW RADAR

We have already seen that CW radar is incapable of measuring range, although ideally suited to measure radial velocity. And we have connected these characteristics with the time and frequency representations [and $R(\tau)$, $\mathcal{H}(v)$] of the emitted signal, as regards their inclusion or not of a sharp point (a mark) on which such measurements can be based. CW radar emits *basically* a constant in time (without the carrier), with no place to base the delay (the shift) of the return echo. So it is no good for measuring range. In frequency, the picture is reversed. The constant has a frequency representation that is an impulse, and a possessor of a finer mark for measuring spectral shifts (velocity) than an impulse is impossible to find. So in conclusion, CW radar's emitted signal is too extremist: too smooth in time, and too sharp in frequency. A compromise is obviously in order, to make it a bit milder in both domains. This compromise comes about by many ways, but the one we are about to discuss uses FM modulation. We already know that such modulation spreads out the spectrum (in fact, at our will) and that such spreading causes shrinkage in the time shape (introducing time *marks*), so this compromise is already apparent, even at this stage of the game.

Let us start by considering the emitted signal with the carrier this time, not without it as we did in the discussion above and in Chapter 4.

The (regular) CW radar emitted signal is

$$m(t) = A \cos \omega_c t \tag{6.2.1}$$

with FT

$$M(\omega) = \pi A[\delta(\omega - \omega_c) + \delta(\omega + \omega_c)] \tag{6.2.2}$$

Since we are including the carrier in our discussion (i.e., we do not consider as our signal the constant A), the emitted signal is not flat; it has some points on which range measurements can be based, say its maxima or minima or zero crossings. In fact, its $R_m(\tau)$ (time ambiguity function), which has the same shape as $m(t)$ (why?), also shows that by exhibiting maxima (or minima, etc.) spaced 2π radians apart. This distance is the distance at which unambiguous range can be measured, a piece of information provided by $R_m(\tau)$, as we already know. Now in view of the fact that the round-trip delay is $T = 2R/c$ and $T = 2\pi/\omega_c$, the unambiguous range is

$$R_{\text{unamb}} = \frac{\pi c}{\omega_c} = \frac{c}{2fc} = \frac{\lambda}{2} \qquad (6.2.3)$$

which is too small for any use in the radar frequency range. Still, range can be measured even with a cosine, as long as its frequency is low enough so that the unambiguous range will cover the radar's R_{max}, although not very well, as the maxima of $R_m(\tau)$ are not very sharp. Nevertheless, let the reader remember this, as we shall need it later. So far, then, we know that the signal of Eq. (6.2.1) cannot measure range very well, whereas it can obviously measure velocity in view of its sharply pointed FT given in Eq. (6.2.2).

The key idea behind the CW-FM radar system is to take another signal $s(t)$ (a sinusoid, a sawtooth, or whatever) with timing marks on it so that it can measure range, and put it in the angle of the carrier of Eq. (6.2.1). The return echo can then be demodulated (FM or PM) and the delay of this new signal can measure the range. Not bad!

Not good, either. Sure, it can measure range. But the FM modulation will play havoc with the spectrum of Eq. (6.2.2). Where will our nice ability to measure velocity go? Well, we have already said that the scheme is a compromise, and we are not going to complain again about the pitfalls of compromises. We are just going to go ahead with our discussion and see where it leads us.

The FM-CW radar, then, sends out a signal of the form

$$m(t) = A \cos(\omega_c t + k_f g(t)) \qquad (6.2.4)$$

which can be thought of as FM with $s(t) = \dot{g}(t)$, or PM with $s(t) = g(t)$. We will think of it as FM. This being the case, its instantaneous frequency $\omega_i(t)$ will be

$$\omega_i(t) = \omega_c + k_f s(t) \qquad (6.2.5)$$

and

$$g(t) = \int_0^t s(t) \, dt \qquad (6.2.6)$$

The number k_f is the usual device constant, and $s(t)$ is a signal that can measure range.

Now, let us talk about this $s(t)$ again. We can use a periodic signal with sharp points (a sawtooth, for example) and an $R_s(\tau)$ that gives a good-sized unambiguous range. But the analysis of the FM signal of Eq. (6.2.4) can be pretty messy with such $s(t)$'s. Here is where we sort of luck out. We mentioned earlier in this section that even a cosine can measure range as long as its frequency is low enough. But if $s(t)$ is a cosine, we know how to work on Eq. (6.2.4) and find its FT. Let us then look at this case and generalize the results to other $s(t)$'s later. In practice, the cosine $s(t)$ is not often used,[1] as a sawtooth type is preferred. But since that, too, is periodic, our generalization will not be very hard to swallow.

To begin with we must choose a cosine

$$s(t) = a \cos \omega_m t \tag{6.2.7}$$

whose frequency ω_m is low enough so that Eq. (6.2.3) gives a reasonably long unambiguous range (say 100 to 1000 Hz).

The emitted radar signal will have the form

$$m(t) = A \cos (\omega_c t + \beta \sin \omega_m t) \tag{6.2.8}$$

where as we know from Chapter 5, $\beta = ak_f/\omega_m$ is the FM modulation index.

Let us assume that this signal meets a target, and denote the return echo by $m_r(t)$. If the total round-trip delay due to the distance of the target is T, then

$$m_r(t) = A_r \cos [\omega_c(t - T) + \beta \sin \omega_m(t - T)] \tag{6.2.9}$$

and after FM demodulation, the originally sent $s(t)$ will have the form

$$s_r(t) = K_1 \cos \omega_m(t - T) \tag{6.2.10}$$

where K_1 is a constant of the FM demodulator. Comparing $s(t)$ with $s_r(t)$, we note that they differ by a phase shift equal to $\omega_m t$. This phase shift can be easily brought out by a mixer (multiplication by a cosine followed by a filter), often called a *phase detector*,[2] which will keep the difference term, that is, whose output will be

$$o(t) = K_1 \cos \omega_m T \tag{6.2.11}$$

from whose value we can easily calculate T and consequently the range of the target. In practical systems this mixing can be done before the demodulation (see Skolnik, 1962, p. 89) and the result will be a beat frequency whose value is related to the target's distance R.

So range can indeed be measured in the CW-FM system, even though not so cleanly as with a PR system. But what about the Doppler shift, and therefore the radial velocity of a moving target?

To see this we must find the FT of $m(t)$ of Eq. (6.2.8). Actually, we already know it from the wideband FM analysis of Chapter 5. It was shown there [see Eq. (5.4.13)] that $m(t)$ can be written in the form

$$m(t) = A \sum_{n=-\infty}^{+\infty} J_n(\beta) \cos (\omega_c \pm n\omega_m)t \tag{6.2.12}$$

where $J_n(\beta)$ are the first-order Bessel functions evaluated at β. In fact, we know that the seemingly infinite series above is actually finite (up to about $n = \beta + 1$),

---

[1]Radio altimeters use it.

[2]In DSB we called it a product or homodyne detector.

so that the FT is nothing but impulses located at ω_c, and $\omega_c \pm n\omega_m$. So the addition of FM has taken our excellent ability to measure velocity (using the nicely spiked FT of the CW radar) and turned it into a mediocre one. The frequency ambiguity function has many peaks, much like the case of the pulse-Doppler radar system, and velocity ambiguities will enter the picture. That is what you get for wanting range-measurement ability in a CW radar.

So much, then, for the case of the CW-FM system, which uses sinusoidal modulation. We should just add that we cheated a little in the analysis above. We talked about the radar's ability to measure range by talking about $R(\tau)$ of $s(t)$ and not of $m(t)$, which is the actual emitted signal. No matter, though. $R_m(\tau)$ can also be found, and the results will be very much the same as above (see Problem 6.1).

What if the modulating signal is not sinusoidal, but some other periodic signal? We should take a quick look at this, since it represents the more practical case, as we mentioned earlier.

Basically, a general periodic signal still leads to an $m(t)$ of the form (6.2.12), except that the coefficients are not necessarily Bessel functions, but some other numbers dependent on β, usually very hard to find (see Problem 6.2 if you wish to try one). Since

$$m(t) = A \cos (\omega_c t + k_f g(t))$$
$$= A \cos \omega_c t \cos k_f g(t) + \sin \omega_c t \sin k_f g(t) \qquad (6.2.13)$$

the messy terms are $\cos k_f g(t)$ and $\sin k_f g(t)$. But since $s(t)$ is periodic (with period $2\pi/\omega_m$, say), the same is also true for the two troublesome terms. Thus they can be expanded into a FS and will theoretically lead to the form of Eq. (6.2.12) as we mentioned above. Therefore, our general conclusions, and the arguments that lead to them, pretty much hold in all cases where $s(t)$ is periodic. The interested reader is referred to Luck (1949) for additional material on this issue. In any event, let us discuss the actual operation of such a CW-FM below, given that, it does represent the more usual case in practice. Too much discussion involving $R(\tau)$ and $\mathfrak{IC}(v)$ can actually get on people's nerves.

A typical CW-FM system is shown in Fig. 6.2.1, stripped down, of course, to its essential operations as regards only range measurement. The modulating

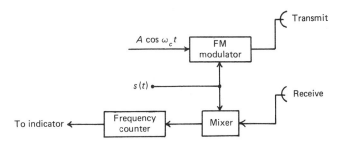

Figure 6.2.1   Block diagram of a CW-FM system.

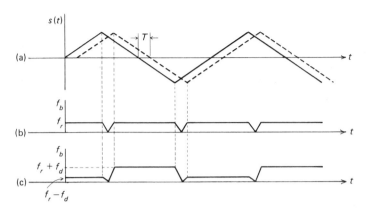

Figure 6.2.2 (a) Linear sawtooth $s(t)$; (b) frequency f_a of the beat note (stationary target); (c) frequency f_b of the beat note (moving target). (After Skolnik, 1962.)

signal $s(t)$ is a triangular wave, as shown in Fig. 6.2.2a, which means that the instantaneous frequency of the carrier is varied linearly from $\omega_c - k_f[s(t)]_{\max}$ to $\omega_c + k_f[s(t)]_{\max}$. The return echo from a target will be delayed by T, which is related to the target's distance R by the expression $T = 2R/c$. This means that the received $s(t)$ (if we think of it after FM demodulation) is a shifted version of the original; or in other words, the emitted signal at the arrival instant of the echo, and the echo itself differ in frequency by an amount related to this shift. A mixer in the receiver can get out this difference in frequency as a beat note much as in the case of sinusoidal modulation that we discussed earlier. The beat note will be of constant frequency for as long as the target is stationary and creates no Doppler frequency which can change the overall location of the spectrum (see Fig. 6.2.2b), save, of course, the turnaround regions of the triangular waveform. The frequency counter is meant to measure this frequency, and after that other circuitry can relate it to the range R and feed the information to an indicator. The operation of the system is substantially as we analyzed it earlier, except that now we see it as a change in frequency between the emitted signal and the echo, rather than a shift in $s(t)$.

The blocks of the system that measure the Doppler shift are not shown in Fig. 6.2.1, but they can be similar to those used in the pulse-Doppler radar system. Incidentally, if the target is moving, the beat frequency will vary in time, as shown in Fig. 6.2.2b. For an approaching target, the beat frequency will be

$$f_b = f_r - f_d$$

on the upswing of the triangular wave and

$$f_b = f_r + f_d$$

on the downswing. To measure the actual value f_r, the two cycles must be averaged out.

We close this section with some remarks on target resolution. CW-FM radar, much like CW radar, is not very useful for range-resolving two or more targets; its $R(\tau)$ is too smooth for that. It could, of course, be changed to accomplish this feat as well, either by changing the modulating signal (a periodic narrow pulse, for example?) or adding circuitry, but such changes either make it a different radar altogether or have not yet paid off in practical results. In any event, it has now become a tradition to call a radar a CW-FM one if the $s(t)$ is also a periodic *continuous* signal (sinusoid, sawtooth, etc.) and not some interrupted type such as a periodic pulse. Still, we wonder what a periodic (or even aperiodic) pulse might do as the FM modulating signal, since if it is narrow enough, it can provide us with satisfactory range resolution, being that its $R(\tau)$ will be sharp and narrow. We let the reader think about this one and go on to the next radar system that we must analyze. The issue will come up again in a later chapter.

6.3 THE CHIRP RADAR SYSTEM

Now it is the PR's turn to be subjected to some FM modulation. Since the regular PR emitted signal already has AM in it (well, the CW one does also, but in a trivial way), the result will be an AM and FM carrier. We have not analyzed such a thing before. And if we recall that FM is a highly nonlinear phenomenon, then combining it with AM may lead us to some surprising results.

There are many types of modulating signals that we can use, but the one that seems most popular is a sloped line, whose form can be written as

$$s(t) = \begin{cases} t & \text{for } |t| < \dfrac{T_p}{2} \\ 0 & \text{elsewhere} \end{cases} \tag{6.3.1}$$

much like the triangular waveform of the preceding section, lasting for only T_p seconds.

Let us consider the (regular) PR emitted signal as just an AM pulse of the form

$$m_e(t) = A \operatorname{rect}\left(\frac{t}{T_p}\right) \cos \omega_c t \tag{6.3.2}$$

and ignore its periodicity, illumination time, and the like, to make our discussion easier to understand. With the signal of Eq. (6.3.1) FM modulating the carrier of Eq. (6.3.2), the instantaneous frequency of the carrier will be

$$\omega_i(t) = \begin{cases} \omega_c + \mu t & \text{for } |t| < \dfrac{T_p}{2} \\ 0 & \text{elsewhere} \end{cases} \tag{6.3.3}$$

(μ is our old k_f), which means that the carrier's frequency changes from $\omega_c - \mu T_p/2$ to $\omega_c + \mu T_p/2$ in a linear fashion, during the interval of the pulse rect (t/T_p). Such a frequency-changing pulse presumably sounds like a chirping bird, and this is the reason for the odd name attached to this radar.[3] Certain animals that employ sonar (bats, for instance) use a chirplike pulse, at least during part of their sound emissions (see Novick, 1971). Nature, you might say, has a way of getting ahead of people's brains. But people's brains are also nature, so this argument reaches the realm of absurdity.

Anyway, integrating $\omega_i(t)$ and ignoring the possible constant phase angle, we obtain the angle of the modulated wave as

$$\theta_i(t) = \omega_c t + \frac{\mu}{2} t^2 \tag{6.3.4}$$

that is, the phase-angle change is quadratic.

The final form of the chirp signal that we shall consider is

$$m(t) = A \operatorname{rect}\left(\frac{t}{T_p}\right) \cos\left(\omega_c t + \frac{\mu}{2} t^2\right) \tag{6.3.5}$$

that is, a cosine lasting T_p seconds, whose frequency is changing in a linear way around ω_c. Figure 6.3.1 shows the frequency variation and the actual form of the $m(t)$ for optical appreciation.

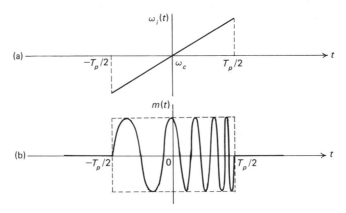

Figure 6.3.1 (a) Frequency variation in chirp; (b) the actual chirp signal.

To be able to say something about the abilities of this signal to measure range, velocity, and to resolve targets, we must find its FT $M(\omega)$ (or amplitude spectrum $|M(\omega)|$), as well as its time and frequency ambiguity functions $R_m(\tau)$ and $\mathcal{K}_m(\nu)$. None of them are very easy, to say the least. The chirp pulse is an

[3]The originator of this appellation was B. M. Oliver, who first used it in a Bell Labs Internal Memorandum entitled "Not with a Bang, But a Chirp" in 1951.

FM signal (with AM), and this is the kind of thing that separates the men from the boys. *Here come the approximations.*

Let us take a first crack at the problem using a *heuristic* point of view. We can think of $m(t)$ as a pulse that has been multiplied by a cosine whose frequency is not constant, but changing. Now the pulse has a $|\sin\omega|/|\omega|$ type of amplitude spectrum. When it is multiplied by a $\cos\omega_c t$, this spectrum gets shifted to $\pm\omega_c$. In essence it is multiplied by an infinite number of cosines with frequencies in $(\omega_c - T_p/2)$ to $(\omega_c + T_p/2)$, so we have a superposition of infinite $(\sin\omega)/\omega$ FTs shifted around ω_c. The result is probably a flat spectrum around ω_c, tapering off at the end point. Let us say that it is completely flat, a pulse-like FT as shown in Fig. 6.3.2a, even though we cannot suppose that many people are convinced at this point. If it is so, what would $\mathcal{H}_m(\nu)$ be? Well, the reader should be able to guess that $\mathcal{H}_m(\nu)$ would be a triangle around $\omega = 0$ of duration $2\mu T_p$, as well as two triangles around $2\omega_c$ of lesser height but of the same duration, something like Fig. 6.3.2b. Very interesting! It does not look too bad for measuring velocities, does it? It has a good peak at the origin with the other peaks pretty far away to introduce serious ambiguities.

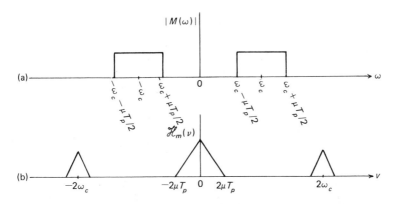

Figure 6.3.2   (a) Approximate spectrum of $m(t)$; (b) approximate $\mathcal{H}_m(\nu)$.

What about $R_m(\tau)$? Well, since we have $|M(\omega)|$, we also have $|M(\omega)|^2$. It has the same shape as $|M(\omega)|$. Furthermore, $|M(\omega)|^2$ is the FT of $R_m(\tau)$, by the Wiener–Khintchine theorem of Section 2.10. Thus $R_m(\tau)$ must have a form

$$R_m(\tau) = \frac{K\sin(\mu T_p\tau/2)}{\mu T_p\tau/2}\cos\omega_c\tau \tag{6.3.6}$$

as can be easily checked by taking its FT and comparing it with Fig. 6.3.2a. This last equation is rather difficult to sketch (a $\cos\omega_c\tau$ amplitude-modulated by the other term), but we have done so, anyway, in Fig. 6.3.3b. Figure 6.3.3a shows the amplitude of the PR pulse for the purpose of comparisons in what follows.

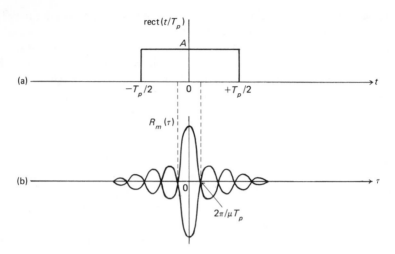

Figure 6.3.3   (a) Original pulse amplitude; (b) the $R_m(\tau)$.

What does this $R_m(\tau)$ say for range measurements? Not bad, not bad at all! It has a nice, though rounded peak at $\tau = 0$, with other peaks of *lesser* magnitude, which should not create too many problems with ambiguities. *In fact, it is better than we would have had without the FM linear modulation*, as the reader can easily verify. Not only that, but the shape of $R_m(\tau)$ can be controlled by us (by varying μ), meaning that we can make the first peak pretty narrow if we desire, although the lesser peaks may also rise.

These results are very interesting. They can even become astonishing if we consider the following. We know that the time ambiguity function $R_m(\tau)$ often has the same shape as the output of a matched filter to the emitted signal $m(t)$. (Why?) Looking over the $R_m(\tau)$, we note that its first main lobe is actually much shorter in duration than the original pulse; in fact, we can make it as short as we like (within reason). This is surprising. We are used to seeing outputs of matched filters which are wider than the inputs; that is why it is surprising. In fact, it is so surprising that the whole idea has its own name (pulse compression), and a great deal of research is still going on in the area (following the declassification of the field; see Klauder et al., 1960). The chirp radar is pretty good in range and velocity measurements, as well as resolution. At the same time, the original AM pulse can be very wide [and still have a narrow $R_m(\tau)$ which is controlled by μ] and so pack a lot of power (longer $R_{\max}$) without damaging the devices. It is a good all-around type of radar, exhibiting advantages that we cannot even discuss at this stage of noiseless analyses.

Now, you say, are we not going a bit too far on the basis of a rather skimpy development? We are happy to announce that these conclusions hold even if the analysis becomes more mathematically rigorous. Lest we be accused of always beating around the bush, we undertake such a task immediately below.

Let us go back and try to find the FT of $m(t)$ of Eq. (6.3.5), following Cook (1960). Ignoring the constant A, we have

$$M(\omega) = \int_{-T_p/2}^{T_p/2} \cos(\omega_c t + \tfrac{1}{2}\mu t^2) e^{-j\omega t}\, dt$$

$$= \tfrac{1}{2}\left(\int_{-T_p/2}^{T_p/2} \exp\{j[(\omega_c - \omega)t + \tfrac{1}{2}\mu t^2]\}\, dt \right. \tag{6.3.7}$$

$$\left. + \int_{-T_p/2}^{T_p/2} \exp\{-j[(\omega_c + \omega)t + \tfrac{1}{2}\mu t^2]\}\, dt \right)$$

by expanding the cosine into exponential terms, of course.

At this point Cook performs some rather intimidating algebraic acrobatics, which bring the above to the form

$$M(\omega) = \frac{1}{2}\sqrt{\frac{\pi}{\mu}} e^{-j(\omega_c - \omega)^2/2\mu}$$

$$\times \left[C\!\left(\frac{(\mu T_p/2) + (\omega_c - \omega)}{\sqrt{\pi\mu}}\right) + jS\!\left(\frac{(\mu T_p/2) + (\omega_c - \omega)}{\sqrt{\pi\mu}}\right) \right. \tag{6.3.8}$$

$$\left. + C\!\left(\frac{(\mu T_p/2) - (\omega_c - \omega)}{\sqrt{\pi\mu}}\right) + jS\!\left(\frac{(\mu T_p/2) - (\omega_c - \omega)}{\sqrt{\pi\mu}}\right) \right]$$

where

$$C(x) = \int_0^x \cos\frac{\pi}{2}a^2\, da \tag{6.3.9a}$$

and

$$S(x) = \int_0^x \sin\frac{\pi}{2}a^2\, da \tag{6.3.9b}$$

are called Fresnel integrals, and cannot be calculated out in closed form. However, we are in luck again, as both of the above happen quite often in engineering applications and have been tabulated or plotted (see Fig. 6.3.4). They are plotted only for $x > 0$ in this figure, but they are well known to be odd functions of x, anyway. What we are interested in is $|M(\omega)|$. Since the exponential

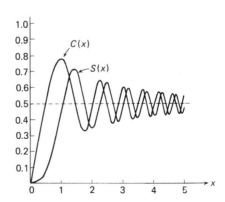

Figure 6.3.4   Two Fresnel integrals.

has absolute value unity,

$$|M(\omega)| = \frac{1}{2}\sqrt{\frac{\pi}{\mu}}\{[C(x_1) + C(x_2)]^2 + [S(x_1) + S(x_2)]^2\}^{1/2} \qquad (6.3.10)$$

where

$$x_1 = \frac{(\mu T_p/2) + (\omega_c - \omega)}{\sqrt{\pi\mu}}, \qquad x_2 = \frac{(\mu T_p/2) - (\omega_c - \omega)}{\sqrt{\pi\mu}} \qquad (6.3.11)$$

Equation (6.3.10) is even messier than our previous FM analysis with the Bessel functions. At least there, the Bessel functions did not include the ω in their argument. Here both $C(x)$ and $S(x)$ are functions of ω and must be sketched out. What must be done is to set the value of ω and then calculate the value of $|M(\omega)|$ and repeat (numerically) until you get the entire spectrum. This is a very sad endeavor indeed—and all this for specific values of μ and T_p that define the radar to start with.

This time we go to Burdic (1968) for help, and give a typical spectrum in Fig. 6.3.5, which the reader must admit is not too far from our original heuristic

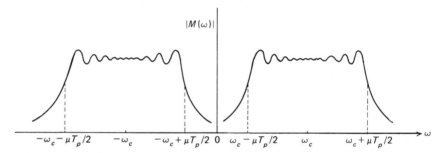

Figure 6.3.5 Typical spectrum of a chirp signal. (Reprinted by permission of Prentice-Hall, Inc., Englewood Cliffs, N.J.)

guess of Fig. 6.3.2a. We told you so.

The $R_m(\tau)$ of the actual signal can also be shown to be (Cook et al., 1965)

$$R_m(\tau) = K\frac{\sin{(\mu\tau/2)(T_p - |\tau|)}}{\mu\tau T_p/2}\cos\omega_c\tau$$

$$\approx K\frac{\sin{(\mu T_p/2)}}{\mu T_p/2}\cos\omega_c\tau \qquad \text{for } |\tau| \le T_p \qquad (6.3.12)$$

so our guess, Eq. (6.3.6) was very close to the target, at least for values $|\tau| \le T_p$, where it counts.

The $R_m(\tau)$ of Eq. (6.3.12) is a function with a fairly narrow peak at $\tau = 0$. If we measure the width of this peak between the first two zeros (right and left of $\tau = 0$), we obtain

$$\text{width of peak} = \frac{2\pi}{\mu T_p} = \frac{2\pi}{\mu T_p^2} T_p \qquad (6.3.13)$$

which for $\mu T_p^2 > 2\pi$ is smaller than T_p. As we mentioned earlier, this entire idea is referred to as **pulse compression**. The **pulse compression ratio** can be defined as

$$\frac{\text{width of original envelope}}{\text{width of peak of } R_m(\tau)} = \frac{\mu T_p^2}{2\pi} \qquad (6.3.14)$$

and is controlled by μ and T_p (i.e., by the radar designer). Compression ratios of 100 or even more are possible and are used in practice.

Lest the reader think that this radar is some kind of panacea, we must hurry to say that this is not so. There are cases of Doppler shift–range combinations for which the chirped signal behaves like a regular PR pulse (Helstrom, 1968). To see this, one needs the general ambiguity function, which we have not yet discussed here. $R_m(\tau)$ and $\mathcal{K}_m(\nu)$ cannot provide us with all the information we want, because the first one $[R_m(\tau)]$ gives information about range (and range resolution) for Doppler shifts $\omega_d = 0$, and the second about velocities, for $\tau = 0$.

The ideas behind the chirp radar form the foundation for many other kinds of radar presently used or still on the design boards. The envelope, for example, need not be a rectangular pulse, nor the $\omega_i(t)$ linearly changing, to get pulse compression. Other envelope shapes (Gaussian, for example), or nonlinear-type frequency modulation lead to similar and sometimes better results. The interested reader is referred to the literature (Nathanson, 1969; Barton, 1975; etc.) for further reading. *We will also see this topic again, after our discussion of PCM.*

We have to assume that the reader has noticed that this section does not contain a block diagram and analysis of a typical chirp radar, as all such sections have done in the past. Sad to say, we will not include one. We shall leave it for the reader to contemplate such a diagram as an exercise. When the reader gets to the matched filter, he or she should stop and think about it a lot. Matched filters for chirp signals are not easy to construct (see Problem 6.9).

One more comment about the name. Chirp radar takes its name from the specific angle modulation it uses, as we have already noted. We covered it in this chapter, because this modulation *is* of the angle type. Still, chirp radar uses double modulation in its signal (AM and FM), and as such it could have been covered under mixed systems in Chapter 10. Not only that, but the angle modulation used widens the spectrum of the original pulse signal, and such systems may be called *spread-spectrum systems*. We will see such systems both in communication and radar in the following chapters, so the chirp radar could have been included there. Then again, it is a pulse compression system and. . . .

It all goes to show you how arbitrary the grouping of systems can be. We chose to cover it in the place of greatest pedagogical advantage. Now do you want another name? The chirp signal is sometimes called LIFMOP (linear frequency-modulated pulse).

PROBLEMS, QUESTIONS, AND EXTENSIONS

6.1. Find the $R_m(\tau)$ for the signal of Eq. (6.2.12) using the Wiener–Khintchine theorem. Is it periodic? If yes, with what period? Do the peaks occur at the same points, as in the signal $\cos \omega_m t$?

6.2. Consider the $m(t)$ given by Eq. (6.2.13). Assume that $s(t)$ [which is in $g(t)$] is a periodic function with period T. Find a general expression for $m(t)$, that is, an expression involving an infinite series of sinusoidal terms. (See also Problem 5.11.)

6.3. Show that the beat note f_b of a CW–FM system (refer to Fig. 6.2.1 and the analysis of it) is given by

$$f_b = f_r = \frac{4 R f_m \,\Delta f}{c}$$

where R the target distance, $1/f_m$ the period of the sawtooth wave, and Δf the frequency deviation (maximum minus minimum) of the carrier.

6.4. We saw that even a plain CW radar system can measure range, but that the unambiguous range [see Eq. (6.2.3)] is too small for practical radar applications. To increase the unambiguous range, we can design a CW radar that uses two frequencies which are very close together to each other in value. Prepare a paper on this type of radar and show how R_{unamb} is increased at the expense of system complexity [you can get help from Skolnik (1962, p. 107)]. Can this idea be extended to three, four, or more frequencies?

6.5. Try to prove that Eq. (6.3.8) is correct [get help from Klauder et al. (1960) or Burdic (1968)].

6.6. Section 6.3 considered the emitted chirp signal as a single pulse. Contemplate heuristically the shapes of $R_m(\tau)$, $\mathfrak{IC}_m(\nu)$, and $|M(\omega)|$ if $m(t)$ is (a) a periodic train of chirp pulses, (b) a group of such pulses (illumination time T_i).

6.7. (a) Consider the chirp pulse and sketch the impulse response of a filter matched to it.

(b) Now find the output of this matched filter if the input was the chirped pulse.

6.8. The name "chirp" pertains to the type of frequency modulation used on the carrier. The AM can be other things than the pulse we discussed in this chapter. So:

(a) Consider the signal

$$m(t) = e^{-at^2} \cos\left(\omega_c t + \frac{\mu}{2} t^2\right) \qquad \text{for } |t| \le T_p$$

where T_p is large enough to include most of the signal e^{-at^2}. Find, heuristically, $|M(\omega)|$, $R_m(\tau)$, and $\mathfrak{IC}_m(\nu)$.

(b) Do the same for

$$m(t) = i(t) \cos\left(\omega_c t + \frac{\mu}{2} t^2\right) \qquad \text{for } |t| \le T_p$$

if $i(t)$ is an isosceles triangle from $-T_p/2$ to $+T_p/2$.

6.9. The practical generation of a chirp signal is no easy matter, nor for that matter is the design of a matched filter. Figure P6.9 shows a system that can generate (approximately) a chirped signal using a tapped delay line. Each tap is spaced Δt seconds from the previous one. The input to the whole system is an impulse.

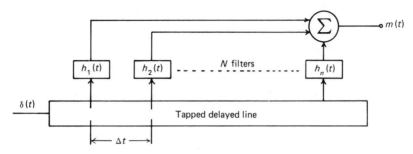

Figure P6.9   Generation (and matched filter) of a chirp pulse.

Each filter is such that its system function is

$$|H(\omega)| = \text{rect}\left(\frac{f - f_j}{\Delta f}\right), \qquad j = 1, 2, \ldots, n$$

where f_j is the center frequency (in hertz) of the jth filter. Now let $f_n = j \, \Delta f$. Find the signal $m(t)$. Then show that the system of Fig. P6.9 is also a matched filter for $m(t)$ if input and output are reversed.

6.10. We saw in Section 6.3 that the output of a matched filter to a chirp signal behaves like a $(\sin x)/x$ function. (Where?) Now the sidelobes are often undesirable. To eliminate them, one may settle for a mismatched filter, which can be produced by a weighting function acting on the matched filter. Search the literature for this interesting topic and prepare a lecture on it. Discuss the "price" one must pay (in some other performance criterion) for reducing the sidelobe effect.

=====7=====

ANALOG PULSE MODULATION: COMMUNICATION SYSTEMS

"They must have known about telephony in ancient China," remarked the Chinese archaeologist to his Greek colleague. "We just unearthed some wires in one of our digging sites."

"That may well be true," he answered. "But the ancient Greeks knew about wireless *transmission. None of our digging sites has ever produced a single wire."*

7.1 INTRODUCTION

We are about to leave the air (space channel, that is) and enter the world of cables (wires, fibers, etc.). The air, as we have noted, behaves like a HPF filter, so a carrier must be used to shift the signal's spectrum to the right-frequency region. Cables, on the other hand, behave like a LPF, so no carrier is needed. The signal (voice or other low-frequency data) can go through it nicely, without the need of a vehicle of any kind. So why bother to discuss it, then, if it is so trivial?

There is no question that you can send *one* signal through the wire channel, but how about *two* or *more*, that is, how about multiplexing? The type of multiplexing we know takes each signal and shifts it to a place in the spectrum unique of its own. And this shifting requires multiplication by a cosine, so you need a carrier. How, then, do you send more than one signal through a wire, without cosines? Use more wires? Not bad! But we meant through the same wire, of course. Sure you could use more wires, more easily than you could use more *spaces* in the previous case, but then we would be comparing apples with oranges.

Pulse modulation systems are meant to show us another way of multiplexing signals, useful when the channel is like an LPF, without shifting their spectra around, without even caring whether their spectra are on top of each other or not (and they will be).

This chapter is about **analog pulse modulation systems**. The word "analog" is used to stress that the transmitted signals are still analog (can take any value in a continuum) even though they are first turned into discrete-time signals, as we will see. Digital signal transmission will be dealt with in a later chapter.

The reader will note by leafing through this book that there is no chapter on analog pulse modulation radar systems. This is sort of obvious, since there is no radar that uses wires as a channel medium. Still, there are radars that make use of analog pulse modulation for *tracking* purposes, although they have it superimposed on an already pulse-modulated cosine, and as such they will be discussed later under mixed systems.

7.2 SAMPLING THEOREMS

The key idea behind all pulse modulation systems is sampling. Without it, these systems would not exist, nor for that matter would the pulse-code modulation systems discussed in Chapter 8. Sampling, then, is what we shall discuss in this section, in a manner dictated by the length of the book. The whole subject is so huge that it could occupy a book of its own. We will try to do the most with the room we have. Needless to say, the ideas of sampling are background-type material and could have been included in Chapter 2. We chose to cover it with a section here rather than burying it there, because it is that important—not just as a help in discussing a chapter or two, but as a precursor to all modern data transmission systems.

Fundamental to all sampling (theorems) is the notion of a bandlimited signal. We have actually seen this before in Problem 2.25. Here, however, we will elevate its status and give it a definition all its own.

Definition (Bandlimited Signal).   We shall call a signal $f(t)$ bandlimited if its Fourier transform $F(\omega)$ vanishes for ω larger than a constant, that is, if

$$F(\omega) = 0 \qquad \text{for } |\omega| > \omega_m \qquad (7.2.1)$$

This definition sets the stage for the **sampling theorem**, a result so remarkable that it deserves a standing ovation.

Theorem 7.2.1 (The Uniform Sampling Theorem).   If a signal $f(t)$ is bandlimited as in (7.2.1), then

$$f(t) = \sum_{n=-\infty}^{+\infty} f(nT)\frac{\sin \omega_m(t - nT)}{\omega_m(t - nT)} \qquad (7.2.2)$$

where $T = \pi/\omega_m$ and $f(nT)$ represents the values of the signal at the points nT.

Proof. There are all kinds of methods for proving this theorem. We will use one that matches our knowledge of expansion theory presented in Chapter 2.

Let us rewrite the above as

$$f(t) = \sum_{n=-\infty}^{+\infty} \left(f(nT)\sqrt{\frac{\pi}{\omega_m}} \right)\left[\sqrt{\frac{\omega_m}{\pi}} \frac{\sin \omega_m(t - nT)}{\omega_m(t - nT)} \right] \qquad (7.2.3)$$

(We have multiplied numerator and denominator by $\sqrt{\pi/\omega_m}$, and then grouped the terms cleverly because we know the result.)

Now any child (and even some adults) can show that the bracketed functions are an orthonormal set; that is, their inner products (IPs) are zero and their norms are unity. This being the case, the terms in parentheses in Eq. (7.2.3) must be the IPs of $f(t)$ and the bracketed terms, the *basis* functions. Let us show this, at least for $n = 0$. We must show that

$$\left(f(t), \sqrt{\frac{\omega_m}{\pi}}\frac{\sin \omega_m t}{\omega_m t} \right) = \sqrt{\frac{\pi}{\omega_m}} f(0) \tag{7.2.4}$$

The left-hand side of Eq. (7.2.4) yields

$$\sqrt{\frac{\omega_m}{\pi}} \int_{-\infty}^{+\infty} f(t)\frac{\sin \omega_m t}{\omega_m t}\, dt \tag{7.2.5}$$

Now there exists a relationship, known as the **generalized Parseval identity**, which states that

$$\int_{-\infty}^{+\infty} f(t)g(t)\, dt = \frac{1}{2\pi}\int_{-\infty}^{+\infty} F(\omega)G(\omega)\, d\omega \tag{7.2.6}$$

which is nothing but a generalization of Problem 2.15 to FTs. Note also that if $f(t) = g(t)$, we obtain Eq. (2.6.8), that is, the usual Parseval identity. In view of Eq. (7.2.6), expression (7.2.5) becomes

$$\sqrt{\frac{\omega_m}{\pi}}\frac{1}{2\pi}\int_{-\omega_m}^{+\omega_m} F(\omega)\frac{\pi}{\omega_m}\, d\omega \tag{7.2.7}$$

where π/ω_m is the FT of $\sin \omega_m t/\omega_m t$ and the limits have been changed to reflect that both $F(\omega)$ and the mentioned FT vanish outside $|\omega| \leq \omega_m$.

The above is obviously

$$\left(\sqrt{\frac{\pi}{\omega_m}} \right)\frac{1}{2\pi}\int_{-\omega_m}^{+\omega_m} F(\omega)\, d\omega = \sqrt{\frac{\pi}{\omega_m}} f(0) \tag{7.2.8}$$

since the integral term is $f(0)$ (the inverse FT evaluated at $t = 0$). So we showed it, at least for $n = 0$. The other coefficients also come out correctly, as the reader can easily verify. The only thing that remains is to prove that the *basis* is complete so that Eq. (7.2.2) is an equality and not an approximation. We will not bother with this, but the reader can take our word for it.

Now let us discuss this theorem and get a better feeling for it. First, note the nature of the coefficients of this expansion [Eq. (7.2.2)], because they are unique. They are actually the *values* of $f(t)$ at equal distances nT (i.e., $-T, 0, T, 2T$, etc.), where T, the **sampling period**, has a value dictated by ω_m. Well, we have seen a lot of expansions in our time, but this is the only one yet with such coefficients. We have often stated in the past (see our discussion of discrete spectra, for example) that once the basis is agreed on, the coefficients *completely specify the signal*. So here we have the remarkable phenomenon that values of a band-limited $f(t)$, at specific points, completely specify this $f(t)$; that is, a countable number of values can determine the rest, which are infinite (uncountable). This is quite an expansion, and it does not even have a name.[1] Somebody has been wronged by history, here.

So if $f(t)$ is bandlimited, we can just keep its values at nT and throw the rest away (they are redundant), that is, *sample it* at these points, and the samples are as good as all of it.

This act of sampling can be visualized with the system shown in Fig. 7.2.1., that is, as a product of $f(t)$ with a comb_T, where, we recall,

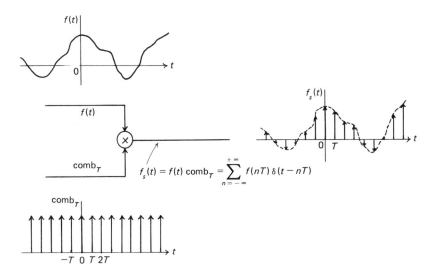

Figure 7.2.1   System for ideal sampling.

$$\text{comb}_T = \sum_{n=-\infty}^{+\infty} \delta(t - nT) \tag{7.2.9}$$

The result—the signal that has values at nT—has been denoted by $f_s(t)$. The whole thing [product of $f(t)$ with the comb_T] is named **ideal sampling** since the comb_T exists only in people's minds.

It will give us a much better feeling for this theorem if we found the FT of the sampled signal $f_s(t)$. Let us assume that $F(\omega)$ is as shown in Fig. 7.2.2a, so that we have some pictures to go with the development. Since

$$f_s(t) = f(t)\,\text{comb}_T \tag{7.2.10}$$

a product in time, $F_s(\omega)$ will be $1/2\pi$ times the convolution of $F(\omega)$ with $2\pi/T \sum \delta(\omega - n\omega_m)$, which is the FT of the comb_T (see Table 2.8.1). Thus

$$F_s(\omega) = \frac{1}{2\pi} F(\omega) * \frac{2\pi}{T} \sum_{n=-\infty}^{+\infty} \delta(\omega - n\omega_m) \tag{7.2.11}$$

[1]Sometimes it is referred to as the Whittaker (J. M. or E. T.) expansion, because both of them (and others) stated and proved this theorem.

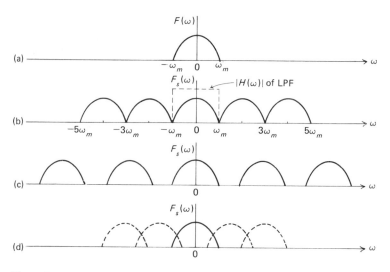

Figure 7.2.2   (a) The FT of the unsampled analog signal $f(t)$; (b) the FT of the sampled signal $f_s(t)$, with the samples as dictated by the theorem; (c) $F_s(\omega)$ with samples closer than T; (d) $F_s(\omega)$ with samples farther than T.

and recalling that the convolution of a signal (in ω this time) with a delta function $\delta(\omega - \omega_m)$ gives back the signal shifted to ω_m (where the delta function sits), we have

$$F_s(\omega) = \frac{1}{T} \sum_{n=-\infty}^{+\infty} F(\omega - n\omega_m) \qquad (7.2.12)$$

which has been plotted in Fig. 7.2.2b.

   This last figure actually contains the *essence* of the **Shannon sampling theorem**, as it is sometimes called. First look at $F(\omega)$ and then $F_s(\omega)$. Is it not obvious that $F_s(\omega)$ contains $F(\omega)$? Of course, it has other terms as well, which are shifted versions of $F(\omega)$. But the sampling of $f(t)$ with the correct $T = 2\pi/\omega_m$ has kept $F(\omega)$ intact, so to speak, inside $F_s(\omega)$, and any time that we wish to recover $f(t)$ we can do it by filtering $f_s(t)$ with an ideal LPF (also shown in the figure). What we did with mathematics is also shown here with an "engineering" type of approach.

   This approach can also tell us what happens if we sample with a period smaller or larger than T, or *equivalently*, if we sample at a *rate* (rate is frequency, i.e., $f = 1/T$ or $\omega = 2\pi/T$) faster or smaller than the rate $f_s = \omega_m/\pi = 2f_m$ or $\omega_s = 2\omega_m$. The result of sampling with a faster rate (more samples than actually needed) is shown in Fig. 7.2.2c. Now, you can get $F(\omega)$ [and thus $f(t)$] even with a sloppy LPF. The result of a smaller rate is shown in Fig. 7.2.2c, and it is somewhat disastrous. There is overlapping (called **folding** or **aliasing**) of $F(\omega)$ with its shifted versions, which causes it to distort. This $F(\omega)$ has been lost, never to be recovered again.

Of course, the whole theorem rests on several important ideas, which as far as the mathematics goes are okay but which must be examined from a practical point of view to see the applicability of this theorem. The first one is that of bandlimitedness. Theoretically, bandlimited signals exist. A Sa (t) or Sa$^2(t)$ are theoretically such signals. Both, however, last forever in the time domain. In fact, it can be shown (Papoulis, 1962, p. 70) that if a signal is time-limited (i.e., it does not last forever; see also Problem 7.1), then it *cannot* be bandlimited. Now we know that in practice we just do not have the time (a lifetime is not enough) to record a signal that lasts forever, so practical bandlimited signals cannot exist. But wait a minute! All practical signals are observed as the output of some device, and no device can be made that will pass all frequencies. So all practical signals must be bandlimited. But we just said that they cannot be unless they are observed forever. As a matter of fact, it would take a scientific conference to resolve this issue. Just assume that they are bandlimited when their FT gets close to zero. You can sample speech as if it had a maximum frequency of 5 kHz (or even less) and still reconstruct it with a sloppy LPF without much effect in intelligibility. So much for this problem.

Now to this business of sampling. It is getting harder and harder to find people who will sit down and record the values $f(nT)$; the job simply does not pay enough. So we came up with electronic sampling, just like that shown in Fig. 7.2.1. But comb$_T$ also does not exist in practice. So what do we do? Does the theorem still hold if we use little pulses or something else instead of comb$_T$? The answer is yes, and we will verify it in the next section.

Finally, a word about the reconstruction of $f(t)$ from its samples. In practice the infinite series is worthless—we use an LPF. But ideal LPFs do not exist, and those that do, do not have the sharp cutoff at $\omega = \omega_m$ that is required to recover $F(\omega)$ properly. To solve this, we simply sample at a faster rate than the one dictated by the theorem.[2] A sampling period of $0.7T$ seems to produce acceptable results in practice.

In closing this section we should like to make some remarks about possible extensions of this theorem, of which there are many. There are versions of sampling theorems that deal with nonuniform sampling periods, versions for signals of other types (nearly bandlimited, bandpass, stochastic), and so on. The only other we shall mention here is the case of a bandpass signal.

A bandpass signal is one whose FT is nonzero in a certain frequency band only. Such a case is shown in Fig. 7.2.3. Quite obviously this type of signal is also bandlimited,[3] as its $f(t)$ is zero for $\omega > \omega_c + \omega_m$ and $\omega < -\omega_c - \omega_m$, and as such it could be sampled with a period

$$T = \frac{\pi}{\omega_m + \omega_c} \qquad (7.2.13)$$

[2]The rate of the theorem is often called the Nyquist rate.

[3]To distinguish between the two even further, the regular bandlimited ones are often called *strictly* bandlimited (they have no gaps up to ω_m).

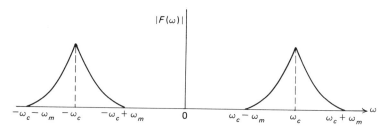

Figure 7.2.3 Bandpass signal's FT.

However, the sampling period of a bandpass signal can actually be increased by taking advantage of the fact that $F(\omega) = 0$ for $|\omega| < \omega_c - \omega_m$. It can be shown (see Panter, 1965, pp. 524–527) that for bandpass signals the sampling period can be

$$T = \frac{\pi}{2\omega_m} \tag{7.2.14}$$

assuming that the frequency band occupied by the signal is located between adjacent multiples of $2\omega_m$. In other words, if it is bandpass, and this last condition is met, we can treat it as if its highest frequency were its bandwidth $2\omega_m$. It is good to remember this, because we might have occasion to use it in the future. See also Problem 7.4 in connection with this theorem.

Finally, let us look at an example.

Example 7.2.1

The signal $f(t)$ is bandlimited up to $\omega = 2\pi \times 10^3$. This signal enters the processing system shown in Fig. 7.2.4. Find the Nyquist sampling periods for the signals $f(t)$, $g(t)$, $l(t)$, and $o(t)$.

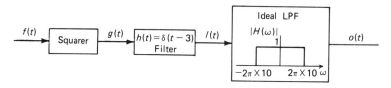

Figure 7.2.4 Processing of $f(t)$.

Solution.   First, direct application of the sampling theorem gives the sampling period for $f(t)$ as

$$T_f = \frac{\pi}{2\pi \times 10^3} = \frac{1}{2} \times 10^{-3} \text{ s}$$

Now if we square a signal, the resulting $g(t)$ is bandlimited up to twice the original band (see Problem 2.25). Thus, for $g(t)$,

$$T_g = \frac{\pi}{4\pi \times 10^3} = \frac{1}{4} \times 10^{-3} \text{ s}$$

What about $l(t)$? This signal is nothing but a shifted version of $g(t)$. Shifts add phase angles to the FTs, but do not change the point at which the FTs are zero (check this). Thus

$$T_l = T_g$$

The output $o(t)$ cannot have any frequencies present that are higher than $\omega = 2\pi \times 10$. The ideal filter will not allow them through. So

$$T_o = \frac{\pi}{2\pi \times 10} = 50 \text{ ms}$$

The key idea behind solving such problems is to keep track of the maximum frequency of the signal. Do not bother finding the FT unless you have to.

7.3 PRACTICAL SAMPLING METHODS

We have already seen that sampling can be visualized as a product of $f(t)$ with a comb_T, and this type of sampling we have called ideal. It is the one that matches the sampling theorem exactly. But theory is one thing and practice another, and comb_T is simply not realizable. In what follows we shall examine a couple of practical methods of sampling, with our eye on whether the sampling theorem is still obeyed or not. Our examination will be based on observing the sampled signal's FT and seeing whether an LPF can reconstruct the original version. Both types will be useful later when analyzing pulse modulation systems.

7.3.1 Natural Sampling

Let us visualize this type of sampling with the aid of Fig. 7.3.1. The switch rotates so that it completes a revolution, in time equal to the sampling period (or less). Every time it touches contact A, it takes a sample. Needless to say, such contact is made in a finite instant of time, so the end result is as if $f(t)$ were multiplied by a periodic pulse train $p(t)$ of unity amplitude. Thus

$$f_s(t) = f(t)p(t) \tag{7.3.1}$$

and this type of sampling is called **natural**. Figure 7.3.2a shows the $f(t)$, Fig. 7.3.2c the $p(t)$, and Fig. 7.3.2e the $f_s(t)$. It should be noted that in natural sampling we do not actually extract *one* value of the signal at the points nT, but a whole range of them around these points, depending on the width of the pulse.

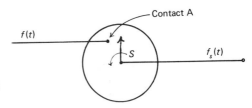

Figure 7.3.1 Visualization of natural sampling.

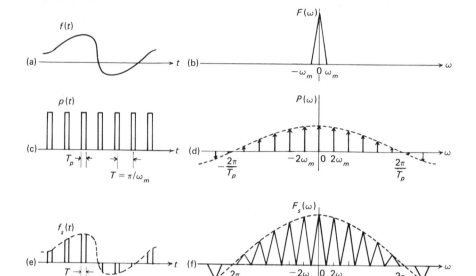

Figure 7.3.2 (a) $f(t)$; (b) $F(\omega)$; (c) $p(t)$; (d) $P(\omega)$; (e) $f_s(t)$; (f) $F_s(\omega)$.

There is not much you can do about it since you cannot realize pulses with zero width (impulses).

To find $F_s(\omega)$ we proceed as follows. We recall that $p(t)$ is periodic, so it can be expanded in a Fourier series

$$p(t) = \sum_{n=-\infty}^{+\infty} C_n e^{jn\omega_m t} \tag{7.3.2}$$

The value of the coefficients is

$$C_n = \frac{T_p}{T} \frac{\sin(n\omega_m T_p/2)}{n\omega_m T_p/2} \tag{7.3.3}$$

where T_p is the duration of the pulse, T the period of the train (equal to the sampling period), and $\omega_m = 2\pi/T$. We now go after the FT of $f_s(t)$ by brute force:

$$F_s(\omega) = \int_{-\infty}^{+\infty} f(t) \sum C_n e^{+jn\omega_m} e^{-j\omega t} \, dt$$

Interchanging summation and integration (can you do this?), we end up with

$$F_s(\omega) = \sum_{n=-\infty}^{+\infty} C_n F(\omega - n\omega_m) \tag{7.3.4}$$

which is shown in Fig. 7.3.2f. It is obvious that the original $F(\omega)$ is still around $\omega = 0$, and can be recovered with an LPF, although with a change of amplitude C_0. Note also that the shifted versions are now weighted (multiplied by C_n) and are no longer equal in amplitude as in ideal sampling. The conclusion is obvious. Natural sampling is feasible.

7.3.2 Flat-Top Sampling

This type of sampling can be understood with the use of Fig. 7.3.3. Unlike natural sampling, in which the pulses are contoured to follow changes in the signal, this sampling has what it says—flat tops—whose value is equal to the value of the signal at nT. Figure 7.3.3 shows the value of the signal at the middle of the pulse width. This need not be so; they could have the value that the signal has at the start of the pulse, or the end, and so on. A variation of this is called **sample and hold**, but we do not need it in this text, so we omit it.

This sampling can be thought of theoretically as an operation in two steps (see Fig. 7.3.3b). The first step $[f_{s'}(t)]$ is ideal sampling, after which we have a series of impulses whose values are equal to $f(nT)$. The second step is the passage of the result through an ideal linear system with the characteristic shown in Fig. 7.3.3c. This system takes each impulse and stretches it out (both to the right and to the left), turning it into a flat pulse.

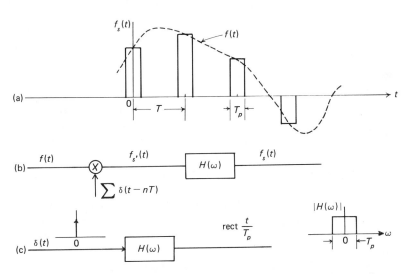

Figure 7.3.3   (a) Flat-top sampling of $f(t)$; (b) block diagram of the sampling; (c) linear system causing flat tops.

The impulse response of such a system (shown as its output in Fig. 7.3.3c) is obviously

$$h(t) = \text{rect}\left(\frac{t}{T_p}\right) \tag{7.3.5}$$

and its transfer function [FT of $h(t)$] is

$$H(\omega) = T_p \frac{\sin(\omega T_p/2)}{\omega T_p/2} \tag{7.3.6}$$

Let us keep these two in mind.

Now, after the first step, ideal sampling, the FT of the signal $f_s(t)$ is

$$F_s(\omega) = \frac{1}{T} \sum_{n=-\infty}^{+\infty} F(\omega - n\omega_m) \tag{7.3.7}$$

as we already know from the preceding section [Eq. (7.2.12)]. The FT of the flat-top sampled signal $f_s(t)$ will obviously be the product of Eqs. (7.3.7) and (7.3.6)—the convolution property of the FT. Therefore,

$$F_s(\omega) = T_p \frac{\sin(\omega T_p/2)}{\omega T_p/2} \frac{1}{T} \sum_{n=-\infty}^{+\infty} F(\omega - n\omega_m) \tag{7.3.8}$$

so it is our old familiar FT of an ideally sampled signal, weighted by the $(\sin \omega)/\omega$ term. The term we are interested in is the one corresponding to $n = 0$, since this is the one that the LPF will catch during reconstruction. This term is

$$\begin{cases} \dfrac{F(\omega)}{T} T_p \dfrac{\sin(\omega T_p/2)}{\omega T_p/2} & \text{for } |\omega| \leq \omega_m \\ \\ 0 & \text{elsewhere} \end{cases} \tag{7.3.9}$$

which is not exactly gorgeous. Flat-topped sampling does not lead to immediate reconstruction with an LPF as was the case up to now. What is needed after the LPF is another filter with $H(\omega)$ equal to the inverse of the $H(\omega)$ of Eq. (7.3.6) (at least in $|\omega| \leq \omega_m$), so its output will be $F(\omega)$—two filters, then, during the process of reconstruction of $f(t)$ after flat-top sampling.

Of course, the analysis above was theoretical—meant to show only that this type of sampling works. The filter with $h(t)$ like the one shown in Eq. (7.3.5) cannot be realized in practice with lumped elements; it has an output before the input $\delta(t)$ has been inserted (not causal). Who cares! Practical flat-top sampling is not performed by the method discussed here, but however it is performed, our analysis holds. That is a satisfying feeling.

Now we have reached the stage of background knowledge that makes the analysis of pulse modulation systems easy. To verify this, continue reading.

7.4 PULSE AMPLITUDE MODULATION COMMUNICATION SYSTEMS

All pulse modulation systems to be discussed in this chapter owe their existence to the sampling theorem. Having established that only the values $f(nT)$ are necessary for the complete specification of a bandlimited signal $f(t)$, the next obvious step is to design systems that will transmit just these values. Systems of this sort are called *pulse modulation systems* (for reasons soon to be discovered), and the one we are about to discuss goes by the name **pulse amplitude modulation (PAM)**.

Such a system is shown schematically in Fig. 7.4.1a. The information signal $f(t)$ (speech, etc.) is first passed through an LPF that ensures that it

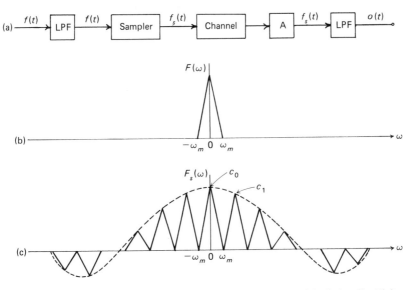

Figure 7.4.1   (a) Block diagram of a PAM system; (b) original signal's $F(\omega)$; (c) sampled signal's $F_s(\omega)$.

becomes bandlimited (up to ω_m, say). Next comes a sampler, one of the types we discussed, or some other kind, whose output $f_s(t)$ is equal to $f(nT)$, with T as the sampling theorem dictates. The signal $f_s(t)$ is ready for passing the channel (wires, etc.), which behaves ideally like an LPF with maximum frequency, say ω_c, such that $f_s(t)$ passes unperturbed. At the receiver, after amplification (block A), the detector is simply an LPF that recovers the original $f(t)$, as explained in the discussion of the sampling theorem. Of course, if the sampling is not natural, but say flat-top, the LPF will have to be augmented by another filter, as explained under flat-top sampling.

It is instructive to find the bandwidth of $f_s(t)$, as the channel must have at least that for its own bandwidth if $f_s(t)$ is to pass undistorted.

Let us assume that $f(t)$ has the FT $F(\omega)$ shown in Fig. 7.4.1b, before and after the LPF. If the sampling is ideal, $F_s(\omega)$ will have a form similar to that of Fig. 7.2.2b; that is, its bandwidth would be infinite, so we are not interested in that. Let us take the case of natural sampling, which is more practical. The output of the sampler is now

$$f_s(t) = f(t)p(t) \tag{7.4.1}$$

where $p(t)$ is a periodic pulse train of period $T = \pi/\omega_m$ (or smaller) and pulse duration T_p. We have already derived the FT of such an $f_s(t)$ in the preceding section [see Eq. (7.3.8)], and this time we even plot it in Fig. 7.4.1c as $F_s(\omega)$. What is the bandwidth of $f_s(t)$?

Well, we could use some fancy definition of bandwidth and get it, but let us not overcomplicate the matter. Let us say that the bandwidth is (approximately) up to the point where the $(\sin \omega)/\omega$ weighting function (the dashed curve) hits the first zero. We could go as far as the second zero if we wanted, but the first zero will be just fine.

With the agreement above, the bandwidth of $f_s(t)$ obviously depends on the width of the pulse of the pulse train $p(t)$. If this width is T_p, the first zero will be hit at $\omega = \omega_c = 2\pi/T_p$, and the bandwidth of $f_s(t)$ will have this value. The same value is the one demanded as the bandwidth of the channel if $f_s(t)$ is to pass *mostly* unperturbed.

The conclusion, then, is that the bandwidth of $f_s(t)$ is approximately equal to $2\pi/T_p$, which is much larger than ω_m. The least it can be is when the pulses are as wide as T (i.e., $T_p = T$, sample-and-hold sampling described in Problem 7.12), and in that case its value is $2\pi/T = 2\omega_m$, reminiscent of the value of the DSB amplitude-modulated signal. But that is *the least*. In most cases where natural sampling is employed, the bandwidth of $f_s(t)$ is much larger, depending on T_p, as we remarked above.

Example 7.4.1

Take speech for $f(t)$, bandlimited up to 5 kHz. Sample it naturally, with pulse width $T_p = \frac{1}{10} T$, where T is the sampling period. Find the bandwidth of the sampled signal $f_s(t)$.

Solution.   First,

$$T = \frac{\pi}{\omega_m} = \frac{\pi}{2\pi \times 5 \times 10^3} = 10^{-4} = 100 \ \mu s$$

Therefore,

$$T_p = 10 \ \mu s$$

and the bandwidth of $f_s(t)$ is approximately

$$\omega_c = \frac{2\pi}{T_p} = 2\pi \times 10^5 = 100 \ \text{kHz}$$

that is, 20 times larger than the bandwidth of $f(t)$.

The above is intriguing. Why bother to use the PAM system if the bandwidth of $f_s(t)$ is many times larger than the original $f(t)$? Why not send $f(t)$ as is, unsampled? Quite obviously, the sampling process spreads out the spectrum—which is undesirable. But spreading a spectrum must have some value, else why would we bother with it? Is it performance in the presence of additive noise? Perhaps. Is it also something else? We will keep the answer secret until a later section.

We have one more thing to take care of before we can close this section, and that is the justification for the name of this system. Why is it called PAM? A look at Eq. (7.4.1) and the answer will become obvious. The idea is a carry-over from sinusoidal modulation and all that business of inserting $f(t)$ in the amplitude, phase, or frequency of the carrier in order to cross the channel. Here the role of the cosine is played by $p(t)$. A proper optical angle at Eq. (7.4.1) reveals that $f(t)$ modulates the amplitude of $p(t)$, $p(t)$ being a carrier of unit amplitude. Of course, $f(t)$ does not need a carrier to cross the LPF channel—it can go through it in fine shape, maybe finer, in terms of bandwidth, anyway. Well, you cannot have everything in interpretations. This one works, but up to a point. Let us keep it in mind, though, because it makes the explanations of the following two systems easier.

One more remark. This system is also called an *analog* pulse modulation system. The reason is that although $f(t)$ is sampled (ideally) at specific points in time, its values can be in a set that is not finite or countable; that is, $f_s(t)$ remains an analog signal as defined in Section 2.2. Ponder this for a while, as we are soon going to be discussing the transmission of signals that are *digital*.

7.5 PULSE DURATION MODULATION SYSTEMS

Let us carry the analogy of the periodic pulse being a carrier of $f(t)$ a little further and the key idea behind this system (and the next) will become obvious. A periodic pulse train, like a periodic cosine, has three parameters that define it, and can therefore be used to carry the information signal $f(t)$. The first is its amplitude, and we finished with that in the preceding section. The other two are pulse duration and period (or pulse position). Attaching $f(t)$ to the duration parameter gives rise to pulse duration modulation (PDM), whereas attaching it to the last one creates a pulse position modulation (PPM) system, discussed in the next section. Both of these systems are analogous to phase and frequency modulation when the carrier is sinusoidal, and we know how tough those were to analyze. These are not any easier. Our discussion here will be qualitative, with a lot of words and nearly no mathematics. The interested reader can search the literature (Schwarz et al., 1966, or Carlson, 1975) if he or she desires a thorough analysis.

The bandlimited $f(t)$ is still first sampled in PDM, so ideally what we must transmit is $f(nT)$, where T is the sampling period. But the $f(nT)$ values will now vary (linearly) the duration of the pulses of the pulse train, not their amplitude. To do this we need a device (a circuit, a function) that will convert amplitude variations to pulse duration variations (or width) in a *linear fashion*. A ramp function can do that, so expect it to play a role in what follows below.

The generation of a PDM signal will be explained with the aid of Fig. 7.5.1. No block diagram will be given, so the reader can think one up. First, the bandlimited information signal is flat-top sampled, and the result is obviously an $f_s(t)$ PAM signal, shown in Fig. 7.5.1a. This signal is next added to the synchronized flipped-over ramp signal shown in Fig. 7.5.1b, and the result is

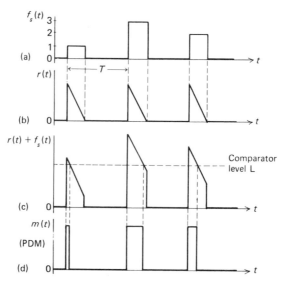

Figure 7.5.1   Generation of a PDM signal: (a) PAM signal; (b) a synchronized inverted ramp; (c) sum of the above; (d) the PDM signal.

passed through a comparator, which gives an output as long as the value of this sum is above level L. It can easily be seen that the resulting signal is wide if $f_s(t)$ is large, and narrow otherwise. The level L is set so that when $f_s(t) = 0$, the output is a certain width τ_0, which can still be decreased for negative values of $f_s(t)$. It is all the work of the ramp, of course, since its values are linearly proportional to time, and this is the conversion we are after in this system.

The demodulation of the PDM signal follows the steps shown in Fig. 7.5.2. The leading edge of the PDM $m(t)$ triggers a ramp function $r(t)$ which increases until the end of the pulse and then stays constant for a specified duration, as shown in Fig. 7.5.2b. Next, this signal, call it $r(t)$, is added to a periodic pulse $p(t)$ (on its flat part), and the result (Fig. 7.5.2c) passes through a comparator. The output is a signal of PAM (flat-top) nature, and a couple of filters (why a couple?) can turn it into the original $f(t)$. It was the work of the ramp again, this time converting duration to amplitude. This is, of course, only one scheme for modulating and demodulating the duration of a periodic pulse carrier. Many others exist, but the ramp (in the form of a sawtooth, triangular wave, or whatever) is always there to do its thing.

What about the $m(t)$'s spectrum or bandwidth? This is difficult (see the references mentioned earlier). Expect a very large bandwidth, probably larger than that of the PAM signal. This system is sometimes called pulse width modulation, for obvious reasons. It enjoys a few applications: for example, in remote control of motor position, and the like.

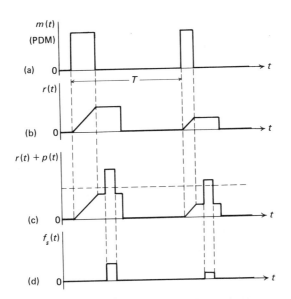

Figure 7.5.2   Demodulation of a PDM signal: (a) the PDM incoming signal; (b) the created $r(t)$; (c) $r(t)$ + a periodic pulse; (d) the PAM signal $f_s(t)$.

7.6 PULSE POSITION MODULATION SYSTEMS

Let us go directly to a mechanism for generating a pulse position modulation (PPM) signal, illustrated in Fig. 7.6.1. No fanfare is necessary here, as we know that this signal must convert amplitude to pulse position. The end product (just as in PDM) is not a periodic signal, and this is mainly what causes the problems in a mathematical analysis. (We had this problem in angle modulation systems as well. Where else?)

Figure 7.6.1a shows a signal that is already in a PDM form, meaning that this initial step is necessary for the creation of a PPM in this scheme. Assume that this signal is first differentiated. The outcome (ideally) is shown in Fig. 7.6.1b. The next step is inversion, followed by half-wave rectification. The results, shown in Fig. 7.6.1c, are impulses, which are located at the end of the pulses of the PDM signal, so their starting position is proportional to width [and thus to the original amplitude of $f(nT)$]. These impulses can be stretched out to narrow pulses and the result is the PPM wave. The demodulation system is left as an exercise for the reader.

This has been a rather quick analysis: no evaluation of the FT, no calculation of the bandwidth required for transmission—they are all too complicated to deal with. Nevertheless, the system works. That being the case, and in view of the fact that "The object of mathematical rigour is to sanction and legitimate

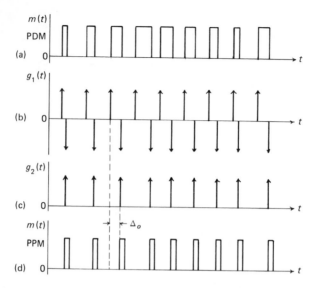

Figure 7.6.1 (a) The PDM signal; (b) differentiation of (a); (c) result of inversion and half-wave rectification; (d) the PPM signal.

the conquests of intuition, and there never was any other object for it" (J. Hadamard), we leave this subject without many regrets or guilty feelings.

7.7 TIME-DIVISION MULTIPLEXING: COMPARISONS WITH FDM

We have left a mystery hanging in the air. Why bother sampling a bandlimited signal $f(t)$ when the resultant $f_s(t)$ has a larger bandwidth than $f(t)$? It is now time to resolve it.

Let us go back to PAM. The information signal $f(t)$ is bandlimited up to ω_m, so the sampling period is at most $T = \pi/\omega_m$. Assume that the sampling is natural and that the width of each pulse $T_p \ll T$. The emitted wave $f_s(t)$ is shown in Fig. 7.7.1 for some imagined signal $f(t)$.

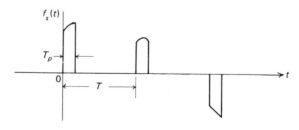

Figure 7.7.1 The sampled signal and the time gap.

The thing to note in this figure is that there is a time gap between the end of each pulse and the begining of the next—a gap equal to $(T - T_p)$ time units. This gap can be utilized to send samples of other signals, samples that are placed in sequence and kept in order. The end result is that more than one signal can be transmitted through the same channel. This method of multiplexing signals is called time-division multiplexing (TDM), the name implying that the time (of the period) T is *divided* into slots, each slot going to one of many band-limited signals.

To clarify this concept we refer the reader to Fig. 7.7.2, which shows a simplified version of time-division multiplexing of n signals. All signals are assumed to be of the same sampling period T, even though that is not absolutely essential for the scheme to operate (see Problems 7.19 and 7.20).

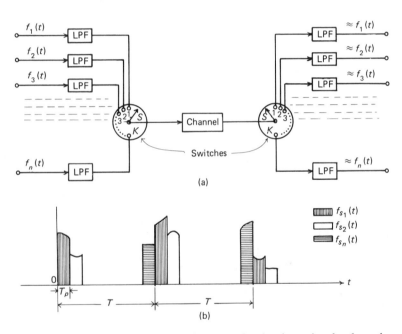

Figure 7.7.2   (a) TDM system; (b) the composite signal entering the channel.

The switches S_1 and S_2 rotate with such a speed that each revolution takes T seconds. Consider first S_1. As it rotates, it takes a first sample from $f_1(t)$, a second sample T seconds later, and so on. In between it takes samples from the rest, also T seconds apart, presumably staying T_p seconds at each contact. Assuming that the sampling is natural, the composite signal sent to the channel is shown in Fig. 7.7.2b. This figure shows the samples touching each other, as if there were no time lapse between contacts of the switch. In practical schemes there is a small gap between samples called a "guard" (time) band, in analogy with the guard band in the FDM system discussed previously. Thus n signals can go through the channel, where

$$n \leq \frac{T}{T_p} \tag{7.7.1}$$

with the equality holding when no guard band is desired (almost never in practice).

In the receiver side the operations are reversed. The switch S_2 (synchronized with S_1) separates out $f_{s_1}(t)$, $f_{s_2}(t)$, and so on, and sends each of them through an LPF to produce $f_1(t), f_2(t), \ldots, f_n(t)$. So, no doubt about it. Many signals can be sent through the same channel by using sampling and the scheme above. But still the original mystery seems to escape us.

To solve it, we proceed as follows. First, let us assume that we send only one signal through the channel. With T the sampling period and T_p the duration of the samples, the bandwidth of the sampled signals $f_{s_1}(t)$ will be

$$B = \frac{2\pi}{T_p} \tag{7.7.2}$$

As we discovered in the section of PAM (Section 7.4), this bandwidth is much larger than the original bandwidth ω_m of $f_1(t)$. But what if we were to send all the sampled signals that can be packed in T, using the TDM described above? The composite signal is actually a new sequence of samples coming from a sum of shifted signals. As such, it can be considered as a new single *signal* sampled with a period of T/n, each sample lasting the same time (no guard band), that is, T/n. This new PAM signal will have the same bandwidth as each $f_s(t)$ separately since it uses the same sample duration—the bandwidth, of course, being measured at the first zero of the dashed curve in Fig. 7.3.2f. This is a key point in this discussion, and the reader is well advised to ponder it. It basically states that the bandwidth of a PAM signal is determined by the width of the samples and *nothing else* (see also Fig. 4.7.2 while you are pondering the issue). The above means that the composite signal entering the channel has a bandwidth of $2\pi/T_p$, so the LPF channel bandwidth must match this to allow it to pass undistorted. Now we recall that with no guard band, the number of signals n is

$$n = \frac{T}{T_p} \tag{7.7.3}$$

or

$$T_p = \frac{T}{n} \tag{7.7.4}$$

Inserting this value into Eq. (7.7.2) and recalling that $T = \pi/\omega_m$, we obtain

$$B = 2n\frac{\pi}{T} = 2n\omega_m \tag{7.7.5}$$

This equation gives the bandwidth required of the channel, irrespective of the number of sampled signals one is sending through it as long as the sample duration is T/n. It is the bandwidth of each sampled signal transmitted alone, or in any combination up to n. So what is the answer to the mystery? Why, do not send just one signal. If you do, you use too much bandwidth unjustifiably. Send n of them simultaneously. The price in bandwidth is the same.

The bandwidth we derived in Eq. (7.7.5), based, we repeat, on the first lobe of the dashed curve of Fig. 7.3.2f or 7.4.1c, comes into conflict with other texts. As recently as 1979 (see Shanmugan, 1979, pp. 553–554), and many times earlier, an argument has been propagated which implies that the PAM-TDM composite signal has a bandwidth of $B = n\omega_m$.

The argument goes something like this. Assume, as we did above, that the signal is bandlimited up to ω_m, which means that its sampling period is $T = \pi/\omega_m$. Conversely, if a signal's Nyquist period is π/ω_m, its bandwidth is ω_m. So far we agree. Now, the argument goes, the composite signal is sampled every T/n seconds, and therefore, by the converse statement above, its bandwidth must be $n\omega_m$. Again we agree.

Where is the fallacy? We think it has to do with what we find! We do not find the bandwidth of the composite sampled signal, but of the continuous one (unknown) from whence the samples came. Similarly, when we first argue that if a signal's Nyquist period is π/ω_m, the signal has bandwidth ω_m, we mean the continuous signal, not the sampled one. What we need here is the bandwidth of the signal that enters the channel, and that signal is not continuous. It could have infinite bandwidth if the samples were ideal, and the argument above would still give $n\omega_m$.

Anyway, there is no need to use roundabout arguments when the answer can be calculated as we did above. Aristotle is reputed to have claimed that women have fewer teeth than men, and many people were convinced at the time.[4] His argument fell through when somebody opened a woman's mouth and counted her teeth.

It is interesting that $n\omega_m$ can be calculated to be the bandwidth of the composite signal if bandwidth is taken at the 3-dB point of the first lobe of the dashed curve in Fig. 7.3.2f. We leave this as an exercise for the reader. One more comment. The bandwidth we calculated is approximate, not exact. This is another complaint that we have with the other argument. It implies that it is $n\omega_m$ *exactly*.

Now we come to some comparisons between FDM and TDM systems. The only legitimate comparison to be made (we feel) is that of bandwidth. A look at Eq. (7.7.5) shows that n time-multiplexed PAM signals need as much bandwidth to be transmitted as n DSB signals. Actually, the B of Eq. (7.7.5) is approximate, and in view of the fact that in FDM you can pack $2n$ signals with $2n\omega_m$ bandwidth (if you use SSB), the FDM may have the edge here. Of course, large bandwidth is not all that bad, as we have repeatedly mentioned—in fact, it is often desirable. It just depends on the application.

---

[4]Even if this rumor is true, we still feel that Aritstotle was one of the greatest philosophers of all time.

Other comparisons between the two are possible, but they usually end up just producing a few laughs. Attempts to compare complexity are such a case, as can easily be ascertained by checking Lathi (1965, p. 488) and Javid and Brenner (1963, p. 202). Technology moves too fast for such comparisons to have any lasting value.

So we close this chapter simply with an observation. In FDM the signals are separated in frequency and they are all on top of each other in time. In TDM the opposite is true. Their sampled versions are separated out in time, but their FTs are not separable. Take your choice. Not really. If the channel is space, you can only use FDM. The choice is there only if the channel behaves as a LPF. (Why?)

Of course, our entire development of TDM was based on PAM. Signals can be time multiplexed even if they are in PDM or PTM form. We shall not discuss these cases here, but the results are similar. The interested reader can look in Taub and Schilling (1971) for an introductory treatment of this issue.

PROBLEMS, QUESTIONS, AND EXTENSIONS

7.1. Let us define a time-limited function $f(t)$ as one for which the following condition holds:
$$f(t) = 0 \qquad \text{for } |t| > T$$
Now try to show this: that an $f(t)$ cannot be time-limited and bandlimited at the same time.

7.2. Prove that under the conditions of Theorem 7.2.1, the expansion elements $\phi_n(t)$ are orthogonal. Then find their norm.

7.3. Prove Eq. (7.2.6).

7.4. Consider a bandpass signal whose FT is shown in Fig. 7.2.3. Show that for such a signal the sampling rate can be
$$f_s = \frac{2(f_c + f_m)}{K}$$
where K is the largest integer not exceeding the value of the ratio $(f_c + f_m)/2f_m$. Then connect the result with the discussion of bandpass signals at the end of Section 7.2.

7.5. (a) Find the Nyquist sampling rate for the following signals (note that they are all strictly bandlimited): (1) Sa (100t); (2) [Sa (100t)]^2; (3) Sa (100t) + Sa (10^3t); (4) Sa (100t)·Sa (200t); (5) Sa (100t)·[Sa (100t)]^2
   (b) As we remarked in Section 7.2, bandlimited signals do not exist in practice. Find the Nyquist rate for the signals (1) $f(t) = e^{-t}u(t)$ and (2) $f(t) = e^{-|t|}$ by assuming that their FT reaches zero when their amplitude spectrum is 5% of its maximum value.

7.6. Prepare a lecture on *quadrature sampling* of bandpass signals. Help can be found in Peebles (1976).

7.7. (Tricky problem). Find the sampling period for $f(t) = a \sin \omega_m t$ by considering it bandlimited up to ω_m. Now that you found it, sample $f(t)$ with this period. What do you get? All zeros, right? How can you reconstruct $f(t)$ from this? Think about this problem. See also Roden (1972, p. 54) for a discussion of this issue. We do not necessarily agree with the argument behind his answer.

7.8. Consider the system shown in Fig. P7.8 and assume that

$$f(t) = \frac{\sin 10^3 t}{10^3 t}$$

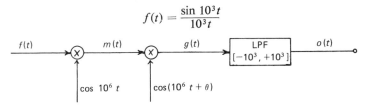

Figure P7.8

(a) Find the Nyquist rates for $f(t)$ and $g(t)$.
(b) Find the Nyquist rate for $o(t)$ for $\theta = \pi/4$ and $\theta = \pi/2$.

7.9. Consider the system shown in Fig. P7.9 with $f(t) = 10 \, \text{Sa}(10^3 t)$ and $g(t) = 20[\text{Sa}(200t)]^2$. Find the Nyquist rate for $m(t)$, $y(t)$, and $o(t)$.

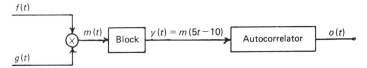

Figure P7.9

7.10. Time- and frequency-domain statements tend to have a duality. Assume that we have a time-limited signal $f(t)$ as defined in Problem 7.1. State and prove a sampling theorem that pertains to sampling $F(\omega)$ [not $f(t)$].

7.11. The signal $m(t) = \cos(100\pi t) \cos(200\pi t)$ is sampled with period 1/400 s. Find the minimum cutoff frequency of an ideal LPF that could recover $m(t)$ from the samples.

7.12. **Sample-and-hold** is a practical method of sampling, akin to flat-top. The sampled value is held flat until the next sample is taken (shown in Fig. P7.12), a feat easily accomplished with a simple circuit like the one shown in Fig. P7.12b. Find the FT of $f_s(t)$ and compare it with Eq. (7.3.8).

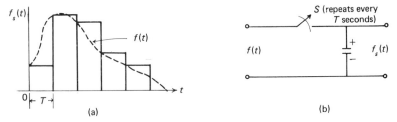

Figure P7.12   (a) Sample-and-hold illustration; (b) sample-and-hold circuit.

7.13. Natural sampling is actually a product in time [see Eq. (7.3.1)]. Derive Eq. (7.3.4) by convolving $F(\omega)$ with $P(\omega)$.

7.14. Consider the signal

$$s(t) = \begin{cases} 1 - \dfrac{|t|}{T_p} & \text{for } |t| \leq T_p \\ 0 & \text{elsewhere} \end{cases}$$

and then the periodic function

$$p(t) = [\text{rep}_T\, s(t)]$$

Next assume that this $p(t)$ is used to sample a bandlimited $f(t)$ (up to $\omega_m = 2\pi/T$) by multiplication. The sampled version $f_s(t)$ is

$$f_s(t) = f(t)p(t)$$

Find the FT of $f_s(t)$ and contemplate whether $f(t)$ can be recovered from $f_s(t)$, and how.

7.15. With reference to Section 7.4, assume that the bandwidth of $f_s(t)$ is determined by the second zero of the $(\sin \omega)/\omega$ curve. Then find the bandwidth of $f_s(t)$ of Example 7.4.1 under this condition. Repeat by defining the bandwidth at the 3-dB point.

7.16. Describe a method for demodulating a PPM wave.

7.17. A serious problem in all TDM systems is **crosstalk**. To see the nature of crosstalk, assume that n equally bandlimited signals, ideally sampled at the sampling frequency, are sent to an ideal LPF channel with maximum frequency ω_c. Find the composite signal at the output of the channel, which should be a series of $(\sin \omega)/\omega$-type functions, carrying the sampled values at their peaks. Do you see that these functions are interfering with each other as regards their maximum values? Can you think of a way to minimize this effect? (The solution lies in the proper spacing between the ideal samples.)

7.18. Consider n voice signals bandlimited up to $f_m = 8$ kHz. How many such signals can be TDM'd (PAM) if the pulse widths are $T_p = T/10$? T is the sampling period and the guard band is $\frac{1}{10}$ of T_p.

7.19. More than one bandlimited signal can be TDM'd, even if they do not have the same sampling period. Consider, for example, $f_1(t)$, with sampling period T_1 and $f_2(t)$ with sampling period $4T_1$. How do you multiplex them? Design a system like the one in Fig. 7.7.2a which can do the job.

7.20. Consider the signals $f(t), f^2(t), f^4(t), \ldots, f^{2k}(t)$ (k an integer). If $f(t)$ has sampling period T (corresponding to an $\omega_m = 5$ kHz) and the width of the samples is $T_p = T/20$, how many of the signals can be TD-multiplexed without a guard band? How many with a guard band of $T_p/20$?

7.21. What about TDM for PDM or PPM? Search the literature and prepare a paper on this issue.

=== 8 ===

PULSE CODE MODULATION: COMMUNICATION SYSTEMS

8.1 INTRODUCTION

We have seen how to transmit an information signal $f(t)$ if it happens to be the *analog* type—continuous time or discrete time. We have seen this for two types of channels, HPF types (space) and LPF (wires, etc.). The time has come to consider digital signals: signals whose values are a countable set and (in practice) usually a finite one (up to n values). This and the next chapter are concerned with such signals, for the LPF and BPF channels, respectively. Needless to say, radar systems will not be considered in conjunction with an LPF channel, as the channel they use is space.

Now digital signals exist in practice all right—of their own accord, that is. If you have to transmit the number of ships that pass the Suez Canal every day to a receiving station in the Canary Islands, the signal you have to send is digital; it can only take one of n possible values. Still, a lot of the information signals that must be transmitted are analog. To use the systems we are about to cover, these signals must first be *converted to digital*, an operation called *quantization*. This operation, then, is the first business at hand, so that the signals to be considered will always be digital, whether they are that way by nature or not.

Strictly speaking, the signals to be transmitted through the systems under discussion will be also discrete-time type; that is, they will have experienced sampling. This is not—theoretically speaking—mandatory, but in practice it is needed for multiplexing. Thus the signals that use these systems are tacitly assumed to possess some sort of bandlimitedness, enough so that the sampling theorem can be used without completely destroying them. Onward, then, with quantization and then PCM systems.

8.2 QUANTIZATION

Quantization is philosophically the sort of thing everybody does when asked to measure something. It is rounding off the outcome of the measurement to its nearest visible digit. If you asked somebody to measure the width of a room and he came back with the number 3.2713572327785858123572432 (and so on) in meters, you would think him crazy. There is no instrument that can do that, and even if there were, it had better be a digital one, because his eye could not possibly have read such a number.

The type of quantization we are concerned with here is, of course, of the same nature, but more *forced* than the natural one. We wish to round off the values of an analog $f(t)$ to a smaller set than that dictated by the inability of an instrument or the human eye. We wish to limit it to a small set of values, so that our transmission process is simplified. The number of values in this case will be dependent on the nature of the signal (we do not want to wreck it in the process) and the system that is to process it for eventual transmission.

The process of quantization can be visualized with the aid of Fig. 8.2.1. In part (a) of this figure we have set the block form of the quantizer and the notation—$f_q(t)$ is the quantized form of $f(t)$.

Figure 8.2.1b shows a typical input/output characteristic of a **quantizer** in terms of values. It is a staircase function whose effect is to turn the analog signal into a digital one. Any value of the input signal in the region $(-\frac{1}{2}, +\frac{1}{2})$ becomes zero, any value in $(\frac{1}{2}, \frac{3}{2})$ turns into unity, and so on. The output will have only integer values, a countable set, and therefore it will be a digital signal. If, before it is inserted into the quantizer, it suffers sampling, the output will be discrete-time digital signal, which can be visualized as a series of impulses of specific heights (ideal sampling), or as a series of pulses of the same heights (practical sampling). Such a case as the last one mentioned is shown in Fig. 8.2.1c, where the practical sampling is assumed natural. Figure 8.2.1d shows the difference between $f_s(t)$ and $f_q(t)$, which is usually referred to as the **quantization error** (or **noise**). With this example behind us, let us discuss quantization in general and see some of its other possible forms.

To begin with, the staircase input/output characteristic does not keep on going forever in a practical situation—it is not possible. Usually, the input is limited to a specific $f(t)_{(max)}$ and $f(t)_{(min)}$, and the number of positive and negative steps of the staircase are finite.

The difference between adjacent discrete output values of the quantizer is called a (quantum) **step**. In the example above (Fig. 8.2.1) the step is the same between all adjacent values, and such a quantizer is called **uniform**. This is not

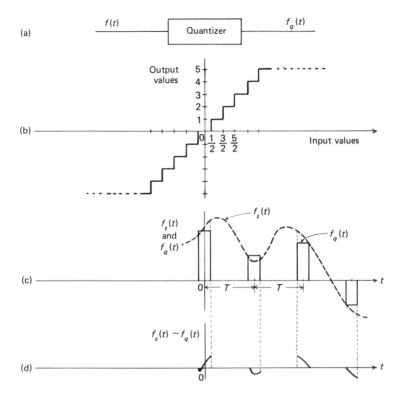

Figure 8.2.1 (a) Block diagram of quantization; (b) input/output relation of a quantizer; (c) a sampled signal $f_s(t)$, and its quantized form $f_q(t)$; (d) quantization error.

always the case in practical quantizers. There are applications where the steps around the origin are small, becoming larger as they go away from it. This "favoring" of the weak values of the signals (by quantizing them less) works well with weak signals and has been studied extensively [see Smith (1957) as an example]. Other cases exist where the quantizer changes its step structure to suit the input signal, and so on, cases that will be ignored here, as they will not be needed later.

Perhaps the most important theoretical remark we can make about quantization is that it is, in general, an *irreversible* process. You can sample a signal, and reconstruct it any time you wish, and you have the sampling theorem behind you to support you theoretically. No such theorem exists to back up your quantization maneuvers. Once a signal is quantized, that is it. Its original, unquantized form is lost forever. There have been some efforts to reconstruct the original version of quantized signals, but they hold only in very specific cases, or the reconstruction process yields an "optimum" original version, not necessarily *the* original version [see Avgeris and Tzannes (1978) for more on this issue]. This being the case, the following question arises: How do we design an optimum quantizer, optimum in the sense of minimizing signal distortion without having too many steps that we cannot deal with? Well, although there are a lot of papers that deal with this subject [typically, see Max (1960)], the answer seems to be empirical—you try a few things and see the results. In color TV, for example, 512 levels seem to produce an excellent picture (nothing much is lost from quantization), whereas 54 levels produce only a bearably good one.

So much, then, for the process of quantization. The key question, of course, is why we should bother with it, particularly since the signal gets distorted, and we run the risk of losing it altogether. What is it that we gain by doing it? What is so important about turning an uncountable infinity of values into a finite set?

Well, we shall not give the full answer to this at this point; it will come in the section that follows. We shall, however, point out one advantage of quantization, an advantage that can be fully exploited in the systems we have already covered: PAM, PDM, and PCM. After all, nothing forbids us from using quantization in those systems either, as long as this process keeps the signal to tolerable limits of distortion.

The advantage has to do with additive noise. To say that by keeping the signal unquantized we reserve its purity, is, of course, nonsense. As soon as we let it go through a device or a channel, noise immediately gets added on (not always *added*, either) and its *real* values are gone, anyway. However, if we first quantize it, we can actually limit the effects of this additive noise. The values of the quantized signal are a known finite set. We can ship it out through a noisy channel, but add quantizers along the way (repeaters) which erase the added noise and return it to its original emitted self. All we have to do is study the noise values, and put the repeaters close enough, so that the noise is not larger than half a quantization step, resulting in a restoration to the wrong value. Think about this one, a little, and make sure that the purported advantage is well understood.

8.3 CODING

Here is the main beauty of quantization. Once a signal's values have been limited to a finite number, these values can be transmitted with a code. A **code** is basically a function that converts the finite values of the signal (or sequences of these values) into sequences of other values. To see this better, and then be in a position to explain the advantages of coding, we consider a simple case of converting the values of a quantized signal into a binary arithmetic code.

Let us assume that we have a signal $f(t)$ bandlimited up to ω_m. This signal is first sampled at the Nyquist rate, say ideally, and then quantized to four values (0, 1, 2, 3). These numbers in the decimal system can be converted to a binary system (with base the number 2) by the formula

$$N = K_1 2^1 + K_0 2^0 \tag{8.3.1}$$

A binary arithmetic code is the mapping of the signal values of $f(t)$ to pairs of values of the coefficient (K_1, K_0) of N. Thus a binary code based on Eq. (8.3.1) for the signal values will be

Signal Values	Code Words	
0	00	(8.3.2)
1	01	
2	10	
3	11	

Let us say that we adopt this code. Then at each instant of time that we are to send a value of the quantized signal, we will send the pair of values of the code instead. This is illustrated in Fig. 8.3.1. Assuming that the first three sampled and quantized values of $f_q(t)$ are (3, 2, 1), as shown in Fig. 8.3.1a, the coded version $f_c(t)$ will be as shown in Fig. 8.3.1b. Quite obviously this last signal is the one that will be sent through the channel, and this conversion to a coded form must presumably provide us with some advantages. What are they?

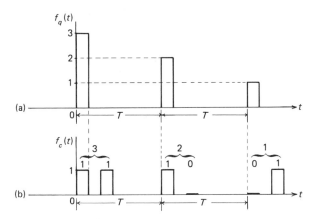

Figure 8.3.1   (a) Naturally sampled-quantized signal; (b) coded version of the signal.

To begin with, the advantage of quantization mentioned at the end of the preceding section (additive noise, repeaters and so on) is kept intact. The values we are transmitting are still few (fewer, in fact, than before), and properly situated repeater-quantizers can erase the noise before it gets too big and converts a "zero" to a "one," or vice versa. Advantage one.

Now think of the receiver. If the transmitted signal is the uncoded one, the receiver must decide which of *four* values are at each sampling moment. If the *coded* signal is sent instead, the transmitter must only decide which of *two* values are at each bit instant. (Zeros and ones are *binary digits*, or bits.)

This makes life a lot easier for the receiver, and it can be designed to perform much more successfully with this simpler task. Advantage two.

There are many other advantages as well. The theory of coding has been advancing by leaps and bounds, providing us with excellent codes that can detect and even correct errors made in the bits, a case always likely to happen no matter how many repeaters you might have in the channel. Then there is the question of secrecy, whether this is for military purposes, company secrets, or safeguarding citizens' right to privacy. This field, called cryptography, also developing rapidly, could not exist without the ability to code, and this in turn would not exist without quantization. For there is no point in trying to code a signal with infinite values. Each code word would have an infinite string of code symbols, and all you could do would be to send one value only. Not much advantage in this, of course.

Quantization and coding are not without disadvantages. Nothing is, as we have often remarked. Quantization distorts the signal irreversibly. The bandwidth increases immensely, as we shall see in the next section. *Complexity* is obviously increased, as each system must have a coder in the transmitter and a decoder in the receiver. Still, the advantages seem to far outweigh the disadvantages, and most communications systems are rapidly being converted to the coded versions about to be discussed in the sections that follow.

8.4 THE PCM SYSTEM

With the background of quantization and coding, the analysis of a typical PCM system becomes trivial. We will do it, anyway.

With reference to Fig. 8.4.1, the original analog signal is consecutively sampled, quantized, and encoded, before it enters the channel. The channel itself is like an LPF filter, and repeater-quantizers are spaced in predecided intervals to reduce the effects of additive noise. At the receiving end, the first block labeled "detector" is the device that accepts the coded bits and (in the presence of noise) must decide whether a "one" is present or a "zero." Usually, a matched filter is used with a set threshold, which, if exceeded, designates the existence of a "one." However, behind this detector there is plenty of theory (statistical theory of signal detection); thus we can say no more about it at this time. Its purpose is to extract the signal from the noise and to hand the original signal $f_c(t)$ to the decoder. The decoder's output is the quantized signal $f_q(t)$, which is approximately equal to $f_s(t)$, and so an LPF will produce a reasonable facsimile of the original analog signal $f(t)$. So much for the trivial part. Let us talk next about some things that are not trivial.

First bandwidth: How much bandwidth is required for the transmission of a PCM signal? We had better be careful here, because questions on bandwidth are quite tricky to answer. Let us take a single signal that has been PCM'd. Can we say, as we did in PAM (Section 7.7), that the bandwidth depends only on the

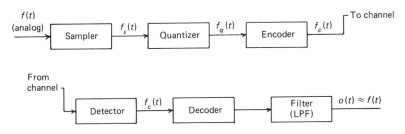

Figure 8.4.1   Block diagram of a PCM system.

width of the pulse (bit) used, and if this width is T_p, then the bandwidth is $2\pi/T_p$? We have seen this to be true when the signal is a finite number of pulses from a periodic train (see Fig. 4.7.2), and then again when the signal is a PAM one. Here we have a series of pulses of equal width and magnitude but spaced at random, so to speak, with no periodicity anywhere. Does it still hold?

Let us go back even further and ask another question. Is the amplitude spectrum of a sum of signals $f_1(t) + f_2(t)$ equal to the sum of their amplitude spectra $|F_1(\omega)| + |F_2(\omega)|$? The answer here is no. From complex variable theory we know that

$$|F_1(\omega) + F_2(\omega)|^2 = |F_1(\omega)|^2 + |F_2(\omega)|^2 + 2\,\mathrm{Re}\,(F_1(\omega)F_2^*(\omega))$$

so it cannot hold unless the third term above is zero, and it is zero only when $F_1(\omega)$ and $F_2(\omega)$ have no overlap. The relationship above holds, of course, even if $|F_1(\omega)| = |F_2(\omega)|$, that is, even when $f_1(t)$ and $f_2(t)$ are exactly the same but shifted with respect to each other, $f_2(t) = f_1(t-5)$ say.

This is actually what we have in PCM. We had it in PAM, too, but there was an easier avenue there to argue our point, since there was some periodicity in the signals. In PCM our signal is a series of pulses equal in magnitude (we are assuming "ones" and "zeros," but "ones" and "minus ones" can also be argued the same way), but shifted randomly along the time axis. What is the effect on the amplitude spectrum of the time shifts? Can we say that the bandwidth of the sum is the same as the bandwidth of each?

Let us take a specific case and then generalize. Say that we have two pulses, $g(t) = f_1(t) + f_1(t-T)$, each 2 seconds in duration. The FT of this sum is obviously

$$G(\omega) = F_1(\omega) + F_1(\omega)e^{-j\omega T} = F_1(\omega)[1 + e^{-j\omega T}]$$

and if $f_1(t) = \mathrm{rect}\,(t/2)$, $F_1(\omega)$ is of the form $(\sin\omega)/\omega$. Thus the above is

$$G(\omega) = K\frac{\sin\omega}{\omega}(1 + e^{-j\omega T})$$

and finally

$$|G(\omega)| = \left|\frac{K\sin\omega}{\omega}\right|\sqrt{(1 + \cos\omega T)^2 + \sin^2\omega T}$$

where K a constant. This last expression shows what is happening. No matter what the second term is (the one with the square root), it is multiplied by the $|(\sin \omega)/\omega|$ term. If we take the bandwidth to be dictated by the first lobe of this curve (the dashed curve in Figs. 7.4.1c and 4.7.2), then the bandwidth is still the same, dictated by the width of the single pulse.

Now we can generalize. No matter how many pulses there are in a PCM signal, the bandwidth will be the same as that of the single pulse, and that, of course, depends strictly on its width. The conclusion is the same as in PAM, and could have been obtained there with the same method, allowing the differences in pulse magnitudes and the like. There was no need to use it there, though, because a much simpler method was on hand.

So, after all this, we can finally state that if we have a *single* PCM signal with the pulse width of its "one" bits equal to T_p, its bandwidth will be

$$B_{\text{PCM}} = \frac{2\pi}{T_p} \qquad (8.4.1)$$

How about the bandwidth of n multiplexed PCM signals? Good question, since the answer can then be compared with FDM or PAM-TDM, for which we know it to be

$$\text{FDM} \longrightarrow B = 2n\omega_m \qquad \text{(DSB)} \qquad (8.4.2\text{a})$$

$$\text{PAM} - \text{TDM} \longrightarrow B = 2n\omega_m \qquad (8.4.2\text{b})$$

Well, in PCM it depends. On what? On the width of the pulses, of course. But now this width depends on the length of the code words, which in turn depends on the number of quantizing levels, and so on. Let us see if we can understand this confusing situation a little bit better.

Let us assume that we have n signals with the same sampling period $T = \pi/\omega_m$ (ω_m their maximum frequency). If we have no coding, and we ignore guard bands, the width of the sampling pulses in a PAM system will be T/n. But we want to code the values of the samples and still pack everything inside a T period. This means that each code word must last T/n seconds. If the length of each code word (assume them equal to simplify the argument; if they are not, think of the maximum length) is now m bits (an m-bit PCM, as it is called), then the width of each bit pulse must be T/mn, resulting in a bandwidth of

$$B_{\text{PCM}} = \frac{2\pi}{T_p} = \frac{2\pi mn}{T} = 2mn\omega_m \qquad (8.4.3)$$

that is, m times what is needed for the other two cases mentioned above. Now m is the number of bits per word. If the code is binary arithmetic, this corresponds to

$$L = 2^m \qquad (8.4.4)$$

levels in the quantization process. Solving for m and inserting the result in (8.4.3), we have the bandwidth in terms of quantization levels L as

$$B_{\text{PCM}} = 2(\log_2 L)n\omega_m \qquad (8.4.5)$$

a very large bandwidth, indeed, no matter which equation you look at.

Example 8.4.1

Find the bandwidth required to transmit 24 speech signals, each bandlimited up to 8 kHz, all PCM time-multiplexed with an 8-bit PCM.

Solution. This example is actually real. Something like it is used in metropolitan areas (short distances) in the United States, designed by AT&T and called the TI system.

The 8-bit PCM assumes that voice has been quantized up to 256 levels, which guarantees very acceptable speech intelligibility. Anyway, the bandwidth asked is

$$B_{PCM} = (2)(8)(24)(8) \text{ kHz}$$
$$= 3.072 \text{ MHz}$$

an awful lot. But large bandwidth is not always detrimental, as we have often remarked, for it can be exchanged with performance in the presence of additive noise. If you cannot use so much bandwidth, increase the width of the pulses, suffering a decrease in the number of signals you can multiplex. You just cannot have everything. Actually, this large bandwidth is not unique to PCM. PDM or PTM have very large bandwidths in multiplexing, as their pulse widths must also be small, to accommodate n signals. We did not discuss this, of course, but we are throwing it out here, since we have the opportunity.

It should also be remarked that there is another way to shorten the bandwidth if one is that keen on it. This is by having a code with more symbols than two (i.e., a trinary, etc.). If we have an m-ary code, the code words are shorter, the pulses wider, and so on, and under the same conditions that led to Eq. (8.4.3), we would end up (do it) with

$$B_{PCM} = 2(\log_m L)n\omega_m \tag{8.4.6}$$

which decreases as m increases. Even so, binary PCM with the largest possible bandwidth is also the most popular, for the obvious advantages (easier detector decision, more coding theory advances, etc.), and because large bandwidth is not always anathema—a remark made for the nth time.

The entire sample-quantize-code business in the transmitter is called analog-to-digital conversion (ADC) and it is often packaged as one unit. The same is true for the receiver's functions after the detector, which are called digital-to-analog conversion (DAC). There are plenty of ADC and DAC devices around, constantly being improved and miniaturized. We will not bother to discuss any of them here; we refer the interested reader to Stark and Tuteur (1979) for a brief discussion or to Schmidt (1970) for a detailed one.

We close this section with a brief discussion of the quantizer-repeaters that presumably are placed every so often in the channel to minimize the effects of noise. They are, of course, quite important. If they work right, the effects of channel noise in PCM can be completely negligible, leaving only the quantization noise (or error) as the main cause for concern.

These devices are often called *regenerative repeaters*, and they are placed at sufficiently close distances along the transmission path to safeguard against accumulation of too much noise, which might render them inoperative. If the code is binary (ones and zeros, say), each repeater must decide whether a "one" is present (or a zero) and alert another device to emit a fresh "one" (or a zero), that is, to regenerate the pulse code sequence. This decision is made much the same way as in the detector at the receiver of the PCM system, say with a matched filter and the exceeding of a set threshold (or not) at the instant of maximum output. For a pulse, the impulse response of its matched filter is also (theoretically) a pulse, and its output a triangle. The maximum value of the triangle is in its middle, so that is the instant at which the device checks whether the threshold is exceeded or not, alerting the next device to regenerate a "one" or a "zero," respectively. Timing circuits are added to the conglomerate so that everything takes place at the right time, and if all goes well, the output of the regenerative repeater is as good as new. All never goes well, of course, and that is where coding theory comes in handy, providing us with proper codes whose errors can be detected and then corrected. But that is another matter, altogether, good for a book twice as thick as this one.

8.5 VARIATIONS ON THE PCM THEME

There exist a group of systems that *look* like a PCM, *behave* like a PCM, and yet are not exactly a PCM. What are they? Who knows? Read and decide for yourself.

8.5.1 Delta Modulation

The delta modulation (DM) system is shown in block diagram form in Fig. 8.5.1a. Let us see how it works. We will concentrate on Fig. 8.5.1b, the modulator. As can be seen there, the analog $f(t)$ is first subtracted from another

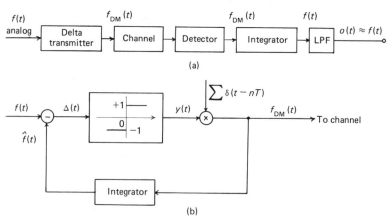

Figure 8.5.1   (a) The DM system in blocks; (b) the DM transmitter-modulator.

signal $\hat{f}(t)$, which, as we will soon see, is a good approximation to $f(t)$. The difference $\Delta(t) = f(t) - \hat{f}(t)$ enters a nonlinear device (a two-step quantizer) whose output is $y(t) = \pm 1$, depending on the sign of $\Delta(t)$. Next, this $y(t)$ is multiplied by a periodic train of impulses (narrow pulses in actual practice) whose period T is the sampling period, and the result $f_{DM}(t)$ is sent through the channel. The signal emitted is obviously a series of equal amplitude positive or negative impulses (pulses in practice), and this is definitely the *look* of a PCM signal. As we will soon see, this $f_{DM}(t)$ carries information about $f(t)$ that can be used to reconstruct it at the receiver.

Let us return to the transmitter to explain the role of the rest of the blocks. The emitted signal $f_{DM}(t)$ is also fed back to an integrator, whose output $\hat{f}(t)$ jumps up or down by an amount equal to the impulse's area (do not forget that the integral of an impulse is a step). This means that $\hat{f}(t)$ is ideally a staircase function, and practically a triangular wave (the integral of a pulse is a ramp). It is also a fairly good approximation to $f(t)$, and this is what we are going to show next, with the aid of Fig. 8.5.2.

In Fig. 8.5.2a a typical analog $f(t)$ is shown, presumably the one that enters the modulator. Since its value at $t = 0$ is positive and $\hat{f}(t)$ is zero (the system is dead), $\Delta(t)$ is positive and so is the impulse of $f_{DM}(t)$. This contributes a positive first step out of the integrator (delayed slightly), as shown in the figure. This story [$\Delta(t)$ positive, positive step for $\hat{f}(t)$] continues until $\hat{f}(t)$ *catches up* with $f(t)$. After that, and if all is well designed, $\hat{f}(t)$ becomes $f(t)$'s staircase shadow.

Since $\hat{f}(t)$ is an approximation to $f(t)$, the receiver's function becomes obvious. After a detector that decides whether the arriving pulses (in practice)

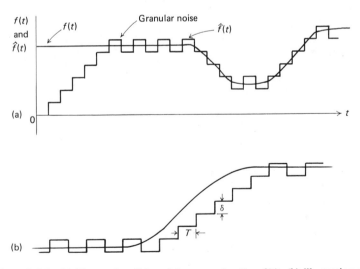

Figure 8.5.2 (a) The analog $f(t)$ and its approximation $f(t)$; (b) illustration of slope overload.

are $+1$ or -1 [i.e., reconstructs $f_{DM}(t)$ from the noise], the integrator provides the staircase approximation $\hat{f}(t)$ and an LPF smooths out the bumps.

It works well—if it all works well, and it does not always. At first, the system has to *warm up*, as they say, to *catch up* with the signal. If the signal has sharp slopes in its variations, $\hat{f}(t)$ may drag behind. Theoretically, of course, sharp slopes and abrupt variations in $f(t)$ mean bigger ω_m (the frequency on which the sampling period is based) and therefore smaller T. But that is not enough. Keeping up with $f(t)$ requires the right T and the right step as well. When this combination is not right, we have the problem of **slope overload**, as it is called, which is illustrated in Fig. 8.5.2b. It is obvious that slope overload will be prevented if

$$\left|\frac{df(t)}{dt}\right|_{\max} \leq \frac{\delta}{T} \tag{8.5.1}$$

since then the slope of the staircase [right-hand side of (8.5.1)] will be sufficient to keep up with all changes of $f(t)$. As an example we mention the case of a sinusoidal $f(t)$, that is,

$$f(t) = A \sin \omega_m t$$

for which Eq. (8.5.1) amounts to

$$\delta \geq A\omega_m T \tag{8.5.2}$$

Of course, in practice the signals are not simple sinusoids, and their maximum slope may not be known. Then the only avenue is empiricism— experimentation. It has been found, for example, that if the relation (8.5.2) holds for the frequency $f = 800$ Hz, the DM system can process speech quite well, even though its maximum frequency could be 10 times higher.

Of course, you cannot make the step δ too high, either; otherwise, you end up with high **granular noise** (or distortion) when the signal is flat, or slowly changing, a phenomenon illustrated in Fig. 8.5.2a. "Everything in moderation," as Socrates is reputed to have claimed.

The delta system has been studied and developed mainly in Europe. It is used primarily for speech transmission in telephone systems. Recent research has come up with an *adaptive* version, which minimizes overload and granular problems. Generally speaking, it is an inexpensive system compared to regular PCM, of somewhat lower quality. It does not much qualify for secret transmissions, as its "code" is obvious.

Talking about its code, what is it, and how much does this system qualify as a PCM? Well, the output is a string of "ones" and "minus ones" (i.e., binary), carrying with it advantages of easy detection, possibility of repeaters, and the like. This string is actually information about the changes of the signal, about its derivative, and that is why an integrator is needed to recover the signal. All these things make it like a PCM. Even so. It does not have the versatility of a regular PCM, where the code is at our choice. We are afraid that its identity crisis may be with us for a while.

There is a variation of the DM system called the **delta-sigma system**, proposed by Inose and Yasuda (1963), and meant to smooth out some of the rough points. Theoretically, it amounts to integrating the signal before it enters the DM system, and therefore differentiating the signal received at the end. The integration smooths out the abrupt slopes of the signal (divides the amplitude spectrum by $1/\omega$, thereby reducing its frequency content) and helps the overload, as well as some other problems presumably caused by a possible dc component. We shall not consider it any further here.

8.5.2 Differential PCM

Differential PCM is a variation of the delta system (that is why it is often called delta-PCM), but such a variation that it can actually be considered a bona fide PCM. It has the last element missing from the delta—the ability to use a code of our choice, with all the advantages that this has to offer.

Let us look at the transmitter only, and the nature of the receiver will become obvious. A look at Fig. 8.5.3a will reveal that there are two differences between the delta transmitter and the present one. The first one is that the two-step quantizer of the error $\Delta(t)$ is now a multistep one. After the multiplier, the signal can have many levels, not ± 1 as it had in the delta. And $\hat{f}(t)$ can jump two or even more steps if the error is big, and thus catch up faster with $f(t)$ if

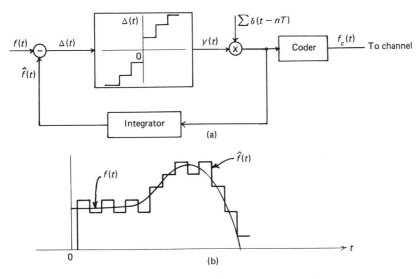

Figure 8.5.3   (a) DPCM transmitter; (b) $f(t)$ and $\hat{f}(t)$.

the latter changes values abruptly. A little thought will also reveal (see also Fig. 8.5.3b) that both the *overload* and *granular* problems can be minimized by decreasing the number of steps. In essence, this system is a PCM, but we are sampling and quantizing the error signal $\Delta(t)$, not $f(t)$ itself.

The other difference is the encoder, which intervenes before emission to the channel and makes this a real PCM system, although not directly on $f(t)$. Its necessity is obvious. The values to be transmitted are now n [it is a PAM signal on $\Delta(t)$]. They can be coded with a code of our choice, usually binary, but not always so. The receiver will have a detector, a decoder, and then the integrator–LPF combination which will first produce $\hat{f}(t)$ and then a smooth approximation of $f(t)$. As we said, the receiver is obvious.

The bandwidth of DPCM signal depends on the width of the pulses emitted by the coder, as usual. For TDM systems employing DPCM, the bandwidth is larger than in DM, since the multilever quantization requires multibit coding. It is comparable to regular PCM, as the reader can show as an exercise.

With this we close our discussion of the various PCM-like systems. It is by no means exhaustive. There are variations on the variations, adding a device here and subtracting one elsewhere. The interested reader can look up Jayant (1974) or Bayless et al. (1973) for further material on this issue.

PROBLEMS, QUESTIONS, AND EXTENSIONS

8.1. We mentioned in Section 8.2 that quantizers often have nonuniform steps. This is equivalent to passing the signal through a **compressor** and then applying the result to a uniform quantizer. The compressor can be modeled mathematically as a device with input/output (normalized) characteristics given by

$$|v_2| = \frac{\log\left(1 + \mu\,|v_1|\right)}{\log\left(1 + \mu\right)}$$

where v_1 is the input, v_2 the output, and μ a positive constant. Plot some input/output curves of the compressor device for $\mu = 0$, $\mu = 10$, and $\mu = 100$. What do you observe? This whole idea is called **μ-law quantization** (see Smith, 1957).

8.2. Find a binary arithmetic code for a signal of eight quantized values.

8.3. Each string of bits (or symbols) of a code that corresponds to a single value is called a **code word**. The number of bits it contains is called the **length** of the code word. What would be the length of each code word of a binary arithmetic code for a signal with 1024 levels?

8.4. A code is called **distinct** if all its words are different from each other. Is the code of Eq. (8.3.2) distinct?

8.5. A distinct code is called **uniquely decodable** if each code word is identifiable when immersed in a sequence of code words so that any sequence of code words will lead to a unique sequence of original signal values. Is the code of (8.3.2) uniquely decodable? Check also codes A and B below.

Signal Values	A	B
0	10	10
1	100	01
2	1000	100
3	1100	1100

8.6. A uniquely decodable code is called **instantaneous** if the end of every code word is recognizable without the need to inspect the succeeding symbol. Check the codes of Eq. (8.3.2) and A and B of Problem 8.5 to see if they are instantaneous. Try to design an instantaneous code of your own (using only ones and zeros). Is an instantaneous code always uniquely decodable?

8.7. Find the Morse code (for English) and determine whether it is distinct, uniquely decodable, or instantaneous. The Morse code is used for coding alphabets or languages.

8.8. Derive Eq. (8.4.6) by using arguments similar to those used in the derivation of Eq. (8.4.5).

8.9. The bandwidth given in Eq. (8.4.6) is based on the bandwidth of a single pulse. More specifically, since the FT of a pulse is a $(\sin \omega)/\omega$ curve, this bandwidth is based on the point of the first zero of this curve. Derive the bandwidth of n PCM-TDM signals using the 3-dB point (not too easy).

8.10. In color TV the quantization levels are $L = 512$. Find the bandwidth of 12 PCM-TDM such signals if all 12 are bandlimited up to $\omega_m = 4.5$ MHz and the guard band is zero. Repeat with a guard band of $T_p/10$, where T_p is the width of the code bits without the guard band. The codes are assumed to be binary-arithmetic types.

8.11. Repeat Problem 8.10 with a code that uses four symbols.

8.12. Prepare a paper on ADCs and DACs by searching the literature for the state of the art.

8.13. Discuss the bandwidth of the DM emitted pulse. What about time multiplexing many DM signals? How does the bandwidth of a DM-TDM compare with that of PCM-TDM?

8.14. Consider a DPCM-TDM system for n speech signals bandlimited up to 8 kHz with 256 quantization levels. Find the needed bandwidth in the channel (with no time guard band).

8.15. (*Intersymbol Interference*) Analog pulse modulation and PCM systems emit pulses. Pulses have an FT of the $(\sin \omega)/\omega$ form which never "ends" (so to speak), so the channel should have infinite bandwidth to pass them undistorted. Practical channels have finite bandwidths, and the result is pulse spreading. Pulse spreading interferes with the shape of neighboring pulses. This problem is called **crosstalk**

in PM systems (see also Problem 7.17) and **intersymbol interference** in PCM systems. Prepare a paper on these issues. Various books on communication systems have sections on them. You may also look up the fundamental work of Nyquist (1928) and the more recent work of Gibby and Smith (1965).

=9=

MIXED MODULATION: COMMUNICATION SYSTEMS

9.1 INTRODUCTION

When the channel does not behave like an LPF (is not a baseband, as they say), the spectrum of the pulses used in PM and PCM must be shifted to match its bandpass. We already know how to do that—with the aid of a sinusoidal carrier. The systems we are about to analyze are **mixed** in the sense that they use two types of modulation on the information signal $f(t)$. First they turn it into a PCM signal (this is the case we will discuss) and then AM, FM, or PM (phase modulation) is used to carry them over the passband channel (air, the cosmos, etc.). For this reason they are often called **digital carrier modulation** systems. Of course, these systems can be used even in cases of baseband channels if frequency-division multiplexing is desired. Data pulses, for example, are often so sent over telephone wires using modulator–demodulator devices called **modems**.

The reason we have chosen the name "mixed" is because we wish to include **spread-spectrum** (SS) systems in this chapter as well. These systems are relatively new, at least in implementation, and a great deal of work is still going on to improve them. They are mixed in the sense that they involve more than one type of modulation, so they can be legitimately included in a chapter such as this one. They are tough systems to implement, but they promise such advantages that no one is willing to let go. Among their most important "hidden cards" are *privacy* and *immunity* to jamming, so their applications up to now have been mainly military. But privacy is beginning to assert itself in other fields as well, and commercial SS systems are rapidly coming into their own. It would be an omission if we did not give a brief introduction to such systems, particularly since we include a chapter on electronic warfare at the end of the book. Actually, *we have already seen an example of a SS system*—not in communications, but in radar. We challenge the reader to ponder this issue after reading the section on spread-spectrum systems, and then state his or her conclusions as part of the answer to the last problem of this chapter.

Incidentally, this will be the last chapter in which communication systems are studied without taking noise into consideration. After this any communication system we analyze will have noise in it, and, in fact, *its effects* will be our main concern. The operation of the noiseless system will be assumed known.

9.2 THE AMPLITUDE-SHIFT KEYING SYSTEM

Taking pulses and sticking them on the amplitude of a sinusoid is old hat for us; we have seen it done repeatedly in radar systems. The only difference in the amplitude-shift keying (ASK) system is that the pulses are close together in time and not periodic in any way. This means that the spectrum of the emitted signal is not going to be discrete, but even this is not absolutely unique, not really. If you include illumination time in radar, the periodicity is only for a short burst, and the spectra become continuous.

Let us first get a mental image of the time "face" of the signal emitted by an ASK system, by having a look at Fig. 9.2.1. In part (a) of this figure, a string of "ones" and "zeros" are shown—the output of the binary PCM system (output of the analog-to-digital converter in Fig. 9.2.1c). This coded version of $f(t)$ enters an amplitude modulator whose output in time will look something like the signal in Fig. 9.2.1b. It is assumed that the binary code is [0, 1] and not [1, −1], and for this reason, the system is often called **on–off keying** (OOK). The [1, −1] type of coding of $f(t)$ results in a PSK system, to be discussed later. It is also assumed that the carrier and the code are synchronized, so that every time a "one" is sent, the carrier covers two exact complete periods. This need not be the case in practice, as it is a lot easier to have a free-running carrier multiplied by the string of binary bits.

The key idea in the ASK system is that the information is carried in the amplitude of the carrier. Furthermore, since the information signal, the code, is never negative, it can be demodulated either synchronously, or asynchronously, by a peak detector. In fact, all variations of AM systems could be utilized (SSB, etc.), depending on bandwidth allocations and the like. In practice, the most common case is asynchronous ASK, as it is simply and cheaply realized.

In asynchronous ASK the signal emitted can be written as

$$m_{\text{ASK}}(t) = f_c(t) \cos \omega_c t \qquad (9.2.1)$$

assuming, of course, that a product modulator (DSB) is in effect. The signal

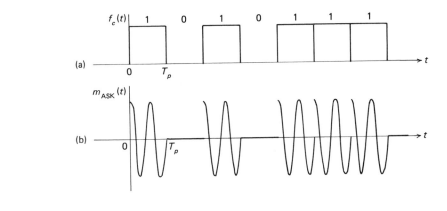

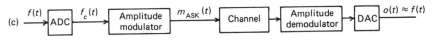

Figure 9.2.1   (a) Output of a PCM system; (b) the ASK signal; (c) block diagram of an ASK system.

$f_c(t)$ may not be the result of a single information signal $f(t)$ as has been implied above, but also of a composite TDM signal. In fact, it usually is the latter, in practice. But no matter what it represents, it is still a string of ones and zeros, as shown in the figure, with the possible exception of a time guard band, which has been ignored.

What about the FT (or spectrum) of $m_{\text{ASK}}(t)$, and the bandwidth required for transmission through the channel? That is pretty easy to answer also—what with all the background we have accumulated by now.

The FT of Eq. (9.2.1) is

$$m_{\text{ASK}}(t) \longleftrightarrow \tfrac{1}{2}[F_c(\omega - \omega_c) + F_c(\omega + \omega_c)] \qquad (9.2.2)$$

where $F_c(\omega)$ is the FT of the string of pulses, a sum of equal pulses in time and duration, but shifted at random, depending on the code. It cannot be found in general (unless you have the specific string), but our discussion in Section 8.4 has shown that the approximate bandwidth of the PCM signal is $2\pi/T_p$, where T_p is the width of each pulse. We spent a lot of words arguing the case there, and we are not about to repeat them here. Anyway, it is obvious that the bandwidth of the ASK signal will be twice that of baseband PCM, since the spectrum of the latter is shifted out to $\pm\omega_c$. If the system comes from an original TDM-PCM, the bandwidth will be twice the bandwidth of the PCM, assuming, of course, that the AM is DSB and not SSB.

The asynchronous or noncoherent ASK has found various uses in practice, but not a lot. One of the earliest voice-band data transmission systems, the ANITSQ (see Koenig, 1958), was of this type, designed to operate over telephone voice circuits using a 1.5-kHz carrier. The next two systems, FSK and PSK seem to be more popular.

One more comment. The ASK system is theoretically similar to a pulse radar system. The emitted pulses are not, of course, periodic, since their time location is governed by the code, but the modulator and demodulator are essentially the same—in a conceptual manner, of course.

9.3 THE PHASE-SHIFT KEYING SYSTEM

The phase-shift keying (PSK) method of sending binary bits over a bandpass channel makes use of the phase of the carrier. The general phase modulation waveform is

$$m_p(t) = A \cos(\omega_c t + k_p f_c(t)) \tag{9.3.1}$$

as we know from the study of angle modulation. The signal $f_c(t)$ is, of course, a sequence of ones and zeros (or ± 1).

It is possible to fix k_p and the level of the bits so that our PSK signal will be

$$m_{PSK}(t) = \begin{cases} A \cos \omega_c t & \text{if } f_c(t) = 1 \\ A \cos(\omega_c t + \pi) = -A \cos \omega_c t & \text{if } f_c(t) = -1 \text{ (or 0)} \end{cases} \tag{9.3.2}$$

Another way to write it is

$$m_{PSK}(t) = f_c(t)(A \cos \omega_c t) \tag{9.3.3}$$

where $f_c(t)$ takes the binary values $+1$ and -1 in a manner dictated by the code. A typical case of a code for $f_c(t)$ is shown in Fig. 9.3.1a, and right below it (Fig. 9.3.1b) the phase-modulated signal emitted to the channel. Figure 9.3.1c shows the overall PSK system in block diagram form.

Phase modulation is very hard to analyze in general, but quite easy in this specific case. In fact, the nature of the modulating signal is such that it is not phase modulation at all, or rather it reduces to amplitude modulation. Indeed, expression (9.3.3) is exactly a carrier, amplitude modulated by the coded signal $f_c(t)$. This being the case, we do not even need a phase modulator to create the PSK signal. A regular product device (balanced modulator) can do a fine job, as shown in the transmitter side of Fig. 9.3.1c.

After the channel, the signal enters the receiver antenna and after some amplification it resumes the form

$$m_{PSK}(t) = f_c(t) \cos(\omega_c t + \theta) \tag{9.3.4}$$

assuming that $A = 1$. Envelope detection is out in this case, since $f_c(t)$ takes negative values, a point in which the PSK differs from ASK (or OOK), even though they are both AM in their essence. The demodulator here must be a synchronous one, that is, a product detector using a locally generated cosine with the same phase. Often, the incoming signal itself is used to generate the cosine

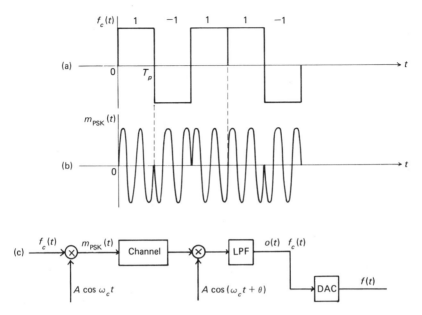

Figure 9.3.1 (a) Coded signal $f_c(t)$; (b) the PSK signal; (c) block diagram of a PSK system.

needed, by a method such as the one described in Problem 3.9. The final operation is the passage through the LPF, which produces a good facsimile of the $+1$ or -1 pulses. But will it really?

Strictly speaking, no. The product detector–LPF combination is meant to work for bandlimited signals, such as speech, music, and the like. Pulses (one, two, or a string) are not a bandlimited signal, as we know from our various discussions of their FT, bandwidth, and so on, so the output of the LPF will be quite distorted. But no matter, the purpose of an PSK is not to send the information signal $f(t)$ directly (in analog form) as it was in the various AM systems. We do not care to have a perfect pulse (positive or negative) at the output of the system. We only care to be able to decide whether it is a $+1$ or a -1, so that we can decode it, and so on. If the LPF cannot do too good a job on this, we can use a matched filter (or a correlator receiver) or other types of filters that will enable us to make an optimum decision in the presence of various types of additive noise. Of course, if this is the case, then after the decision is made, logic circuits can be used to generate the code, which are not shown in Fig. 9.3.1c.

From the discussion above, questions about the FT or bandwidth of $m_{\text{PSK}}(t)$ are easy to answer. After all, the signal emitted is similar to the one used in ASK, and everything ascertained there applies here, with minor modifications due to the existence of the negative pulses. The bandwidth of the $m_{\text{PSK}}(t)$ will be twice that of the bandwidth of the pulses, and their bandwidth depends on their duration. This can be used as a guideline for the bandwidth of the LPF, if an LPF is to be used in the PSK receiver.

9.4 THE FREQUENCY-SHIFT KEYING SYSTEM

As the name implies, in the frequency-shift keying (FSK) system the string of binary bits is latched on the instantaneous frequency of the carrier. We recall that the general FM wave has the form

$$m_f(t) = A \cos(\omega_c t + k_f g(t)) \tag{9.4.1}$$

where

$$g(t) = \int f_c(t)\, dt$$

$f_c(t)$ being our coded signal. The instantaneous frequency of $m_f(t)$ is

$$\omega_i(t) = \omega_c + k_f f_c(t) \tag{9.4.2}$$

and since $f_c(t) = \pm 1$, this can be written as

$$\omega_i(t) = \begin{cases} \omega_c + k_f = \omega_c + \Omega \\ \text{or} \\ \omega_c - k_f = \omega_c - \Omega \end{cases} \tag{9.4.3}$$

and integrating $\omega_i(t)$, we obtain

$$m_{\text{FSK}}(t) = A \cos(\omega_c \pm \Omega)t \tag{9.4.4}$$

as the signal emitted by the FSK transmitter. It is, of course, understood that each one of the two cosines of Eq. (9.4.4) is emitted for a short period, as long as the $+1$ or -1 pulses last, their duration being T_p seconds.

A typical string of binary bits for $f_c(t)$, together with the $m_{\text{FSK}}(t)$ they generate, are shown in Fig. 9.4.1a and b, respectively. The modulation constant k_f is picked so that the frequencies $\omega_c + \Omega$ and $\omega_c - \Omega$ are reasonably far apart to be distinguishable at the receiver, a point that will become clearer as the discussion evolves.

Figure 9.4.1d shows a typical FSK system. Actually, it is nothing but an FM system and could have been omitted altogether. In FSK the modulating signal is so specific—positive or negative pulses—that we can forget entirely that we are dealing with FM, and all its difficulties in modulating, demodulating, and finding FT and spectra. Its transmitter and receiver can be visualized in very simple terms, as we will see later.

Before we look at simplified transmitter and receivers, however, it is worthwhile to take a closer look at the emitted signal and find its frequency representation. Equation (9.4.4) is a representation of an $m_{\text{FSK}}(t)$ signal, but not a complete one—it does not represent the signal of Fig. 9.4.1b, for example. The equation only gives the form that the emitted signal has, in each T_p time segment. The actual emitted pulse is more correctly a sum of shifted such pulses, shifted,

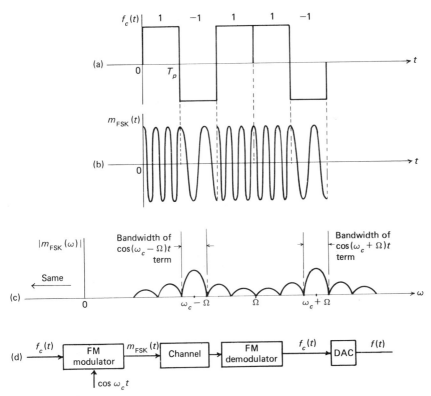

Figure 9.4.1 (a) Digital signal $f_c(t)$; (b) the FSK emitted signal; (c) envelope of the amplitude spectrum of the emitted signal; (d) simplified block diagram of an FSK system.

of course, by T_p. More precisely, each term of this sum is a cosine with frequency $\omega_c + \Omega$ or $\omega_c - \Omega$. We can consider this sum as being split into two parts. One part contains a sum of shifted terms of the form $\cos{(\omega_c + \Omega)t}$, and the other of the form $\cos{(\omega_c - \Omega)t}$. Now the first sum is exactly like an ASK (OOK) type of signal, whose FT is a complicated mess around $\omega_c + \Omega$, but with an envelope of the form $(\sin{\omega})/\omega$, and with bandwidth determined by the width of the interval T_p. The same is true for the other sum, with its FT centered around $\omega_c - \Omega$, so the overall amplitude spectrum resembles two PSK (or ASK) spectra symmetrically displaced by $\pm\Omega$ around ω_c. The distance between $\omega_c - \Omega$ and $\omega_c + \Omega$ is therefore a key factor in determining the overall bandwidth of the FSK signal. If they are far apart, this bandwidth is four times the original bandwidth of the code pulses, assuming, of course, that we ignore the gap between them if they are too far apart. Figure 9.4.1c shows the envelope of the $m_{FSK}(t)$ amplitude spectrum in the case when the two center frequencies are farther apart than the bandwidth of one of the sums mentioned above.

With this picture in mind, we can now take a look at simplified versions of FSK transmitters and receivers, which we promised just before our discussion on the FT and the bandwidth. First, the transmitter. Figure 9.4.2a shows a scheme for generating the signal $m_{FSK}(t)$ without the use of a FM modulator device, at least conceptually. Whenever a positive pulse is emitted, the switch S is alerted to swing upward and cause the $\cos(\omega_c + \Omega)t$ to be emitted. Negative pulses cause $\cos(\omega_c - \Omega)t$ to be emitted.

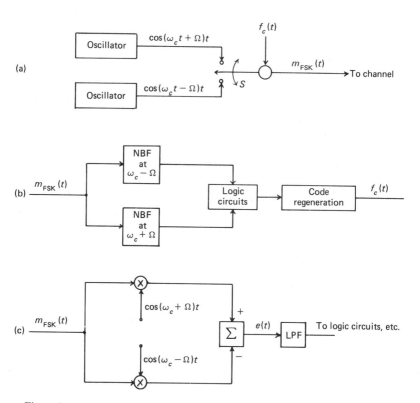

Figure 9.4.2 (a) Simple transmitter for FSK; (b) tuned-filter receiver for FSK; (c) correlation-type receiver for FSK.

Now let us go to the receiver side. One possible simple receiver is shown in Fig. 9.4.2b, and works presumably on the assumption that when $\cos(\omega_c - \Omega)t$ is sent, only the upper filter will have an output, whereas the opposite is true if $\cos(\omega_c + \Omega)t$ is sent. This is why we remarked earlier that the two frequencies should be reasonably far apart. So at each T_p interval and depending on which filter has an output (or exceeds a *threshold* if there is noise), logic circuit can be alerted to produce the appropriate bit and reconstruct the original $f_c(t)$. The highly tuned NBF filters can be replaced by matched filters followed by envelope detectors and the systems will still operate well, in fact, even better. The whole idea here is (as it was in the previous two systems as well) that one need only decide whether a $+1$ or -1 is present, and not extract $f_c(t)$ from the incoming signal as was the case in analog AM and FM systems. Once the decision is made, the $f_c(t)$ can be regenerated quite easily.

Still with this idea in mind, the reader should take a look at a second receiver possibility for the FSK, shown in Fig. 9.4.2c. This receiver operates synchronously with the incoming carriers and can be shown to be quite effective in the presence of additive noise. To see how it works, let us assume that the incoming signal is

$$m_{\mathrm{FSK}}(t) = A \cos{(\omega_c - \Omega)t} \tag{9.4.5}$$

lasting for a time interval of duration T_p.

The signal above is multiplied by both locally generated cosines (in synchronization) and the results are subtracted. The signal $e(t)$ is

$$e(t) = A \cos{(\omega_c - \Omega)t} \cos{(\omega_c + \Omega)t} - A \cos^2{(\omega_c - \Omega)t}$$
$$= -\frac{A}{2} + \text{high-frequency terms} \tag{9.4.6}$$

and only the $-A/2$ term will go through the LPF, alerting the logic circuits to generate a -1. When the incoming "bit" has the form $A \cos{(\omega_c + \Omega)t}$, the output of the LPF will be $+A/2$, as the reader can easily verify, and this means, of course, that the bit received is a $+1$.

One more remark and we close our discussion on FSK. The reader should think about multiplexing coded signal with the three different methods we have described so far. Is it not obvious that the FSK is at a disadvantage on this issue? (Explain why.)

9.5 THE DIFFERENTIAL PHASE-SHIFT KEYING SYSTEM

Although the PSK is a good system, superior in many respects to the other two we have discussed so far, it has a serious flaw in requiring synchronous type of detection in the receiver. To correct this, a modification of the PSK has been developed, the differential phase-shift keying system (DPSK). The coded signal is modified somewhat before phase modulation in this system, and the result is quite nice. The demodulation can be performed at the receiver without the need to generate a synchronous cosine—in fact, without the need to generate anything at all. Of course, a price will have to be paid for such an advantage (we know this well, by now), but we leave this for the end of the discussion.

Let us assume that we have the $f_c(t)$ type of signal of the ± 1 variety (each pulse lasting T_p seconds) shown in Fig. 9.5.1a. This $f_c(t)$ is not the one that is PSK'd and sent over the channel, but another one that gives information about the "transitions" of $f_c(t)$. This other one, a binary signal denoted by $g_c(t)$, is also

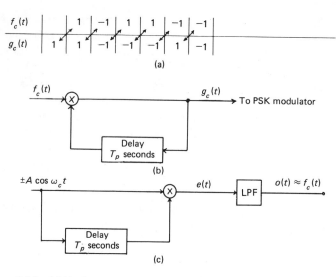

Figure 9.5.1   (a) Explanation of $g_c(t)$; (b) generation of $g_c(t)$; (c) receiver scheme for recovering $f_c(t)$.

shown in Fig. 9.5.1a. To begin with, $g_c(t)$ has some value, say $+1$ as shown in the figure, and this value is emitted as the first bit to the PSK modulator. Then the first value of the $f_c(t)$ shows up. The value of $g_c(t)$ is determined by multiplying the first value of $g_c(t)$ with the first value of $f_c(t)$. The story repeats until $f_c(t)$ has ended. Each value of $g_c(t)$ is the product of its previous value with the newly arrived value of $f_c(t)$. If the reader ponders this scheme, he or she will realize that $g_c(t)$ changes value only after the bit of $f_c(t)$ that shows up is a -1. If it is not, then $g_c(t)$ just stays put. A change in $g_c(t)$ provides information about -1 bits, whereas a lack of change provides information about $+1$ bits. It should be noted that the bit duration of $g_c(t)$ is also T_p, and thus this system, being in all other respects a PSK, has the same amplitude spectrum and bandwidth as the PSK.

A system that can produce this $g_c(t)$ from the incoming $f_c(t)$ is shown in Fig. 9.5.1b, and it operates in exactly the way as described in our explanation of the nature of $g_c(t)$. Let us assume this time that the output of the product device [our $g_c(t)$], prior to the entering of the first bit of $f_c(t)$, has the value -1. This value is also at the output of the delayer, so it gets multiplied by the first bit in the output; if it is a -1, change, and so on. Clever.

The signal $g_c(t)$ is pushed to a PSK signal-generating scheme, enters and leaves the channel, and arrives at the receiver in the form $\pm A \cos \omega_c t$ (phase angle assumed zero), where the $\pm A$ are the bits of $g_c(t)$. At this point, another clever little scheme takes over, shown in Fig. 9.5.1c, which is sort of the inverse of Fig. 9.5.1b, with due changes to take care of the demodulation.

Let us say that the first arrival is $A \cos \omega_c t$ [the first bit of $g_c(t)$ in Fig.

9.5.1c], and so is the second one. These two are multiplied to produce

$$e(t) = A^2 \cos \omega_c t \cos \omega_c(t - T_p)$$

$$= \frac{A^2}{2} \left[\cos \omega_c T_p + \cos 2\omega_c \left(t - \frac{T_p}{2} \right) \right]$$

which turns into

$$o(t) = \frac{A^2}{2} \cos \omega_c T_p$$

after the LPF (approximately, as we know). Aside from the known constant $\cos \omega_c T_p$, the term is positive, indicating the presence of a $+1$, as is the first bit of $f_c(t)$ in Fig. 9.5.1a. If the first bit is a -1, the output will be negative. The system works, without the need for a locally generated cosine in synchronism with the incoming one as in regular PSK. Of course, the LPF (or a matched filter) will not produce the exact pulse. It will give an output from which it can be decided whether $f_c(t)$ is $+1$ or -1, and other devices will regenerate it. But this story is the same in all the digital carrier systems that we have covered. It is not only that the pulses are not bandlimited. It is also the noise that causes problems, distorting their shapes. We will see more on detecting pulses in a later chapter.

What is the price that one must pay for the elimination of synchronous detection? Well, to begin with, you do not only eliminate circuitry, you also add —the $g_c(t)$ generation, and so on. That is part of the price. The biggest price, though, is in performance in the presence of noise. Since each value of the regenerated $f_c(t)$ at the receiver depends on the previous value of $g_c(t)$ as well, an error in $g_c(t)$ due to noise can cause two errors in $f_c(t)$. Errors tend to occur in pairs of bits if the noise is high, and their overall effect is more detrimental than in regular PSK. Of course, one can use an error-correcting code, but then we add complexity, and so on; it never ends.

9.6 MULTILEVEL MIXED SYSTEMS

In all our previous systems the coded form $f_c(t)$ of the information signal $f(t)$ was assumed binary. This covers the majority of baseband systems that need a carrier to traverse a BPF-type channel, but not all of them. Often, the code is multilevel, or the original signal is in a multiple-level PAM form, uncoded. How can we modify the systems described previously to take such eventualities into consideration?

The reader can probably answer this question without help. Nevertheless, in the material of this section, we will run through a quick synopsis of some of the existing modifications. It is a good idea for the reader to set the book aside at this point, conceive of the possible modifications, and then check and compare them with the material below.

A **multiple-level ASK** (MASK) system will emit a signal of the form

$$m_i(t) = A_i \cos \omega_c t \qquad (9.6.1)$$

in each interval T_p, where A_i corresponds to the values of the level. The levels could be the values of a PAM signal (quantized, though uncoded) or the levels of a code. The modulator and demodulator would not differ from the regular ASK, nor would the form of the $m(t)$'s $|M(\omega)|$ or its bandwidth. (Why?) The only problem with such a system is that the receiver is much more complicated and prone to error, since the decision is no longer one of two things (yes or no type), but *which* of the many levels has been emitted. If all levels are positive (they can be made to be, if they are not), a peak detector can do the job, although it will have difficulty in distinguishing the levels in the presence of severe noise. You are better off coding the levels into binary logic and using the simple binary detector, even though you might suffer an increase in bandwidth if the code pulses are narrower.

The **multilevel** (actually **multitone**) **FSK** (MFSK) system will emit a signal of the form

$$m_i(t) = A \cos (\omega_c + \Omega_i)t \qquad (9.6.2)$$

of duration T_p, where Ω_i $(i = 1, 2, \dots, n)$ are frequencies corresponding to the various levels. A system for generating such a system can be an obvious generalization of Fig. 9.4.2a. The bandwidth of the signal emitted in MFSK is approximately $2nB$, where B is the bandwidth of the code pulse, assuming of course, a reasonable separation of the frequencies generated. A tuned-filter type receiver (one for each frequency) such as the one shown in Fig. 9.4.2b will work about as well as before. A generalization of the correlation-type receiver of Fig. 9.4.2c is also possible, and the reader should verify that it works.

Multilevel (or rather **multiphase**) **PSK** (MPSK) must emit one carrier with as many phases as the levels. Four-level MPSK has actually been used in practice. It emits a signal of the form

$$m_i(t) = A \cos (\omega_c t + \varphi_i) \qquad (9.6.3)$$

in each interval T_p, of course. Usually, the phases are taken equidistant. Thus for a four-level PCM, the phases are 0, $\pi/2$, π, and $3\pi/2$, or $\pi/4$, $3\pi/4$, $5\pi/4$, and $7\pi/4$.

The receiver is a synchronous type, using four locally generated coherent cosines, which, however, can be produced by phase shifting a *master* one in coherence. Such a scheme is shown in Fig. 9.6.1. Each product is followed by an LPF, as usual.

Let us assume that the input is the signal

$$m_1(t) = A \cos \omega_c t \qquad (9.6.4)$$

which corresponds to level 1. All LPFs will have some output. As we know from our analyzes of AM (DSB with phase error, check it), the first LPF will have

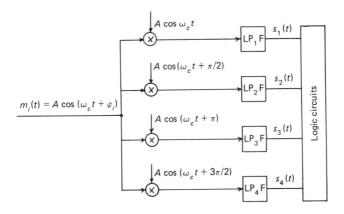

Figure 9.6.1 Coherent receiver for MPSK.

the output

$$s_1(t) = \frac{A^2}{2} \tag{9.6.5}$$

whereas the rest of them will have

$$s_2(t) = \frac{A^2}{2} \cos \frac{\pi}{2} = 0 \tag{9.6.6}$$

$$s_3(t) = \frac{A^2}{2} \cos \pi = -\frac{A^2}{2} \tag{9.6.7}$$

$$s_4(t) = \frac{A^2}{2} \cos \frac{3\pi}{2} = 0 \tag{9.6.8}$$

The *largest positive* output is $s_1(t)$ and the logic is alerted to regenerate level 1. The logic is always alerted to pick the level corresponding to the largest positive output, that is, the one coming from the coherent cosine.

This type of system has been actually applied in practice, as we mentioned a little earlier. Not exactly as we described it here, but the idea is still the same. The application is in the Bell System Model 207 data system, which operates over telephone lines. The four levels are not actually levels, but pairs of binary bits. The system starts with a binary code (ones and zeros) and then sorts them into four pairs (00, 01, 10, 11). These four pairs are the four levels. Each pair causes a cosine of a distinct phase to be emitted, the four phases being 45°, 135°, 225°, and 315°. Since each modulated cosine pulse represents two bits of the code, the signal emitted can last $2T_p$ seconds (T_p is the duration of the code pulse), and thus the bandwidth required is halved. One can play around with such things and come up with some very practical results, indeed. The idea of grouping binary bits and calling them levels is a good one. Whatever you lose in receiver difficulties, you at least gain in bandwidth.

Can we have a multilevel DPSK? Think about that for a while.

9.7 SPREAD-SPECTRUM SYSTEMS

The coverage of spread-spectrum (SS) systems is the reason we called the chapter "mixed" rather than "digital carrier" systems as the previous systems can be called. The word "digital" is associated only with ASK, FSK, PSK, and their variations, and it did not seem quite proper to include the SS systems within such a title. So we ended up with mixed systems, instead. Of course, SS systems have a sinusoidal carrier also, as they are mainly used over the space channel (a BPF). They need not be, however, not all of them anyway, as we shall see later. At least one version (direct sequence) can also be used over an LPF channel of adequate bandwidth. Even so, their forte is over the space channel, since they have the ability to camouflage the information so that a hostile person cannot easily unscramble it, something not yet that important over telephone lines. It may be soon, however.

As indicated by our last statement, SS systems are used primarily for military communications or wherever else secrecy is important. They are a fairly recent discovery, or, more accurately, a fairly recent declassification.[1] Their theory is still being developed, and some problems they have in implementation are still being ironed out. Our treatment here will be elementary—no noise, as usual. The interested reader is referred to Dixon (1975, 1976), or Cook et al. (1982) for more detailed treatment or for recent results on these systems.

A word about their name. It comes from the fact that in these systems, the emitted signal's spectrum is much wider than that of the information signal's it carries. So what, you say. That is nothing new. FM and PM systems are like that, and so are all the variations of pulse modulation and PCM that we have encountered, including the ones covered in this chapter up to now. You are right, of course. Even so, there is a difference. The spreading in the systems up to now was caused by the information signal itself. The spreading in SS systems is caused mainly from another signal that carries no information. Anyway, the difference will become clear by the end of the discussion.

There are many types of SS systems, and we cannot possibly cover all of them in the limited space that we have here. We will simply discuss the principal ones and leave the combinations (the hybrids, as they are called) for additional reading.

9.7.1 Direct-Sequence Systems

Let us go directly to Fig. 9.7.1a, where a direct-sequence spread-spectrum system (DS-SS) is illustrated in block diagram form. The signal $f_c(t)$ represents the PCM version of an information signal $f(t)$; that is, $f(t)$ has been sampled, quantized, and coded into a binary code. A typical such signal is shown in Fig. 9.7.1b, and we shall use it to explain the nature of the rest of the system.

---

[1] This means "no longer secret."

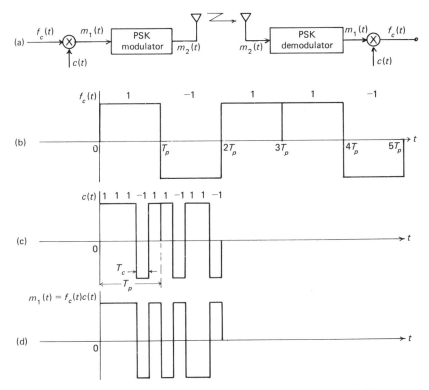

Figure 9.7.1   (a) Typical DS-SS system; (b) typical $f_c(t)$; (c) typical $c(t)$; (d) their product in time $m_1(t)$.

The first operation is a multiplication with another binary digital signal $c(t)$, done in practice with a balanced modulator. *It is this operation* and a similar one at the receiver end that makes this system what it is. Take that away in Fig. 9.7.1a and you end up with a pedestrian PSK system.

What is this $c(t)$, and what effect does it have on the otherwise normal PSK system? To begin with, this $c(t)$ has nothing whatever to do with the information signal $f_c(t)$. In fact, it has nothing to do with anything. It is just a "random" code, as they say. Now we do not know much about the mathematical meaning of the word *random* at this point, so we cannot really explain this concept to the fullest. Still, the reader can conceive of it as follows. A hired seal flips a coin with his nose, and every time a head turns up, he barks, "a one"; otherwise, "a zero." This produces a signal like the $c(t)$ we want for the DS-SS system. In practice, there exist devices that can replace the seal, even though they cannot do as good a job as he can, and so are called pseudorandom generators (they have some periodicity, although a long one).

So, seal or device, we can come up with such a $c(t)$, something like the one shown in Fig. 9.7.1c. Take a good look at it because it (with the product operation) is the key to the whole system.

The first thing we note is that the duration of its bits (denoted as T_c) is much smaller than T_p, the duration of the bits of $f_c(t)$. Second, the $f_c(t)$ and $c(t)$ must be in synchronism as regards their transition. The period T_p must be a multiple of T_c; otherwise, we can loose control of the nature of the emitted signal, and therefore of the overall system. More on this below.

With $c(t)$ explained, we can move on to the operation of multiplication of $f_c(t)$ with this $c(t)$.[2] Looking over Fig. 9.7.1d, we note the following. The resultant signal $m_1(t)$ is again a binary-type signal, resembling $c(t)$ more than $f_c(t)$. This, of course, makes sense, since the pulses of $c(t)$ are of shorter duration than those of $f_c(t)$. The new signal $m_1(t)$ is, therefore, a series of pulses *whose minimum width is T_c*. This is where the synchronism we mentioned above comes in. If they were not synchronized, this minimum width could be anything—much smaller than T_c. This we do not want, as will become obvious later.

Let us run through the rest of the system quickly, and then return later to any hidden meanings of what is happening behind the surface. The new signal $m_1(t)$ advances to a PSK modulator which turns it into a double phase cosine, capable of crossing the BPF (space) channel (actually, a product modulator, as we know from PSK). This step would not be needed if the channel is of LPF form, nor is PSK the only system that can be used to push $m_1(t)$ through the space channel, just the most popular one. Anyway, the original $f_c(t)$ seems to first modulate $c(t)$ [a product is AM modulation, and why not think of $c(t)$ as a digital carrier?], and the result phase modulates a cosine. Add to that that $f_c(t)$ is already a modulated signal [remember the original $f(t)$?] and we actually have three modulations here. It certainly qualifies for discussion in this chapter.

At the receiver, after phase demodulation (amplification, regeneration, etc.), we end up again with

$$m_1(t) = f_c(t)c(t) \qquad (9.7.1)$$

which is multiplied again with a locally generated $c(t)$ identical (and in synchronism) with the one used at the receiver, to produce

$$f_c(t) = [c(t)]^2 f_c(t) = f_c(t) \qquad (9.7.2)$$

since

$$[c(t)]^2 = 1 \qquad (9.7.3)$$

The rest is history.

Now we go back to the transmitter again and discuss the hidden nature and properties of $m_1(t)$. Nothing difficult, of course—our background is quite sufficient to understand everything that we are going to say.

Let us recall that the amplitude spectrum of $f_c(t)$ is of $|(\sin \omega)/\omega|$ form, and that its bandwidth is determined by the width T_p. This bandwidth B_c is (see Section 8.4)

$$B_c = \frac{2\pi}{T_p} \qquad (9.7.4)$$

---

[2]Incidentally, the multiplication of $f_c(t)$ by $c(t)$ could be made after the PSK modulation, with no change in the overall picture of the system.

based, of course, on the first zero (null) of the envelope of the rather messy $|F_c(\omega)|$.

What about the bandwidth of $m_1(t) = f_c(t)c(t)$? Well, $m_1(t)$ is still a series of pulses, albeit of a much shorter duration $T_c < T_p$. Its $|M_1(\omega)|$ will still be of $|(\sin \omega)/\omega|$ form (in terms of the envelope) and its first null bandwidth B_m will be

$$B_m = \frac{2\pi}{T_c} \tag{9.7.5}$$

much larger than before, much larger indeed. In fact, in view of the fact that

$$T_p = nT_c \qquad (n = 4, 5, \ldots) \tag{9.7.6}$$

as we mentioned earlier, we have that

$$B_m = nB_c \tag{9.7.7}$$

that is, the bandwidth of $m_1(t)$ is n times that of $f_c(t)$. This is where the synchronism and Eq. (9.7.7) do their thing. If they were not there, the bandwidth of $m_1(t)$ would be anything, since its shortest pulse could be of any minute width. This way, things are under our definite control.

The moral of the story is that $c(t)$ controls the bandwidth of the emitted signal, since after the PSK operation the bandwidth simply doubles. This random code sequence causes the spectrum of $f_c(t)$ to *spread*, to as wide a bandwidth as we desire [as short as we can make the $c(t)$ pulses]. And this spreading has nothing to do with the information signal $f(t)$ or its coded form, $f_c(t)$. That is a **spread-spectrum** system. Spreading of spectra in all our previous systems (FM, for example) was connected with the information signal, and such cases are not traditionally included in the SS systems group.

Now spreading a spectrum is not always what it is cracked up to be. Sometimes it is detrimental. The use of too much spectrum is expensive. What do we get in return?

Many things (some good, some bad). The most important thing we get in the "good" column is "secrecy" of our transmitted signal. Demodulation of $f_c(t)$ demands precise knowledge of this random pulse sequence $c(t)$. Furthermore, the product operation must be done in synchronization with the incoming $m_1(t)$. [Does this remind you of AM synchronous detection? It should if you think of $c(t)$ as a carrier.] These two things, both of them unknown to an adversary, are what causes this system to have high security in secret communications. Of course, it can also be decoded (by an adversary). The perfect cryptographic code does not yet exist. But it is much, much harder to crack the DS-SS system than the regular PSK system. And even if new, improved cryptographic codes are used on $f_c(t)$, using a SS system makes them even safer.

Actual implementation of such a system was not that easy in practice. Many problems had to be ironed out. We have the all-important aspect that the receiver *must know* $c(t)$ and have a basic clock rate in synchronization. The $c(t)$ knowledge must somehow be passed on to him or her either by another channel (in which case synchronization is easier), or be agreed upon in advance.

In the latter case, there may be many $c(t)$'s available to sender and receiver, and another code that decides which is used on a given moment, and so on, to the nth power.

Direct-sequence systems, also called **direct-spread** or **pseudonoise** systems, owe their specific name to $c(t)$. We have unwittingly implied that they are used only by the military. Far from the truth. There are companies (we will not say which) that make them for commercial use as well. After all, the business world has its secrets also. Maybe the reader does, too.

There is one other issue we would like to discuss, the issue of multiplexing in SS systems, but we leave it for after the discussion of frequency-hopping systems.

9.7.2 Frequency-Hopping Systems

If we gave the reader another name for frequency-hopping (FH) systems, the odds are that he or she will get the overall picture immediately and be able to outline their operation. They can be called **code-selected multiple-frequency FSK**. What do you say now?

Our familiar FSK system takes in binary pulses and produces cosines of two different frequencies, located at equal distances around the carrier frequency ω_c. In other words, if the carrier's frequency is ω_c, the two bits (± 1) are sent out as $\cos(\omega_c \pm \Omega)t$. The FH system changes ω_c, in a manner dictated by a random code sequence. Every time ω_c is changed, a new pair of $\omega_c \pm \Omega$ frequencies are formed to correspond to the ± 1 of the coded information signal $f_c(t)$.

A simplified block diagram for such a system is shown in Fig. 9.7.2. Our binary-coded signal $f_c(t)$ FSKs a carrier whose frequency changes every T_c seconds (the **chip rate**), where T_c is usually equal to or larger than the duration T_p of the $f_c(t)$ pulses. If the carrier's frequency is changed every T_p seconds or.

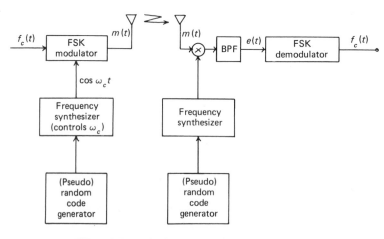

Figure 9.7.2 Block diagram of a FH-SS system.

less, the system is said to have a *fast* chip rate. *Slow* chip rates correspond to changes in ω_c every 2 bits of $f_c(t)$, or more.

The change in the carrier's frequency ω_c is random in this system (pseudorandom in practice), and it is dictated by a pseudorandom number sequence coming from the generator. This generator can be visualized as follows. Let us say that the frequency synthesizer is capable of putting out 10^3 different ω_c's, each of them corresponding to the numbers 1 to 10^3 coming from the generator. Then the previous seal is asked to bark numbers from 1 to 10^3 as he feels like it. Anyway, the code produced by the generator dictates the ω_c of the carrier, and this code must be known to the receiver if he or she is to be able to recover the emitted FH-SS signal. Not only the code, but the precise times of the changes as well (i.e., synchronization)—the same story as in the DN-SS system. Incidentally, the frequency synthesizer may look innocent, but it is not. It is, in fact, guilty of delaying the development of such systems for quite a long time, until recent technology solved most of the problems of meeting the practical stringent specs for its design.

Let us go to the receiver side next. The incoming signal is of the form

$$m(t) = A \cos (\omega_c \pm \Omega)t$$

with ω_c known to the receiver (he or she knows the code that produces it). The receiver frequency synthesizer produces a local cosine, such that the product of the incoming signal with the local cosine has a spectrum in the region where the two filters sit, of the regular FSK receiver—the tuned filter type (see Fig. 9.4.2b). It is the old trick of the variable oscillator and stationary filters used in the superheterodyne AM receiver. Of course, a correlation-type detector can also be used, the type we have already seen in Fig. 9.4.2c.

Where is the spreading of the spectrum in this case, and how is it accomplished? The answer is actually obvious. We already know the shape of the amplitude spectrum of the single ω_c FSK signal; it is a $|(\sin \omega)/\omega|$ shape around $\omega_c \pm \Omega$. Every time you change ω_c, you move the whole thing around. The spread of the values of ω_c plus twice the bandwidth of the pulses determines the overall spectrum of the system. And it can be quite large. In some systems the discrete values of ω_c are as high as 2^{20}, and even though you might be able to live with some overlap of the shifted spectra, the bandwidth can be huge. Is it worth it? Who is to say? If the security of a nation (or of a large company) is dependent on it, "You pays your money, and you buys your bandwidth." In any event, the large bandwidth in this system has nothing to do with the information signal $f_c(t)$. It depends on the synthesizer and the code that controls it, and that is why this system is classified as an SS system.

We mentioned earlier that one of the biggest problems with these systems is the frequency synthesizer. It is worthwhile discussing this issue a little further, although we cannot even come close to fully covering it in a simple presentation such as we have adopted here.

There are many kinds of frequency synthesizers (there are many kinds of everything, it appears), available in theory or practice, usually classified as **direct** or **indirect**. The latter use a phase-locked loop.

A fairly simple direct kind is shown in Fig. 9.7.3, employing three oscillators of frequency ω_1, ω_2, ω_3 and a bunch of switches, filters, and product devices. Depending on how you open up switches and create products and then filter the results, you can get any of the three input frequencies, sums or differences of pairs, and so on. A control signal at the output can open and close filters to produce the desired frequency in accordance with some code which is to be used in a FH-SS system. You can increase the input frequency number to n and get a lot of them out. Of course, other practical specs must be met as well, most important of which is rapid switching from frequency to frequency. But we do not have the space to go into all that, so we leave the subject altogether. It suffices to say that the design of direct-frequency synthesizers can be systematized (see also Problem 9.10).

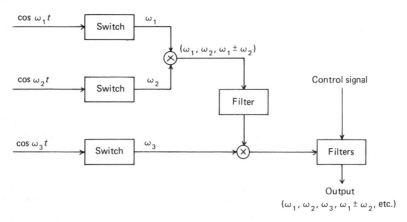

Figure 9.7.3 Simple direct frequency synthesizer.

With this we close our discussion on SS systems, save, of course, the subject of multiplexing, which we shall take up in the next subsection. We discussed only two general types and left the rest of them for additional reading. It is not actually very hard to visualize all kinds of other hybrid possibilities. It just needs a little imagination. This hopping business can be extended to *time*, for example, leading to **time-hopping** systems. In these, the time of transmitting depends on a random code. You can combine time hopping with frequency hopping (**time-frequency hopping systems**), and so on. And all these are not just figments of the imagination of an armchair scientist. Far from it. A time-frequency hopping system has actually been designed (RACEP, by Martin Co., Orlando, Florida). And many other hybrids are on the market, already in use around the world. There is no end to the power of the human mind for creating, and the human spirit for constructing.

One more comment. Are SS systems completely new to us? See Section 9.1 and work on Problem 9.12 to get the answer.

9.7.3 The Code-Division Multiple-Access System

How are we to multiplex more than one (hopefully n) signal of the sort emitted from our SS systems? Let us see if we can use one of the existing multiplexing systems, TDM or FDM.

Try TDM first. TDM can actually be used (at least theoretically) for DS-SS systems if they were to be sent over an LPF-type channel directly without a cosine type of carrier (no PSK). It would be similar to PCM type TDM, except that each signal would be spread "spectrum-wise" even farther, by the effect of the random code. The key to this type of multiplexing is, of course, that all users go through a central control unit, which allots their time slots. If this cannot be done, as in the case of the space channel, TDM is impossible.

Well, if TDM is impossible in the space channel, how about FDM? This is quite possible, theoretically, but not really all that practical. SS-type signals have such huge bandwidths that if you started multiplying them further by n (the number of multiplexed signals), you might end up with more spectrum than is presently in use for everything. Other systems (AM, FM, etc.) would start a protest demonstration. So much for this idea.

Are we stuck, then? Is there no way to multiplex signals that are allowed to overlap in both time and frequency? The answer is certainly no to both questions; otherwise, we would not be using a whole subsection to discuss the issue.

The system we are about to discuss is the code-division multiple-access (CDMA) system. It is also called **spread-spectrum multiple-access** (SSMA) system, but we don't like this name—it is too restrictive. The term CDMA may entice researchers to think that the idea is not good only for SS systems, and find ways to extend it to other systems. We must not put more restrictions on the human mind than is absolutely necessary.

Let us look at DS-SS systems first. Say that n users are using an allowed spectrum simultaneously. One of the receivers would then accept all the emitted signals, or more accurately a sum of all. Let us say that the sum signal looks something like this:

$$r(t) = \sum_{i=1} A_i f_{c_i}(t - \tau_i) c_i(t - \tau_i) \cos(\omega_c t + \theta_i) \qquad (9.7.8)$$

where $f_{c_i}(t)$ are the n emitted signals, $c_i(t)$ the pseudorandom codes for each, θ_i the phases of each arrival, and A_i the respective amplitudes of each signal.

The jth receiver is interested in getting out A_j, the bit from her corresponding transmitter. She is assumed to be synchronized to the transmitter, so for her τ_j and θ_j can be set to zero. She takes the sum signal $r(t)$ and processes it as shown in Fig. 9.7.4 a correlation-type receiver with an integrator that behaves like a LPF (i.e., eliminates the high-frequency terms). The signal at the output

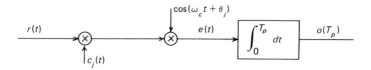

Figure 9.7.4   Illustration of CDMA reception.

of this receiver will be

$$o(T_p) = A_j + \sum_{\substack{i=1 \\ i \neq j}}^{n} A_i \int_0^{T_p} f_{c_i}(t - \tau_i)c_i(t - \tau_i)c_j(t) \cos \theta_i \, dt \qquad (9.7.9)$$

The A_j bit is out, but so is the second term of Eq. (9.7.9), which clouds it. How can we make this term ($n - 1$ terms, actually) equal to zero? Make all the cos θ_i = 0? Good idea, but unworkable. (Why?)

Let us look at a typical term of the sum, say the kth. It has the form

$$k\text{th term} = A_k \cos \theta_k \int_0^{T_p} f_{c_k}(t - \tau_k)c_k(t - \tau_k)c_j(t) \, dt \qquad (9.7.10)$$

where $f_{c_k}(t - \tau_k)$ is the bit (say $+1$) of the kth transmitter. The signal $f_{c_k}(t - \tau_k)$ can be constant over $[0, T_p]$, but it could change sign as well, because the integration period of the jth receiver is not in sync with each transmitter, only with his own. To get an idea of what is happening, let us assume that this synchronism exists with all the transmitters, and then we can come back and drop this assumption later. If $f_{c_k}(t - \tau_k) = 1$, it means that $c(t - \tau_k)$ is also in sync (i.e., $\tau_k = 0$). Therefore,

$$k\text{th term} = A_k \cos \theta_k \int_0^{T} c_k(t)c_i(t) \, dt \qquad (9.7.11)$$

which is nothing but the inner product of the $c_k(t)$ and the $c_i(t)$ random code sequences. If all the sequences used by all the transmitters (and therefore the receivers) are orthogonal to each other, all the terms above will be zero and multiplexing is possible. So, some orthogonality among the $c_i(t)$ codes is in order here. But how does it manifest itself when the synchronization assumption above is lifted?

Things begin to get messy now. If $f_{c_k}(t)$ is not in synch with $c_i(t)$, it can change sign during the interval $[0, T_p]$, at the point $\tau_k \neq 0$. If this is the case, the kth term is

$$k\text{th term} = \pm A_k \cos \theta_k \int_0^{\tau_k} c_i(t)c_k(t - \tau_k) \, dt$$

$$\pm A_k \cos \theta_k \int_{\tau_k}^{T_p} c_i(t)c_k(t - \tau_k) \, dt \qquad (9.7.12)$$

and these are the terms that must be zero in the general case and all like it coming from the other transmitters. Notice that we have made both partial sums with $\pm$ signs, to account for the fact that the $f_{c_k}(t - \tau_k)$ may start at -1 and switch to $+1$, or vice versa.

The two terms above, then, a sort of sum of two partial cross-correlations, rather than one inner product as was the case in Eq. (9.7.11), are the ones that decide whether CDMA can work or not. A lot of work has gone into studying codes for which the sequences have zero (or near zero) for the above two terms, as the reader can verify by checking Sarwate and Pursley (1980), for example. Of course, if the nature of the code sequence is restricted by such requirements, their randomness is decreased, and so must be their privacy capabilities.

Actually, the problem of properly picking the $c_i(t)$ sequences so that Eq. (9.7.12) is zero is not even the most serious problem in DS-SS CDMA (our abbreviations are becoming a sequence also). A more crucial one appears to be the *near–far problem*, caused by the varying differences in the geographical distances between transmitters and receivers. If, for example, the ith receiver is much closer to the kth transmitter than he is to the ith one (the one he wants), the power from the kth one might be so high that it might completely mask the ith one, even if the terms of Eq. (9.7.12) are small. In such cases, the multiplexing scheme fails completely.

What about some kind of CDMA for frequency-hopping systems? It is possible. How? Well, each user can be given a hopping pattern that is orthogonal to all other hopping patterns. This is not easy to do, even for a small number of users, and often two of them are overlapping in frequency (a *hit*). Errors are common and so are error-correcting codes to correct them. More information on these issues can be found in Solomon (1973) or Nettleton and Cooper (1981). More references on every issue pertaining to SS systems are to be had at the end of Dixon (1975) or Pickholtz et al. (1982).

It is obvious that the search for ways to multiplex SS signals is far from over. The field is still new. A lot of the ideas are still on the drawing board, and if anybody has any practical results, he or she probably is not saying.

PROBLEMS, QUESTIONS, AND EXTENSIONS

9.1. Fig. P9.1 shows another possibility for generating the local oscillator needed in the synchronous detector of an ASK system. Show that the output is indeed a cosine with the same phase as the carrier. The local oscillator is of a low frequency, so the output of the first mixer is in a region of frequency where a very narrow BPF can be constructed. Why is the second mixer used?

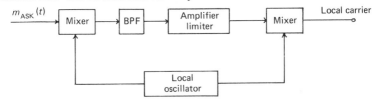

Figure P9.1

9.2. The $f_c(t)$ entering an ASK system has the code $(1, 1, 0, 0, 1)$, each bit lasting 1 μs with a 2-μs guard band. Find the FT and sketch the amplitude spectrum of $m_{ASK}(t)$. Also find its bandwidth.

9.3. The signal $f_c(t)$ described in Problem 9.2 enters a PSK system. Find the amplitude spectrum of $m_{PSK}(t)$ as well as its bandwidth. What should be the bandwidth of the LPF of the synchronous detector of the system?

9.4. Repeat Problems 9.2 and 9.3 with each bit lasting 1 μs but with a 0.1-μs guard band.

9.5. Consider an FSK signal that emits the signal $A \cos(\omega_c \pm \Omega)t$, and assume that the bit stream entering it is $f_c(t) = 1010$. Find the FT of $m_{FSK}(t)$ if $\omega_c = 10^6$ and $\Omega = 10^3$. Sketch the amplitude spectrum $|M_{FSK}(\omega)|$. What is the bandwidth of $m_{FSK}(t)$?

9.6. The code signal $f_c(t) = 10110111001$ enters a DPSK system whose output is first set at 0. Find the signal $g_c(t)$ (see Fig. 9.5.1). Verify that the receiver shown in Fig. 9.5.1c recovers the original $f_c(t)$.

9.7. Multilevel PSK systems usually try to use phase angles that are symmetrical around the circle $(0, 2\pi)$. Can you come up with a general expression for the values of the phase angles of an m-level MPSK system?

9.8. Consider a possible four-level $(1, 2, 3, 4)$ FSK system whose $f_c(t)$ input (bit stream) is $f_c(t) = (1, 3, 4, 2, 2)$. The signal emitted from this system is expressed as

$$m_{MFSK}(t) = A \cos(\omega_c \pm n\Omega)t \qquad (n = 1, 2)$$

Now assume that each pulse of $f_c(t)$ lasts 1×10^{-9} s. What should Ω be so that there is no overlap (with respect to the first null) between the spectra around each emitted frequency?

9.9. A DS-SS system has an input PCM signal $f_c(t)$ whose bits last $T_p = 1$ μs. The code sequence $c(t)$ has bits lasting T_c seconds. Find the bandwidth of $f_c(t)$, and then T_c, so that this bandwidth is spread out by a factor of 100.

9.10. Prepare a lecture on the **mix-and-divide** method of designing frequency synthesizers, proposed by Stone and Hastings (1963). A brief description of the method is also described in Dixon (1975, p. 107). The lecture should include a description of the basic module, the steps used in the design, and an example.

9.11. Consider the problem of multiplexing SS signals (DS-CDMA). Assume that the sequences $c_i(t)$ have four bits (± 1). How many orthogonal sequences can you produce? Can you generalize to $2n$? (*Hint:* Think of the Walsh functions mentioned in Problem 2.16.)

9.12. Can the chirp radar system be considered an SS system? Why?

=10=

MIXED MODULATION: RADAR SYSTEMS

10.1 INTRODUCTION

Here we go again with the word "mixed," an excuse to cover whatever pleases our fancy, this time for radar. We start with what can be called **pulse-compression** or **spread-spectrum** radar. The signal it emits is mixed. It has both AM and angle modulation, but the latter now comes from a digital signal, and for this reason we did not cover it in Chapter 6.

After that, the word "mixed" will become vague, and it will be used to cover some tracking radar systems, systems that perform continuous measurements of a target's location or velocity. Whenever we can justify the appellation "mixed" we will do so somewhere in the section. Otherwise, we will conveniently ignore the problem and hope that the reader will forgive the omission.

Incidentally, this is the last chapter where noise effects are not taken into account. Future analyses will always include them.

10.2 PULSE COMPRESSION RADAR (CODE MODULATION) OR RADAR SPREAD-SPECTRUM SYSTEMS

Both pulse compression and spread-spectrum (SS) concepts have been discussed before, so it is a good idea to start with a brief synopsis of what we know up to now. This way we will better understand the newness (if any) of the systems under discussion.

The SS idea was stressed primarily in Chapter 9 when the SS communication systems were discussed. We saw there that a SS communication system is one whose emitted signal spectrum is very wide (much wider than that of the information signal) and that this widening is caused by a secondary signal, irrelevant to the information signal. So far, so good.

Now the pulse compression idea was first introduced in Chapter 6, during our discussion of the chirp radar system. This name originates in the fact that the pulse emitted by this radar lasts fairly long (so a lot of power can be packed without damaging the peak-power-limited radar devices), and yet this long pulse (which is linearly FM-modulated) can be compressed again at the receiver with a matched filter (or a correlator), to provide good range accuracy *and* resolution. Still, as we noted several times since then, chirp radar qualifies as a SS system, in the sense that the emitted signal's spectrum is spread out to a width much greater than the width of the spectrum of the amplitude-modulating pulse. If we stretch our imagination a little and consider this pulse as the information signal, the spreading is caused by an extraneous signal (the one causing the linear FM), so the chirp radar is an SS system.

So, in terms of radar, we have already covered a system that can be called both an SS and a pulse compression (PC) system. Why, then, do we have a new section on this issue? It all has to do with the angle modulating signal, the one that causes the spectrum spreading. When we were discussing chirp, our coverage of systems was still based on *analog* signals, not *digital* ones. The angle modulation in the chirp comes from a linear frequency modulating signal which is analog (a sloping straight line). With the passage of the chapters, we are now quite familiar with coded (or digital) signals. So here is the proper place to discuss the PC (or SS) systems that use such signals for spectrum spreading, or for compression of the signal at the receiver.

There is, then, a large group of radar systems called PC systems (the SS name has only recently come into widespread use), and chirp radar is one of them. The ones we are mainly interested in for this section are PR systems in their AM nature and they use a coded signal in the angle, to spread the spectrum. This coded signal is a pseudorandom code as it was in the SS communication systems. Now since we can tamper with the angle of a cosine by tampering either with the phase or with the frequency, we had better split up our systems into two kinds.

10.2.1 Phase-Coded PC Systems

The phase-coded PC system is like a direct sequence spread spectrum (DS-SS) communication system except in name. The tampering with the angle is done on its phase. We assume that the regular PR emitted signal is of the form

$$s(t) = A \operatorname{rect}\left(\frac{t}{T_p}\right) \cos \omega_c t \qquad (10.2.1)$$

that is, a single pulse modulating a cosine, lasting T_p seconds. Now we can come in and add phase modulation to the pulse above, much as we did in the direct sequence (DS) SS system, that is, using a binary-coded signal (or a code sequence) whose pulse duration T_c is much shorter than the duration of the rect (t/T_p) pulse (i.e., $T_c \ll T_p$).

Since we know (from PSK) that phase modulation of a cosine with a binary code (± 1 or 1, 0) can end up making a cosine $\pm\cos \omega_c t$, we can conceive of the whole thing as the sequence of operations shown in Fig. 10.2.1a. The rect (t/T_p) signal is first multiplied with the $c(t)$ code sequence, and the result amplitude-modulates the carrier. Typical rect (t/T_p), $c(t)$, and emitted signals from this *phase-coded* PC radar are shown in Fig. 10.2.1b, c, and d, respectively.

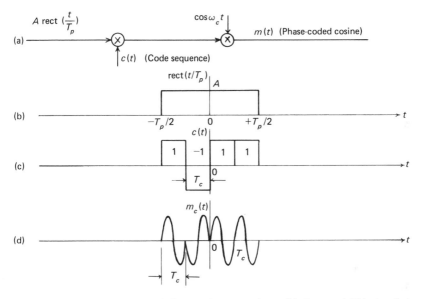

Figure 10.2.1   (a) Phase-coded PC radar transmitter; (b) the rect (t/T_p) signal; (c) the random sequence $c(t)$; (d) the emitted signal.

Note that $4T_c = T_p$ in the figure, but it can be much higher in practice. Note also that the frequency of the cos $\omega_c t$ is assumed such that each code bit causes one complete period in the cosine, but this may not be so either. Anyway, the whole scheme is identical to the DS-SS system transmitter. The only difference is that here the rect (t/T_p) is our information signal (so to speak), and in the actual radar it is periodic (ignoring illumination time), whereas in the DS-SS communication scheme it was $f_c(t)$, another coded signal.

Now in radar we are interested mainly in range and velocity measurements, and in resolution. To be able to say a few words about these things, we must find $R_m(\tau)$ and $\mathcal{K}_m(v)$, the time and frequency ambiguity functions of the emitted signal $m(t)$. What can we do about these? Well, not very much, not at this level anyway. Chirp radar was bad enough, and that one even had some periodicity. Some $R_m(\tau)$'s and $\mathcal{H}_m(\nu)$'s have been found for specific types of pseudorandom codes [see Skolnik (1970, Chap. 3), for example], but they require knowledge of random processes and cannot be done at this stage.

Are we going to give up? Not completely. We will at least look for $R_m(\tau)$ heuristically, to get an overall feeling. To find $R_m(\tau)$ we can look for $M(\omega)$, then $|M(\omega)|^2$, and via the Wiener–Khintchine theorem, invert it to get $R_m(\tau)$. As we have already noted, the present system is very much a DS-SS—only the name is different. Therefore, its $|M(\omega)|$ has got to be similar to the amplitude spectrum of Eq. (9.7.5) (i.e., a mess), with an envelope of $|(\sin \omega/\omega|$ shape. The inverse FT will give us, by the Wiener–Khintchine theorem, something like a *triangle envelope*, which is fine for range measurements. Of course, the argument is heuristic and rather weak. Mathematical results do bear it out approximately, however (see Cook and Bernfeld, 1967, p. 243). The triangle is, in fact, quite narrow, only twice as wide as the pulses of the pseudorandom code. The reception can then be performed with a matched filter (just as in chirp radar), and the output will be quite suitable for measuring ranges. The resolution will not be too bad either, as the triangle is narrow, and can be made even narrower if we increase the number of the pulses within T_p (i.e., make T_c smaller).

What about $\mathcal{H}_m(\omega)$? Even messier, we are afraid. One can use a heuristic argument to get a feeling for it as well, but we leave it for the reader as an exercise (see Problem 10.1). Most of these need stochastic processes to be done more elegantly, anyway.

In conclusion, phase-coded pulse compression radar systems are a good possibility, although their mathematical analyses present problems at an elementary level. They have a lot of advantages (wide spectrum, more power, good accuracy, etc.). Even so, their application has so far been limited. It appears that chirp radar can do all this in a better way overall, even though it, too, has some problems, as we have noted. All of these systems (SS types) are quite important in the presence of noise jamming, something we will not begin to appreciate until a later chapter.

10.2.2 Frequency-Coded PC Systems

In the frequency-coded (FC) PC system, it is the frequency of the carrier that is varied in some fashion, dictated by a code, the radar system remaining basically of PR type. The idea behind the FC-PC is very much the same as in the frequency-hopping communication systems, except that instead of the bits of the coded information signal $f_c(t)$, here we have the pulses of the PR radar. Each pulse (or each group of pulses) multiplies a carrier of different frequency, so the entire spectrum is constantly shifting. The overall bandwidth is much the same as in the FH-SS system, in fact easier to conceive, since here only positive pulses are emitted. As for finding the actual FT, $R_m(\tau)$, or $\mathcal{H}_m(\tau)$ mathematically, they depend on the nature of the code. The reader can think of the entire thing in the context of a single pulse, and then we have the known expressions of the PR system. The only difference is that the carrier's frequency changes from pulse to pulse, and allowances have to be made for this at the receiver. The receiver is, of course, right next to the transmitter and knows the changes. This is again a much simpler case than in the FH-SS communication system, where the frequency-hopping code has to be known in advance, or transmitted separately. Similarly, all synchronization problems are simpler in the radar system, where the transmitter and receiver are usually in the same place. We leave the rest for the problems at the end of the chapter.

We close our discussion of PC (or SS or whatever) systems by pointing out once more that chirp radar holds the leading position in this field, so much so that the word "chirp" has become almost synonymous with pulse compression. Look at the book by Barton (1975). Its title is "pulse compression radar" and its contents are almost entirely devoted to chirp. The ease of its implementation and overall performance beat everything in sight at this stage of radar development.

10.3 TARGET-TRACKING RADAR SYSTEMS: GENERAL REMARKS

Our noiseless analyses of communication and radar systems are drawing to a close. This is our last chance to discuss tracking radars, and we shall take it. A basic knowledge of such systems is necessary not only for an overall appreciation of radar accomplishments, but also for a better understanding of electronic warfare (Chapter 15).

The sections that follow cover a group of radars that can perform "continuous" measurements of a target's relative position in range, azimuth, elevation angle, and even velocity. By keeping track of these parameters, an accurate prediction of the future position of a target can be estimated, at least for the near future. Such prediction of a target's movement is especially important in military applications of radar, particularly in weapons control and missile-range instrumentation. The tracking radar's role in such applications is to keep track of the position of a target and feed this information to the guidance and steering mechanism of a missile before and during its flight. Without TTRs, missile guidance is almost impossible.

Up to now we have been chiefly concerned with the abilities of a radar to measure range and velocity of a target. The key role in these measurements was played by the signal emitted from the radar, since it is the changes in its parameters (time shift, spectrum shift) that enable us to measure these two target properties. In all of our previous radars, the main distinguishing feature was the signal emitted—even their names were based on it. We took pains in our analyses to look at this signal in time and in frequency, and even (whenever we could), to find its time and frequency ambiguity functions, functions that have a role to play in the accuracy of the two measurements. Generally speaking, then, up to now, the key to the radar was the signal emitted. Do target-tracking radar (TTR) systems present anything new in this respect?

No, not really. In terms of signals, the radar systems that perform tracking are not different in any way from those we have already discussed. They differ only in the way in which they use the available information, and possibly in the antenna structure. In all other respects, they are like one of the radar systems already discussed, primarily that old standby, pulse radar.

We shall divide the TTRs into two groups: those that track in range and velocity, and those that track in angle. This division is strictly artificial; it has to do only with our presentation here. Tracking in range will be taken up first, as it is judged more familiar to us with the background accumulated. Small subsystem variations to our familiar radar systems can perform tracking in range and the matter can be disposed of rapidly. Velocity tracking will not be discussed at all; it has been delegated to the Problems. The last two sections of the chapter are devoted to the principal angle-tracking radars, conical-scan and monopulse. They will both be discussed in some detail, as they will make our work in Chapter 15 a lot easier to comprehend.

10.4 RANGE-TRACKING RADAR

The only difference between a simple PR system and range-tracking radar is in the processing of the return echoes. The name makes it sound as if we are dealing with a new radar system, but really, we are not. In fact, most angle-tracking radars (to be discussed later) have the proper subsystems to track range also, not only elevation and azimuth angles, as might be assumed.

We already know that the range is measured by estimating the round-trip time of the emitted pulses. The tracking of the range, which means the continuous estimation of the target distance, is accomplished by the method of **range gating**. Once the distance of a target is initially estimated, a device (such as a gate) is set up to work as a switch and open the receiver at the precise instant when the return echo is expected, and to keep it open for as long as the echo pulse is expected to last. Of course, the target distance may be changing, so the gating time must be continuously readjusted; otherwise, the target may be "lost."

In actuality, most range gating[1] systems use *two* gates (a split-gate system), not just one. The idea of the split-gate system is explained with the aid of Fig. 10.4.1. Figure 10.4.1a shows a typical return echo as it looks approximately after demodulation. Below it we have sketched two pulses, which represent the opening and closing of the two half-gates used in the system. The first, the **early gate**, opens on the arrival of the signal and lasts for half its duration. The other, the **late gate**, opens when the early gate closes and stays open for the same

---

[1]Gating is officially defined as "the process of selecting those portions of a wave that exist during one or more selected time intervals."

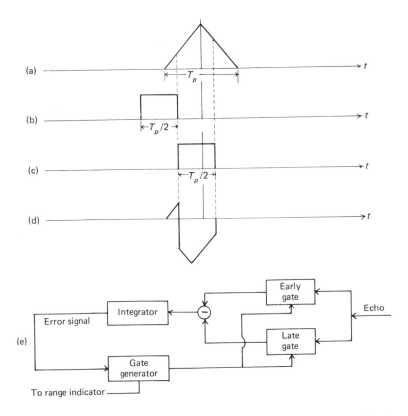

Figure 10.4.1 (a) Return echo; (b) early gate; (c) late gate; (d) difference signal; (e) overall gating system.

duration. The outputs of these two gates are then subtracted and the difference signal integrated, as shown in the overall block diagram subsystem of Fig. 10.4.1e. If the gates are properly timed, the integrator will have zero output. This case is not shown in the figure. What is shown is an incorrect gating time; that is, the early gate does *not* close in the middle of the duration of the echo. The difference signal will be as shown in Fig. 10.4.1d, and the integrator will have a negative output, a sign that the gating time must be changed. The error signal is then introduced to a gate generator, which changes the gating times until the error signal is driven to zero. The whole story goes on continuously and the split gate keeps the target *locked* in range. At any moment, the distance of the target can be obtained from the gate generator as a separate output, possibly through a visual indicator. We should remark that the system has been described in a rather simplified way. Mixers, IF amplifiers, clocks, AGCs, and the like have been omitted, to make the explanation as simple as possible.

Range gating, and in fact, all tracking, can also be done manually by an operator, particularly if the target's speed is low and its position is changing slowly. In most modern radars, however, it is done automatically. Modern target speeds make it quite a tiring endeavor, and sometimes (missile tracking, for example) almost impossible.

So much for tracking one of the coordinates of a target's position. Now we must see what we can do about tracking the other two, the elevation and azimuth angles. They are tracked by angle TTRs, which are discussed in the following two sections. As we have already remarked, velocity tracking will be omitted (see, however, Problem 10.4).

10.5 ANGLE TRACKING : CONICAL-SCAN RADAR

Angle tracking is, of course, continuous measurement of the angles (azimuth and elevation) of the location of a target. Before we can take up such continuous measurements of the angles, we should discuss how angles are measured to begin with. It is a subject that we have largely ignored up to now, other than having stated previously that angles are measured by the position of the antenna pointing to the target.

Well, the central idea is precisely that. If a target is detected, and the antenna is facing it squarely, with its beam, then the angles of the antenna beam axis determine the angles of the location of the target. The important phrase here is *facing it squarely*, for it leads us to the importance in angle measurements of the *beam* of the antenna. The angle-measuring antenna must have a very narrow beam (pencil beam), a clear maximum on its main lobe. A wide beam with a flat maximum cannot adequately do the job of facing the target squarely. It can move a little and still face the target squarely so to speak (i.e., get an echo as strong as before). A pencil-beam antenna cannot move at all (at least theoretically); otherwise, the target is lost (no echo). The principle should be clear, and we will not illustrate it further.

So an angle tracking radar is characterized chiefly by a pencil-beam antenna (or antennas, as we will see). But such a beam covers very little area in the space around the radar location—theoretically, a single point, in fact. A radar using such an antenna can hardly be expected, then, to search for and detect a target, unless it knows approximately where the target is. It would be like looking for a fly in a huge, dark auditorium, using only a narrow-beam flashlight. Angle-tracking radars must cooperate with *search and detect* radars, so that they can get an initial approximate location of a target before they can zero in for tracking. Of course, if their beam is very narrow, finding the target even if its approximate location is given may not be immediate. To cover a small area, the target usually performs some antenna movements that are designed to cover a small neighborhood, as shown in Fig. 10.5.1. They are classic movements, their name a direct consequence of the nature of their path.

Of course, we do not mean to imply that no tracking radar can do both searching and tracking—just that these two functions contradict each other.

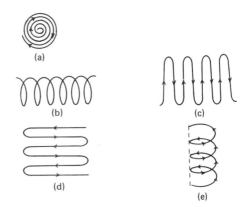

Figure 10.5.1 Antenna movements in angle tracking radars for acquisition: (a) spiral scan; (b) Palmer scan; (c) nodding scan; (d) TV (or raster) scan; (e) Helical scan.

To search and detect, you need a wide beam; to track, a narrow one—compromise, and the result will be mediocre in both. In essence, all radars can do everything to some degree; it is a matter of processing the information, and the accuracy desired. In fact, there exists a radar in actual use that does both, but not in a "continuous" manner as we have been discussing here. This radar uses a wide fan-shaped beam and rotates it very fast in azimuth (full circle in a couple of seconds). All target locations are fed to a computer, which keeps track of not one, but many targets at the same time. This **track-while-scan** (TWS) radar will not be discussed further here, but let it serve as an example that nothing is absolute. Anyway, tracking radars are meant to track just one target, and do so continuously, and these are our primary interests in these sections.

Our first angle-tracking radar is the **conical-scan** radar. Its name and operation are strongly dependent on the antenna system that it uses, and this will be our first concern in the analysis of this system. Up to now we have looked at antennas in a way that restricts their possible nature and activities. Take a look at Fig. 10.5.2, where a typical antenna is portrayed. The parabolic curve represents its mechanical (so to speak) nature, and the beam its electromagnetic (invisible) one; each has its own axis of symmetry. The axis of symmetry of the geometry of the antenna mechanical system we shall call the **tracking axis**; the axis of the beam, the **beam axis**. So far we have taken it for granted that these two axes coincide, and this, of course, may not be true. The case shown in Fig. 10.5.2b is the one of interest in angle-tracking radars, even though we shall not get into a detailed discussion of how this is accomplished in practice. The two axes in this last antenna do not coincide. Not only that; they can also move in various directions almost independently of each other. The geometrical (tracking) axis can be moved all around a 360° solid angle. The beam axis is, of course, bound by the parabolic curve, but it, too, can move in all directions in a solid angle determined by this curve.

After this simplified explanation of how the two axes are interrelated, we are ready to have a look at the antenna of a conical-scan angle-tracking radar. As the name implies, there is a conical scan involved here. This scan is performed

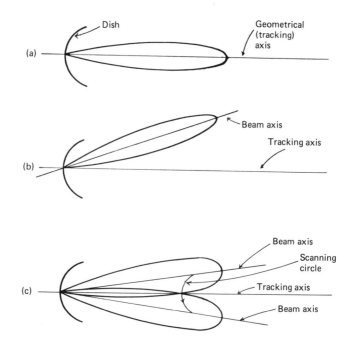

Figure 10.5.2 Antenna nature: (a) tracking and beam axis in the same direction; (b) negation of (a); (c) the conical-scan antenna system.

by the beam around the mechanical axis of the antenna, resulting in tracing out a cone, as shown in Fig. 10.5.2c. In practice, this can be accomplished by an electrically unbalanced rotating dipole feed or by a lens that rotates in front of a fixed feed (or by other methods). The result is a beam "squinted" from the tracking axis, which rotates around the mechanical axis as shown in the figure, with a rate of rotation that is about $\frac{1}{10}$ to $\frac{1}{100}$ the repetition rate (PRF) of the radar, assuming, of course, that the radar is our old friend the PR, and it usually is. If the radar is of the CW type, the scan rate has no limit, except for that determined by the practical means of causing it.

So the conical-scan radar is basically a pulse radar, with the funny antenna system we described above. But how does it perform the angle tracking? Well, it has to do with amplitude modulation (AM) again. To see this, witness the following. To have a continuous measurement of the angles, the tracking axis must be fixed squarely on the target, even while the target is moving. If this is accomplished, the tracking axis direction can be measured, to provide us with the azimuth and elevation angles of the target. So our problem is to get a square fix of the tracking axis right on the target, and keep it there. That is where the conical scan of the beam and the mentioned AM come in.

Consider the situation when the tracking axis is fixed right on the target, as shown in Fig. 10.5.3a (we do not care how it got there). Pulse radar emits RF

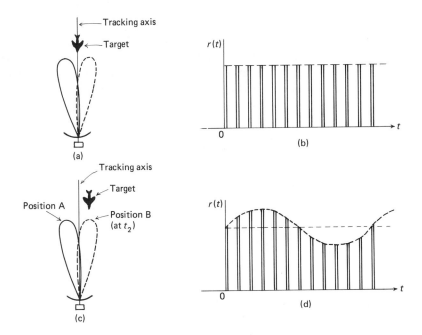

Figure 10.5.3   (a) Target on tracking axis; (b) the received pulse train; (c) target off tracking axis; (d) the received pulse train.

pulses, and it receives their echoes, while the beam axis rotates around the. tracking axis. These pulses do not leave and enter from the peak of the antenna beam; nevertheless, they all leave and enter from the same point of the beam since the beam is rotating in perfect symmetry around the target. Its center remains equidistant from the target. The end result is that the received signal $r(t)$ is made up of echoes that have the same amplitude, as shown in Fig. 10.5.3b, after the carrier has been removed. So far, so good. The only AM modulation involved is the one we already know of, the AM of the carrier by the pulses, which is of no interest here.

Now let us take the case when the target is a little off the tracking axis of the antenna, a case shown in Fig. 10.5.3c. In this case, the echoes in $r(t)$ do not have all the same amplitude; they appear something like we have shown in Fig. 10.5.3d. When the beam is passing through position A, for example, the pulses are weak in amplitude, whereas the opposite is true when it is passing through position B, where it is hitting the target near the maximum of its beam. That is where the AM modulation comes in, the one we were talking about above. When the target is off the tracking axis, the returned echoes have an AM modulation on their amplitude, and this is what shouts out loud that the tracking axis must be shifted. The circular rotation of the beam axis, its presumed symmetry in three dimensions, and so on, cause this AM to be sinusoidal in nature, its frequency equal to the frequency of the beam scan.

Now, knowing that the tracking axis is not on target is important all right, but more important for tracking is to know how to turn it and put it in the right place. This is where synchronization circuits, reference signals, and the like come in to help, and this is what we must try to explain. We will need a new figure, Fig. 10.5.4 and its various subparts. They are all needed for the explanation.

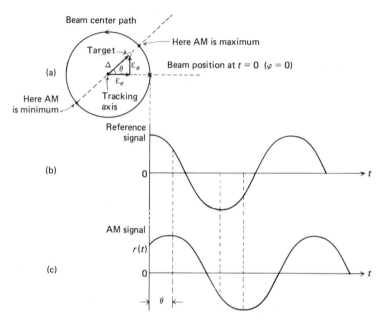

Figure 10.5.4 (a) Axes and target configuration; (b) the reference cosine; (c) the AM signal from the beam scan.

Consider Fig. 10.5.4a, which shows a typical case of target and axes location, looking at it from a top view, a view on a plane perpendicular to the tracking axis. The center of the beam (the beam axis) traces a path around the tracking axis in a counterclockwise direction, while a reference cosine is generated within the radar to keep track of its position around the path. This reference cosine is shown in Fig. 10.5.4b, and its $t = 0$ corresponds to the beam axis position shown in Fig. 10.5.4a, to the right of the target.

As can be seen in Fig. 10.5.4a, the target is at a small distance Δ from the tracking axis (a fraction of the beam width). The azimuth and elevation errors are denoted by ϵ_a and ϵ_e, respectively, and are given by

$$\epsilon_a = \Delta \sin \theta \qquad (10.5.1)$$

$$\epsilon_e = \Delta \cos \theta \qquad (10.5.2)$$

where θ is the angle shown in the figure. It is these errors that the conical-scan radar will presumably drive to zero and then align the tracking axis with the target.

Now, let us look at the AM signal superimposed on the return echoes. It is shown in Fig. 10.5.4c, after it has been shifted down the vertical axis so that it has no dc value. The reader should study Fig. 10.5.4a to verify that this cosine takes its maximum value when the beam center passes through the point closest to the target. This cosine is shifted by θ when compared in phase with the reference signal. Furthermore, the amplitude of this cosine is related (approximately linearly, say) to the distance Δ—if $\Delta = 0$, for example, no AM; if Δ is equal to the "squint" of the beam axis, maximum AM; and so on. When all of the above is taken into account, we can write this AM signal as

$$r(t) = K\Delta \cos(\omega_s t - \theta) \tag{10.5.3}$$

where ω_s is the beam scanning frequency, K a constant relating Δ to the distance (called the error slope), and Δ and θ the quantities already defined.

Now we can expand the cosine in Eq. (10.5.3) and make use of Eqs. (10.5.1) and (10.5.2). The result is

$$r(t) = K\Delta(\cos \omega_s t \cos \theta + \sin \omega_s t \sin \theta)$$
$$= K\epsilon_e \cos \omega_s t + K\epsilon_a \sin \omega_s t \tag{10.5.4}$$

which is the key to the operation of this angle-tracking radar. The AM on the echoes carries both the errors in its two quadrature components. Special circuits can extract their value and drive the appropriate feedback servos to align the target with the tracking axis. This operation is done in a continuous manner, and the tracking axis can presumably follow the movements of the target. The angle coordinates of the tracking axis are continuously measured and translated into angle position of the target. The whole thing is pretty nifty.

We are ready for the discussion of the conical-scan subsystem shown in blocks in Fig. 10.5.5. As we have already noted, the overall system is usually a PR radar and thus most of its components are not shown. Only the subsystem that does the tracking is emphasized in the figure.

The return echoes are received (TR, ATR, FR demodulations, range gates, etc., are not shown) and led to a mixer and IF amplifier, and then to an AM demodulator (envelope detector?) to get $r(t)$ out. The AGC should be noted at this point, as it is necessary to keep the signal $r(t)$ always at the same level, and not have it vary with the distance of target, size of target, and the like. Let us not forget that the amplitude of $r(t)$ is presumably related only to Δ.

The signal $r(t)$ [which has the form of Eq. (10.5.4)] is then shipped to two error detectors (what is their nature?), which also entertain the reference signal as their second input. The reference signal is, of course, generated somewhere in the motor that causes the beam scan. One of the error detectors receives the reference signal *as is* [to get the error in the cos $\omega_s t$ term of $r(t)$], and the other after it has been turned into a sin ω_s t. The outputs of the error detectors represent the azimuth and elevation errors. They serve as inputs to the feedback servo systems, which drive the tracking axis motors for the eventual alignment with the target.

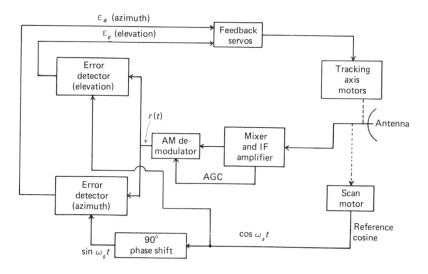

Figure 10.5.5 Conical-scan radar subsystem.

Here is the moral of the story. The present radar uses double AM modulation (the pulses modulate an RF carrier, but then they are themselves modulated by a low-frequency carrier), so it sort of fits in with our present chapter of mixed systems.

Another moral has also just occurred to us. Even though the configurations may look different, the principles behind them seem to be recurring.

This completes our discussion of conical-scan TTR. An earlier (discrete) version of it, called **sequential lobing** tracking radar, has been abandoned almost entirely and will not be covered.

Conical-scan radars are an easy and relatively inexpensive way to angle-track, but they have their problems. To demodulate the AM produced by the scanning, the receiver must wait for quite a few return echo pulses (at least four). This makes the radar a prey to **angular jitter** caused by pulse-to-pulse fading, a result of target scintillation. There is also the possibility of moving parts on the target (propeller?) which may produce an AM component on the return echoes and create difficulties in tracking. Moving parts are often placed there on purpose by an adversary to "break track." Wouldn't it be nice if we could have a radar that could angle-track a target using only *one* return echo?

10.6 ANGLE TRACKING: MONOPULSE RADAR

We saw how a rotating beam with an axis "squinted" off the tracking axis of the antenna can be used to measure and track the angle coordinates of a target. In essence, this beam emits and receives pulses from discrete points around a circle whose center is the tracking axis of the antenna, at least four of them. So to get a measurement of both angles of a target, *we must send out and receive (at least) four pulses from four equidistant locations of the beam around the tracking axis.*

You can do the same thing by having four stationary beams symmetrically located around the tracking axis of the antenna, each squinted off by the same angle from it. This type of antenna configuration and the system that goes with it to process the information is called a **monopulse** (or **simultaneous lobing**) radar. It does have more complexity (four horns, four feeds, etc.) than a conical-scan radar, but that is the price that must be paid for eradicating the main problem of the conical-scan, discussed at the end of the preceding section. Monopulse radar can do what its name implies—it can measure the angles of a target using only one pulse. Pulse-to-pulse variations brought about by physical or intentional causes have no effect on it whatsoever.

There are many types of monopulse radar. The only one we shall describe here is one called **amplitude monopulse** (or just monopulse). **Phase-monopulse** radar, or other combinations of the two, will be left to the reader for further study. The monopulse concept is so important in theory and practice that an entire book has been devoted to it (Rhodes, 1980).

The amplitude monopulse radar system uses four beams in a manner shown schematically in Fig. 10.6.1. They represent the four discrete positions of a conical-scan radar around the scanning circle. The reason we need four is to be able to measure (and track) *both* azimuth and elevation angles. If only one angle is desired, two of them will do the job, say 1 and 2 for azimuth and 2 and 4 for elevation. This being the case, we can take up an explanation of this radar by considering two of its beams and seeing how they measure one of the angles. The generalization to four will then be fairly obvious.

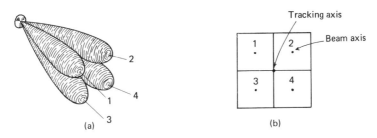

Figure 10.6.1   Beam arrangement in an amplitude monopulse.

Let us isolate beams 1 and 2 and see how they can measure the azimuth angle. They are shown in Fig. 10.6.2 (plane view) in polar coordinates. A target is also shown there located slightly to the right of the tracking axis. When the radar emits a pulse, it does so simultaneously from both beams. The result of this "sum" transmission is equivalent to emitting the pulse from the sum pattern

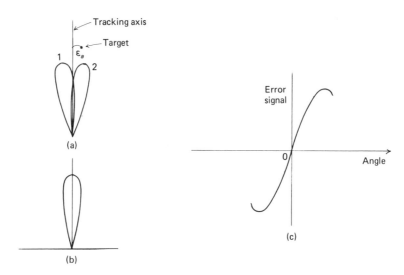

Figure 10.6.2 (a) Beams 1 and 2, target off tracking axis; (b) the sum pattern of the two beams; (c) the error signal.

of the beams, shown in Fig. 10.6.2b, a pencil-type beam as desired in all angle-tracking radars. So much for emission.

Now we consider reception. During reception of the return echo, the outputs of the two beams are subtracted. If the target is on the tracking axis ("azimuth-wise," so to speak) the return echo will hit each beam at a point of equal gain and the output of the subtracter, the error signal, will be zero. If, on the other hand, the target is slightly off-axis by an azimuth error ϵ_a, as shown in Fig. 10.6.2a, the output from beam 2 will be larger than the one from beam 1, and the error will be positive. The error signal from the difference is thus related to azimuth angle error, with a curve like the one shown in Fig. 10.6.2c, and it is this error signal that can be used to drive the servos and realign the target with the tracking axis.

Now, in actuality, the receiver not only receives the difference signal from the two beams, it can also receive the sum signal, and does. It can use it to measure and track in range, and as a reference signal to compare it with the difference signal. A reference signal is needed to find the direction of the azimuth error, as the error signal comes envelope-modulating the carrier and its sign must be ascertained.

A simplified block diagram of a monopulse radar (still with two beams) is shown in Fig. 10.6.3; this time the beams are meant to measure the elevation angle error. The return sum signal (Σ) is mixed with a local oscillator, IF-amplified, and then used for range measurement and tracking by a different subsystem. The same sum signal is also sent to a phase detector to be compared

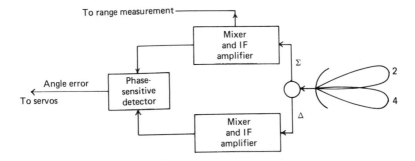

Figure 10.6.3   Block diagram of a monopulse radar.

in phase with the difference signal (Δ), which, of course, would exist only if the target is off-axis in terms of elevation. If it does exist, it can be either $K\epsilon_e \cos \omega_{if}t$ or $-K\epsilon_e \cos \omega_{if}t = K\epsilon_e \cos (\omega_{if}t + \pi)$ and this is checked by the phase sensor. The result is a positive or negative voltage analogous to the error in elevation angle, and it is used to steer the servomotors toward alignment. In other words (or rather in symbols), the error voltage will be

$$\text{error voltage} = c\epsilon_e \tag{10.6.1}$$

but proving it would require a lot of rather difficult mathematics. See Skolnik (1962) if you do not believe it.

This basic idea is easily extended to four beams, so that both angle errors are measured with one return pulse. Now the sum signal is the sum of all four beams (see Fig. 10.6.1); that is,

$$\Sigma = m_1(t) + m_2(t) + m_3(t) + m_4(t) \tag{10.6.2}$$

where $m_1(t), m_2(t)$, and so on, are the signals from each beam. This is the signal used to measure and track range, and as a reference for the error signals. The error signals are

$$\Delta_{el} = [m_1(t) + m_2(t)] - [m_3(t) + m_4(t)] \tag{10.6.3}$$

for the error in elevation and

$$\Delta_{az} = [m_1(t) + m_3(t)] - [m_2(t) + m_4(t)] \tag{10.6.4}$$

for the error in azimuth. Of course, the radar could have used only $[m_1(t) - m_3(t)]$ or $[m_2(t) - m_4(t)]$ for Δ_{el}, but doing as shown in Eq. (10.6.3) is even better. It is doing both differences and adding them up for stronger Δ_{el}. Similarly, in Eq. (10.6.4) advantage is taken of the four existing beams by taking the sum of the pertinent differences, $[m_1(t) - m_2(t)]$ and $[m_3(t) - m_4(t)]$, for stronger Δ_{az}. The rest of the system is double that of Fig. 10.6.3, as regards the error processing, phase detection, servos, and the like. Of course, if the same antennas are used for both transmission and reception (they usually are, as this system is basically a PR one), then TR, ATR, and so on, are also used (not shown in the figure we presented).

The monopulse radar does demand increased complexity and cost (compared to conical-scan), as it requires a fixed but multihorn feed assembly and three receiver channels (Σ, Δ_{el}, Δ_{az}) matched in gain and phase. But the added complexity and cost are well worth the price. Aside from the fact that it can track *with only one pulse* (a great advantage in an electronic warfare regime), it has even better performance in the presence of additive noise, which extends the range by 30 to 40%. It is also very difficult to try to jam with intentional noise. It can turn around and actually use this noise to track the jamming culprit. What more can you ask of a tracking radar?

Why did we cover monopulse radar in this chapter of mixed radar systems? Hard to say. Maybe because it can be thought of as an extension of conical-scan, and we have already argued that that system belongs in this chapter.

PROBLEMS, QUESTIONS, AND EXTENSIONS

10.1. Think of a heuristic argument to find $\mathfrak{K}_m(\nu)$ of a phase-coded PC radar system. [*Hint:* $\mathfrak{K}_m(\omega)$ is a correlation function, which is like a convolution, which in time is a product.]

10.2. In a frequency-coded PC system, the emitted signal is

$$m(t) = s(t + 5T_p) \cos \omega_1 t + s(t) \cos \omega_2 t + s(t - 5T_p) \cos \omega_3 t$$

where $s(t) = A \text{ rect}(t/T_p)$. Find $M(\omega)$, $R_m(\tau)$, and $\mathfrak{K}_m(\nu)$ if ω_1, ω_2, and ω_3 are spaced so that their distance is three times the bandwidth of $s(t)$.

10.3. Consider the system portrayed by the block diagram of Fig. P10.3. It is a radar subsystem meant to track a target's velocity, that is, meant to sense changes in target velocity and adjust to them. Try to analyze its operation. [*Hints:* The frequency discriminator emits a signal (the error signal) when the IF frequency is different from its center frequency. The VCO has as an output a cosine whose frequency depends on the value of the error signal. When the error signal is zero, at what point of the system can we have a reading of the target velocity?]

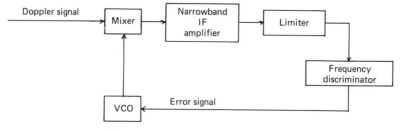

Figure P10.3   Velocity tracking subsystem.

10.4. Consider a conical-scan angle-tracking radar. If its IPP is 10^{-6} s, how fast should the beam rotate so that at least six pulses are sent out in one revolution?

10.5. A conical-scan's beam rotates around the tracking axis and makes a complete revolution in 2 s, and during this time 10 pulses must be emitted and received. Find the PRF of the radar. How far can this radar detect a target without ambiguity?

10.6. For the purposes of a conical-scan radar, a cassegrain antenna is often used. Prepare a paper on the principles and operation of this type of antenna [see Skolnik (1962), for example].

10.7. Repeat Problem 10.6 for a lens antenna.

10.8. Search the literature and prepare a paper on means to attain the beam arrangement of an amplitude-monopulse radar.

10.9. As mentioned in Section 10.6, there exist also phase-monopulse systems. Prepare a short paper on the issue. Note carefully the beam arrangement. Discuss its differences with amplitude monopulse. Are there any hybrid forms? [You can find help in Skolnik (1962, 1970), Rhodes (1980), and Barton (1975, Vol. 1).]

=11=

PROBABILITY AND RANDOM VARIABLES

11 .1 INTRODUCTION

We have been avoiding noise up to now, citing pedagogical reasons for this act. We have always felt that noise, a difficult quantity to model mathematically, clouds the issues of the analyses—it steals the scene, so to speak, to the detriment of an understanding of the operation of the systems. At this point, however, it is time to consider it. We know how the various systems work, and our mind is now clear to accept some new mathematical background, necessary to model noise. For as the reader must have deduced, the existing background is incapable of handling it.

What is so unusual about noise that demands new methematics to model it? Well, it is kind of unpredictable, in ways that make our background in signal analysis defunct. Take resistor thermal noise, for example, a voltage developed across a resistor when it gets hot. You can heat a resistor up to some temperature, T_0 say, and observe this noise all right, as you do for any other signal. You can even model this waveform mathematically as best you can (graph, table, expansion on a series, etc.), and design a communication system that will minimize its effect on the information signal. But all this work will have been in vain. The next time you heat up the *same* resistor to the same temperature T_0, the noise waveform will not be the same as before, nor would a third time produce either of the previous two waveforms. Thermal noise in a resistor is not just one waveform—a **deterministic** signal, so to speak—but a gigantic family of waveforms, maybe uncountably infinite in number.

Unpredictable phenomena such as these can be handled with the theory of **random** (or **stochastic**) **processes**. This theory is based on the theory of random variables, which in turn owes its existence to the theory of basic (abstract) probability. So we must present the necessary background in the reverse order: probability, random variables, stochastic processes. This chapter covers the first two topics, and the next one covers the stochastic processes. Of course, all the material presented here has been condensed to suit our future purposes. Each topic is worthy of a book in its own right—more than one, in fact. Our presentation here will stress concepts and ideas, and consider the needs of future chapters. "Epsilonics"[1] will, by necessity, be kept to a minimum.

Stochastic processes are not meant only to model noises—far from it. Any phenomenon that produces a family of waveforms can be modeled using this theory, and this includes many useful information-type waveforms. Human speech is also best modeled as a stochastic process; it, too, is unpredictable in the same general way. After all, if every time someone spoke we knew the waveform in advance, there would be no need to design communication systems to send it anywhere. Of course, we have been analyzing communication systems up to now, sort of implying that the speech signal was a deterministic one. Well, we are sorry about that; it was a dirty trick, we agree. Let pedagogy again serve as the excuse, and it will be for the last time.

Incidentally, the word "noise" usually implies something undesirable, something detrimental to the communication or radar system on hand. But even this is not absolute. One person's noise is another person's useful signal, in many cases. A ship trying to detect a submarine by using the noise emitted from the submarine's engines while a "flock" of whales is singing nearby considers the whales' song to be noise. A nearby marine biologist trying to record the whales' song thinks of the submarine engine's rumblings as noise. A pilot tries to jam a conical-scan radar with noise, and a nearby monopulse radar *uses* the noise as an information signal to track the pilot's plane. So the word "noise" may have a bad connotation, and we will use it as such in what follows, but let us also keep in mind that it, itself, can serve as an information signal in certain applications. But all of this is actually more pertinent to our discussion in Chapter 12 and the chapters that follow. Our immediate task is the theory of probability and random variables and we will go to them now without further elaboration.

11.2 BASIC PROBABILITY THEORY

This field is concerned with the problem of finding a mathematical description (**a model**) of phenomena that do not always produce the same outcome. The result of flipping a coin, of measuring the length of people's noses, the phase

---

[1] This term, meaning mathematical rigor, was overheard at the graduate school of The Johns Hopkins University during the early 1960s.

angle of a received AM wave, the number of cars crossing a bridge per day—all of these are such phenomena—and without probability, we would not know what to do with them.

What does probability theory suggest we should do with them? It suggests the following.

11.2.1 Single Experiment

First, such a phenomenon, one that has more than one outcome, should be called a **probabilistic phenomenon** or an **experiment**. Then each such experiment has as a **probabilistic model** two things, Ω and P.[2] We define them and explain their meaning below.

The Ω.   This is the *set* of all outcomes of the experiment, outcomes that can be observed or even imagined. It is called the **universe**, **sample space**, or **certain event** of the experiment. It can have n points, countably infinite points, or uncountably infinite points.

If the experiment is the flip of a coin, then

$$\Omega = \{\text{heads, tails}\} \tag{11.2.1}$$

or maybe a third result—standing on edge.

If the experiment is "heights of human beings," then

$$\Omega = 0 \text{ to } 3 \text{ meters} \tag{11.2.2}$$

or something similar, depending on how you feel about the term "human beings." We can imagine people as tall as 3 meters and as short as zero, the latter being the height immediately after an egg is fertilized.

There is something personal in this Ω—something unscientific, so to speak —but we will not let it bother us. Most scientists usually agree about the Ω on hand, in practical problems, anyway. Those that do not are usually condemned to a life of isolation.

So much for the Ω. Now we come to the P—the toughy. We plunge into it a bit more formally, assuming that the reader is familiar with basic set theory, unions ($\cup$), intersections ($\cap$), subsets ($\subset$), the empty set ($\varnothing$), mutual exclusiveness ($A \cap B = \varnothing$), and the like. The explanation of the meaning of P will be taken up after the formal definition.

The P. P, called the **probability measure** of the experiment, is a numerically valued function (a mapping) of all subsets of Ω which have the following axiomatic properties.

(1) $\qquad\qquad 0 \le P(E) \le 1 \qquad \text{for any } E \subset \Omega \qquad (11.2.3a)$

(2) $\qquad\qquad P(\Omega) = 1 \qquad\qquad\qquad\qquad\qquad\qquad (11.2.3b)$

(3) $\qquad\qquad$ If $E_1, E_2, \ldots$ are mutually exclusive, then

---

[2]Actually three things, but the third (the σ-field of events) will not be dealt with here.

$$P \left\{ \bigcup_{i=1}^{\infty} \right\} E_i = \sum_{i=1}^{\infty} P(E_i) \tag{11.2.3.c}$$

P, then, takes every subset of Ω (which could be abstract: oranges, hippopatami, seaweed, etc.) and assigns to it a probability [$P(E)$ is read as the probability of the subset E], a number. This number must be between zero and unity, as property (11.2.3a) dictates. The only set with the right to the number unity is the sample space Ω [property (11.2.3b)]. If two or more sets are mutually exclusive, the probability of their union is the sum of their respective probabilities [property (11.2.3c)].

That's all fine, but what does it mean? How do we find P for a given experiment so that we can complete its probability model, (Ω, P)? Well, the best answer we can probably give to these questions is that we do not know, and that "yours is not to reason why, yours is but to do or die." Since no one is going to be satisfied with this, we attempt some answers.

To begin with, P is obviously the most important part of the probability model (Ω, P). It basically reflects, in mathematical jargon, what *we all feel* about probability—the likelihood of something taking place. This is very hard to define. Many a night has been spent on defining it, justifying it, convincing others that it serves science, and so on, by a lot of people. Laplace thought he had something when he said, "If Ω is made up of n outcomes, all of which are *equally likely* to occur, then

$$P(E) = \frac{n(E)}{n} \tag{11.2.4}$$

where $n(E)$ is the number of points in the subset E." Actually there is not much here. The phrase "equally likely" restricts us to specific types of phenomena, and it probably makes the definition circular, as well. Even if accepted, it would hold only for Ω with n outcomes, a small fraction of the practical probabilistic phenomena that we wish to model.

Other people have also tried. Von Mises even caused the creation of a "school" of people who *believed* in his definition, called "relative frequency." It goes like this: "The probability of a subset $P(E)$ is given by

$$P(E) = \lim_{n \to \infty} \frac{n^*(E)}{n} \tag{11.2.5}$$

where $n^*(E)$ is the number of times the subset (event) E occurred in n repetitions of the experiment." Not excellent, but not bad either. The definition talks about the experiment being repeated. It gives the probability as a limit. Does the limit exist? How large n should be to find it approximately? The problems are obvious.

To make the story short, Kolmogoroff (1933) and others finally suggested the axiomatic approach presented here, where P is defined as we stated earlier, ignoring what it means and how to find it. Ignoring these problems does not mean, of course, that they go away. So we will settle this conflict as follows.

We accept the common-sense feeling that we all have, the one that suggests the "odds in favor of something happening." To find P in a given problem, we can use Laplace's idea or the relative frequency definition (if the occasion warrants it), or whatever else will convince the world.

We will come back later to this problem of finding the probability. We are not completely done with it.

Example 11.2.1

Find a probability model for the experiment "throw an honest die."

Solution. We must find the doublet (Ω, P). Now $\Omega = \{1, 2, 3, 4, 5, 6\}$; no one can argue with that. What about P? Well, the die is honest. Why not use Laplace's idea? So for any E,

$$P(E) = \frac{\text{number of points in } E}{6}$$

For example, $P(\text{even number}) = \frac{3}{6} = \frac{1}{2}$, since $E = \{2, 4, 6\}$.

Note that the problem asks us to find *a* probability model, not *the* probability model. This is because we do not really have solid arguments to prove that the model we found is the correct and only model. Can you think of others?

Incidentally, readers should convince themselves that the P we found does meet the definition of P, including the properties (11.2.3).

Example 11.2.2

Find a probability model for the experiment "heights of human beings."

Solution. Ω for this experiment could be

$$\Omega = \{0, 3\} \quad \text{meters}$$

a continuum. What about P? We do not really know how to find it with what we have said so far. Laplace is out—the outcomes are uncountably infinite, and there is no reason to believe that they are equally likely to occur—whatever that means. Relative frequency seems to be applicable, but we do not have the time to take every subset and use the definition with the limit. So forget it for the time being.

We hope that the reader is still with us. Probability theory may have some problems; everything does in its foundational arguments. Even Zeno's paradoxes,[3] put forth more than 2000 years ago, still cause difficulties. Set theory, a fundamental branch of math starts with a definition that may be circular. (Check it. What is a set?) Probability theory is deadly serious stuff, and quite useful, as we will see. We are just pointing out some of the difficulties, hoping that the reader will develop a mature attitude toward it.

Let us go back and do some more math. It is always comforting, as it gives the *illusion of action.*

P has other properties aside from the axiomatic ones that we have presented. They are all, of course, proven by use of the axiomatic ones, and the rules

---

[3]If you do not know what they are, look them up in an encyclopedia. We think you will enjoy them.

of set theory. Two of them are:

(1) $$P(\varnothing) = 0 \qquad (11.2.6)$$

(2) $$P(E_1 \cup E_2) = P(E_1) + P(E_2) - P(E_1 \cap E_2) \qquad (11.2.7)$$

and several more are given in Problem 11.1. We will prove only (1) here, leaving the rest for the reader.

We can start with

$$P(\Omega) = 1$$

which is one of the axiomatic properties. Now from set theory,

$$\Omega = \Omega \cup \varnothing \qquad (11.2.8)$$

so

$$P(\Omega \cup \varnothing) = 1 = P(\Omega) + P(\varnothing) \qquad (11.2.9)$$

the last step from property (11.2.3c), since Ω and $\varnothing$ are mutually exclusive. Thus

$$P(\varnothing) = 1 - 1 = 0$$

and we have proved it.

Now, a few more questions. We know that the probability model of an experiment is (Ω, P), and we appreciate the difficulties in finding it. But let us say that we found it in a given case. What good is it? How do we use it? What problems can it solve?

Once a probability model is found, it represents the experiment in a mathematical way. P is its important part, of course, for it gives us the probability of occurrence of any set we want. In a sense, to appreciate the model you have to change your way of thinking. Up to now, models have been deterministic. If you had a signal $f(t)$, you plugged in a value of t and you got a number for $f(t)$. This is not true with probabilistic models. You can only find the "probability of something happening"—never what exactly will happen—and you have to be satisfied with that. From now on, we will always move in vague steps, steps that are likely to happen with large or small probabilities. It is a new way of thinking, and it takes time getting used to it. Anyway, a better answer to the questions above, one *likely* to satisfy a high percentage of readers, will come by and by, as the material evolves. After all, eventually we want to model noise. If this theory will lead us there, then it has a lot of usefulness.

We close this subsection with an interesting observation. Let us take the case of an experiment with a finite number of points n in its Ω (i.e., $\Omega = \{\omega_1, \ldots, \omega_n\}$). To complete the model you must find $P(E)$ for all subsets of $E \subset \Omega$—at least that is what the probability theory says you must do. In actual practice, though—and this is the important thing—*you only have to find the values of P for each point (outcome) of Ω*. If you have that, you can find the probability of any other subset by using the properties of P. This means that $P(\omega_i)$ for all $\omega_i \in \Omega$ is as good as having the $P(E)$ that the theory specifies.

Example 11.2.3

The experiment is the "throw of a die." Its $\Omega = \{1, 2, 3, 4, 5, 6\}$. Say that we found $P(\omega_i) = \frac{1}{6}$ for all ω_i, that is, an honest die (à la Laplace). Find the probability of the subset E = odd number = $\{1, 3, 5\}$.

Solution. $E = \{1, 2, 3\}$, but it can also be written as $E = \{1 \cup 2 \cup 3\}$, where 1, 2, and 3 are subsets disjoint (mutually exclusive) to each other. Thus by property (11.2.3c),

$$P(E) = \tfrac{1}{6} + \tfrac{1}{6} + \tfrac{1}{6} = \tfrac{1}{2}$$

As we said above, this is a very important observation and it holds as well even when Ω has a countably infinite set of points. You just have to find the P of each point. This idea puts some order in $P(E)$, which was somewhat chaotic before.

11.2.2 Two or More Experiments

Often we are interested in modeling two (or more) experiments, as they occur simultaneously, and we would like to answer questions that pertain to both. In backgammon, for example, we throw *two* dice at the same time, and a great part of the time we are interested in both of their outcomes. In communications and radar we might have noise and speech together somehow (both probabilistic in nature) and might be interested in modeling them *jointly*. What now? The concepts do not change much in this case, just the math.

Let us assume that we have two experiments with individual universes Ω and Z. We will assume that the points in each are of a finite number n and m, respectively. This is the easiest case and the one that will best illustrate our concepts. Thus

$$\Omega = \{\omega_1, \omega_2, \ldots, \omega_n\} \tag{11.2.10}$$

$$Z = \{Z_1, Z_2, \ldots, Z_m\} \tag{11.2.11}$$

Their **joint probability model** is again a doublet, a universe and a P. But now, since you are considering them simultaneously, the universe is $\Omega \times Z$ (the Cartesian product of Ω and Z) with points

$$\Omega \times Z = \{(\omega_1, z_1), \ldots, (\omega_1, z_m), (\omega_2, z_1), \ldots, (\omega_2, z_m), (\omega_n, z_1), \ldots, (\omega_n, z_m)\} \tag{11.2.12}$$

The P of the joint experiment specifies the probabilities of subsets of this new joint sample space. Its form is $P(A, B)$ with $A \subset \Omega$ and $B \subset Z$. It certainly obeys the three axiomatic properties, and every other property derived from them, in this new form, of course. It also suffers from the same problems—how to find it, and so on.

The model above is valid for the case when the points in Ω and Z are not finite in number. Let us stick to the finite (or countably infinite) case, however, as we have something important to remark.

Just as in the single experiment case, if we can find $P(\omega_i, z_j)$ for all i and j, we are in business—it, and the properties of P, can give us the probability of any joint set (A, B). Not only that, but witness also the following. Any point $\omega_i \in \Omega$ can be thought of as a subset of $\Omega \times Z$ given by

$$\omega_i = \{(\omega_i, z_1), (\omega_i, z_2), \ldots, (\omega_i, z_m)\}$$
$$= \{(\omega_i, z_1) \cup (\omega_i, z_2) \cup \ldots \cup (\omega_i, z_m)\} \tag{11.2.13}$$

which are all disjoint, and property (11.2.3c) gives us

$$P(\omega_i) = \sum_{j=1}^{n} P(\omega_i, z_j) \tag{11.2.14a}$$

But as we already remarked earlier, this $P(\omega_i)$ is actually the individual P (the marginal P) of the experiment with universe Ω. Similar reasoning gives us

$$P(z_j) = \sum_{i=1}^{n} P(\omega_i, z_j) \tag{11.2.14b}$$

the P of the other experiment. Therefore, if we know the joint $P(\omega_i, z_j)$, *we can find from it* the **marginal P's** of each experiment and, since we know the individual universes Ω and Z, their **marginal probability models**. A joint model then is more general—it has the marginals in it. The idea does not work the other way, however. In general, having the marginals $(\Omega, P(\omega_i))$ and $(Z, P(z_j))$ does not help in finding the joint model. There is one situation in which it does, and this case will be explained in the next subsection.

Example 11.2.4
The experiment: the throw of two dice in a backgammon game. Assume them honest (à la Laplace). Find the probability of throwing at least one 3. Also find the marginal model of one of the dice.

Solution.   Let us assume that one die is green (G), the other red (R), so that we do not confuse the experiments. The universes are

$$\Omega = \{G_1, G_2, G_3, \ldots, G_6\}$$
$$Z = \{R_1, R_2, R_3, \ldots, R_6\}$$
$$\Omega \times Z = \begin{Bmatrix} (G_1, R_1), \ldots, (G_1, R_6) \\ \cdots\cdots\cdots\cdots\cdots \\ \cdots\cdots\cdots\cdots\cdots \\ (G_6, R_1), \ldots, (G_6, R_6) \end{Bmatrix}$$

Using the Laplacian interpretation of probability (we have got to use something), we assign

$$P(\omega_i, z_j) = \tfrac{1}{36}$$

all equally likely outcomes in $\Omega \times Z$.

So now we have the joint model. Next, the set (event) "at least one 3" is made up of the joint points (G_3, R_j), where R_j means anything for red, as well as (G_i, R_3), where G_i means all green values except G_3, since (G_3, R_3) was included in (G_3, R_j). Thus

$$P\{\text{at least a three}\} = \tfrac{11}{36}$$

Now for the marginal $P(G_i)$. Say that $G_i = G_1$. Using Eq. (11.2.14), we have

$$P(G_1) = \sum_{j=1}^{6} P(G_1, R_j) = \tfrac{6}{36} = \tfrac{1}{6}$$

and in general, for any G_i,

$$P(G_i) = \tfrac{1}{6}$$

that is, if they are simultaneously honest, they are also so individually. This is nice to know.

Everything we have said in this section can generalize to three or more experiments. Take three, with a joint universe $\Omega \times Z \times Y$ of a finite (or countably infinite) number of points, denoted as (ω_i, z_j, y_k). Then the universe and $P(\omega_i, z_j, y_k)$ make up the joint model. From this, if we have it, we can find the two-dimensional joint models as well as the individual (marginal ones), but not vice versa. Thus

$$P(\omega_i, z_j) = \sum_{k} P(\omega_i, z_j, y_k) \tag{11.2.15a}$$

$$P(\omega_i, y_k) = \sum_{j} P(\omega_i, z_j, y_k) \tag{11.2.15b}$$

$$P(z_j, y_k) = \sum_{i} P(\omega_i, z_j, y_k) \tag{11.2.15c}$$

and

$$P(\omega_i) = \sum_{j} \sum_{k} P(\omega_i, z_j, y_k) \tag{11.2.16}$$

and similarly for $P(z_j)$ and $P(y_k)$. Prove them all as an exercise.

Moral: If (big if) we have the n-dimensional probability model of n experiments, we have everything—marginals, two at a time, three at a time, and so on.

You may be wondering at this point what all of this has to do with modeling noise. We must have patience.

11.2.3 Conditional Models: Independence

In all our probability models so far, the assumption is that if an experiment is performed, its Ω is sure to take place. This is because no matter what the outcome (point), the outcome is in Ω, and therefore Ω will take place. The P that we assign to each subset of Ω is assigned under the implication—the *condition*, so to speak—that Ω has occurred. In other words, all probability assignments can be thought of as **conditional probabilities** and even denoted as $P(E|\Omega)$ [rather than $P(E)$], the symbol being read "the probability that E occurred given that Ω occurred," or the **conditional probability of E given Ω**. In view of the known property $P(\Omega) = 1$ and that $E \cap \Omega = E$, we can write

$$P(E) = P(E|\Omega) = \frac{P(E \cap \Omega)}{P(\Omega)} \tag{11.2.17}$$

Equation (11.2.17) is fundamental to an understanding of **conditional probability models**. Let us see how.

Let us assume that we have an experiment with *known* model (Ω, P). Next we assume that the experiment was performed, and we learned (somehow) that some set B *has taken place*. This knowledge can be used to change the original P of the experiment, to assign a new P to all subsets of Ω, *conditioned* on the knowledge that B has occurred.

The new probability measure, denoted $P(E \mid B)$, *can be found* by

$$P(E \mid B) = \frac{P(E \cap B)}{P(B)} \tag{11.2.18}$$

and together with Ω they make up a **conditional probability model** of the experiment, conditional on B. We will not prove Eq. (11.2.18), but the reader can easily see that it is a generalization of Eq. (11.2.17), with B taking the place of Ω, since B is the new *certain* event (set). It is quite clear that each probability model (Ω, P) has many conditional ones (any of its sets can be learned to have occurred), and all these conditional models *can be found* from the original model.

Example 11.2.5
Take the experiment "throw a die" with $\Omega = \{1, 2, \ldots, 6\}$ and $P(\omega_i) = \frac{1}{6}$, that is, with a known model. Now assume that the die was thrown and we learned that the outcome was $B = \{\text{even number}\} = \{2, 4, 6\}$. Find the new conditional model.

Solution. We must find $P(\omega_i \mid B)$. This is sufficient since the experiment has finite number of points n. Using (11.2.18) and the fact that $P(B) = \frac{1}{2}$,

$$P(1 \mid B) = \frac{P(1 \cap B)}{P(B)} = \frac{P(\varnothing)}{\frac{1}{2}} = 0$$

and the same is true for $P(3 \mid B)$ and $P(5 \mid B)$. Now

$$P(2 \mid B) = \frac{P(2 \cap B)}{P(B)} = \frac{P(2)}{\frac{1}{2}} = \frac{1}{3}$$

Similarly,

$$P(4 \mid B) = P(6 \mid B) = \tfrac{1}{3}$$

It should be noted that the new model is different from the old. Knowledge of the occurrence of "even number" changed the probabilities drastically. The original P was based only on the knowledge that Ω had occurred.

Now we are ready for the notion of statistical independence. We say that two sets (events) E_1 and E_2 of a model (Ω, P) are **statistically independent** (S.I.) if and only if (iff)

$$P(E_1 \cap E_2) = P(E_1)P(E_2) \tag{11.2.19}$$

This definition does not make much sense the way it is given, but note the following. Say that Eq. (11.2.19) holds for E_1 and E_2. Now assume that we learn that E_2 has occurred. Then from Eq. (11.2.19),

$$P(E_1 \mid E_2) = \frac{P(E_1 \cap E_2)}{P(E_2)} = \frac{P(E_1)P(E_2)}{P(E_2)} = P(E_1) \tag{11.2.20}$$

and this makes plenty of sense. Even though the definition does not say it directly, it implies that $P(E_1 \mid E_2) = P(E_1)$ and $P(E_2 \mid E_1) = P(E_2)$ (prove it);

that is, knowledge that one of the two has occurred *does not change the probability of the other*. The concept of S.I. is of immense importance for what we have in store in the future. Remember: Eq. (11.2.19) defines it, but Eq. (11.2.20) states its meaning more clearly.

So much for conditional models and S.I. in a single experiment. The ideas can be generalized to a two (or more)-experiment model. Only the notation changes slightly, as we shall see.

Let us assume that we have a two-dimensional model $(\Omega \times Z, P(\omega_i, z_j))$. Let us also assume that sets of Ω are denoted by E_i and sets in Z by F_j. Joint sets are denoted by (E_i, F_j), but it should be noted that in terms of set theory,

$$(E_i, F_j) = E_i \cap F_j \tag{11.2.21}$$

and this is what causes the change in notation that we mentioned above.

In a two-dimensional model we are interested in how the marginal probability model of one of the two experiments, say $P(\omega_i)$, changes with the knowledge that a set from the other experiment took place. In other words, we want to find $P(E_i | F)$ [or $P(\omega_i | F)$]. The formula that provides us with the result is

$$P(E_i | F) = \frac{P(E_2 \cap F)}{P(F)} = \frac{P(E_i, F)}{P(F)} \tag{11.2.22}$$

in view, of course, of Eq. (11.2.21).

Thus we can write that the conditional model $P(\omega_i | F)$ is found from

$$P(\omega_i | F) = \frac{P(\omega_i, F)}{P(F)} \tag{11.2.23}$$

Do not forget that since the joint model is known, so are the original marginals, and $P(F)$ is known.

The concept of S.I. generalizes here immediately. In practice, we are interested in whether sets in Ω are S.I., or not, with sets in Z. So two sets $E \subset \Omega$ and $F \subset Z$ are S.I. iff

$$P(E \cap F) = P(E, F) = P(E)P(F) \tag{11.2.24}$$

and if Eq. (11.2.24) holds, it can be easily shown that

$$P(E | F) = P(E) \tag{11.2.25a}$$

$$P(F | E) = P(F) \tag{11.2.25b}$$

which give the meaning of S.I.

In fact, in a two-experiment model we have a new concept, that of **statistical independence (S.I.) of the two experiments**. This holds iff

$$P(E, F) = P(E)P(F) \tag{11.2.26}$$

for all $E \cap \Omega$ and $F \cap Z$, and since the models are actually specified by $P(\omega_i, z_j)$ if the universes are not uncountably infinite, the above can be written as

$$P(\omega_i, z_j) = P(\omega_i)P(z_j) \qquad \forall \, i, j \tag{11.2.27}$$

Very interesting, indeed. Why? Well, we have stated earlier that if you know the joint model of two experiments, you can find the marginals, *but not*, *in general*, *vice versa*. The reason we said *in general* is because there is the foregoing exception. If the two experiments are S.I., you obviously can find the joint from the marginals, by using Eq. (11.2.27).

Everything that we have said about two experiments generalizes to three or more. We will not bother with it here, not for all the possible expressions, anyway. Let us just state that three experiments are S.I. iff

$$P(\omega_i, z_j, y_k) = P(\omega_i)P(z_j)P(y_k) \qquad (11.2.28)$$

and Eq. (11.2.27) holds in all three possible pairs.

We close this subsection (and the section) with an interesting and useful relationship, usually called **Bayes' theorem**. In a single-experiment model, it has the form

$$P(E_1 | E_2) = \frac{P(E_1)P(E_2 | E_1)}{P(E_2)} \qquad (11.2.29)$$

that is, it gives $P(E_1 | E_2)$ from $P(E_2 | E_1)$ *and* $P(E_1)$, $P(E_2)$. Its proof is easy. Start with Eq. (11.2.18),

$$P(E_1 | E_2) = \frac{P(E_1 \cap E_2)}{P(E_2)} \qquad (11.2.30)$$

and note that the same equation can give you

$$P(E_2 | E_1) = \frac{P(E_1 \cap E_2)}{P(E_1)} \qquad (11.2.31)$$

or

$$P(E_1 \cap E_2) = P(E_2 | E_1)P(E_1) \qquad (11.2.32)$$

Substitution of Eq. (11.2.32) into Eq. (11.2.30) gives the result immediately.

In a two-experiment model, Bayes' theorem has the form

$$P(E | F) = \frac{P(E)P(F | E)}{P(F)} \qquad (11.2.33)$$

where $E \subset \Omega$ and $F \subset Z$. Same thing. For an interesting application of Bayes' theorem, see Problem 11.5.

This completes our discussion of **basic probability theory**, where the outcomes are not necessarily numerical. A little more can be found in the Problems. Let us summarize the key ideas:

1. *Single experiment.* The complete probability model is $(\Omega, P(E))$. It is hard to find. Help can be had from Laplace's ideas or the relative frequency interpretation. If the complete model is known, all conditional models can be found.

2. *Two (or more) experiments.* The complete model is $(\Omega \times Z, P(E, F))$— also hard to find, with help as in 1. If it is known, all marginal and

conditional models can be found. If marginal *and* conditional models *are* known, the joint model can be found. (How?) If only marginals are known, the joint model cannot be found, except in cases of S.I.

11.3 SINGLE RANDOM VARIABLES

The basic probability models that we have covered so far have abstract universes; that is, their outcomes (points) are not necessarily numerical. In this section we wish to restrict attention to experiments with numerical outcomes, and such experiments are called random variables. Random variables are a lot easier to deal with; their model can be a lot simpler than that of the abstract model. They have a lot of other advantages, too, which we will stress as the material evolves, not the least of which is the fact that they will lead us to the subject of stochastic processes. That is really what we are mainly after—we need it to model noises and nondeterministic information signals—and everything we do until then is mostly a vehicle to get us there.

Our main concern will be with the development of the probability model of a random variable (r.v.). Let us assume that we have an abstract probability model $(\Omega, P(E))$. The points in Ω are ω_i and they can be of finite number, countable, or of uncountable infinity. The probability measure $P(E)$ is presumably defined on all subsets $E \subset \Omega$. Now let us consider taking every point $\omega_i \in \Omega$ and by some rule (a mapping, a function) assigning to it a value on the real value. This rule $X(\omega_i)$, or simply X, is called a (real) **random variable**.[4] Once we do that, we end up with a probabilistic phenomenon that has numerical outcomes only. We shall call this new, derived experiment the **random variable X**, and it is our task here to discuss its probability model. In actuality, we have seen random variables before (the throw of a die is one), but in this section we are looking at them officially.

Before we take up the model of a random variable, we should like to make some general comments about it. The notion seems to cause a lot of grief to many engineering students. It is usually the mathematics behind it that causes the grief.

Although engineers call every phenomenon with more than one outcome a random variable, mathematicians insist that you prove that it is. To prove it, one must show that the numerical outcomes are assignments to some abstract outcomes of an abstract probability model. In most engineering problems you

---

[4]Actually, if you want to be mathematically exact, you must include an additional condition in the definition, a condition that deals with σ-algebras. But we promised in the introduction to omit such "epsilonics." It should also be noted that the values of the r.v.'s are real. One can actually define a complex r.v. Z as well, by assigning complex numbers to the outcomes ω_i, so that

$$Z(\omega_i) = X(\omega_i) + jY(\omega_i)$$

where X and Y are real r.v.'s. All our r.v.'s in what follows will be assumed real unless noted otherwise.

can do that—it just takes a little ingenuity. For example, take the throw of a die. As engineers, we claim immediately that we are dealing with a random variable. To satisfy the mathematicians you can say that the experiment "throw a die" is abstract. The die may not have dots on it, but each face may be painted a different color. The random variable comes about by assigning to each color the numbers 1 through 6. That should make the mathematicians happy. Similarly, behind the random variables "heights of people," "weights of people's ears," and so on, are abstract outcomes, the people themselves. Keep this in mind, even though we will not bother to express it from now on.

Next, let us deal with the probability model of the single r.v. X. First, its universe Ω, which is obviously the real line $(-\infty, +\infty)$. Sometimes it may be a portion of it, but we can take it to be the entire line every time, without any problem. If it is not the entire line, it will show up in the other part of the model, anyway. So *all* r.v.'s have the same universe Ω. We can keep that knowledge in the back of our heads and otherwise ignore it from now on. The model of a r.v. will be characterized primarily by the "probability measure" $P(E)$, $E \subset (-\infty, +\infty)$, and it is this part of the model that we shall discuss next. The beauty of the r.v. surfaces at this point, so pay heed.

The probability measure $P(E)$ is a function defined on all subsets of Ω. It is a *set function* that takes every set of Ω and gives it a number from zero to one. Set functions are tough to deal with, and for this reason we have essentially ignored them during our discussion of the abstract models, when Ω is a continuum. If Ω is finite (or denumerable), we were able to boil it down to $P(\omega_i)$, but if Ω is a continuum, specifying $P(E)$ for all sets $E \subset \Omega$ is chaotic; we cannot put a handle on it. In random variables, we can deal with it. In fact, by a clever series of steps we can change it from a set function to a point function, the type we all know how to work with.

Here are the steps. Let us think about the universe Ω, the real line. It goes from left $(-\infty)$ to right $(+\infty)$. Concentrate on sets of the form $(-\infty, x]$, that is, sets that all have $-\infty$ as their left end, and some real value x as their right end. Let us now define the probability measure P on only these sets of the real line. The result, the probability that the r.v. X takes values in sets of the form $(-\infty, x]$, that is,

$$P\{X \in (-\infty, x]\} = P(X \leq x) = F(x) \tag{11.3.1}$$

is called **the cumulative distribution function** or **the probability distribution function** of the r.v. X. It is our first clever step that we promised above. Why is that so clever? For one thing, it looks like a point function; it is even denoted as one, by $F(x)$. It is still a probability of *sets*, but since all the sets have $-\infty$ as the left side, we can remember it, and ignore it. When you write $F(3)$, you actually mean the probability that X is in $(-\infty, 3]$. So $F(x)$ looks very much like, and is, a function of points x, the right end of the sets $(-\infty, x]$. In fact, if you had it, you could even plot it, and this, too, is important, because it provides you with a useful optical image.

Aside from that, $F(x)$ is important because it can actually *replace* the $P(E)$, since it can provide us with the probability of any subset of Ω, as $P(E)$ was meant to do. We will see this in a minute, but first, a word about its properties.

Properties of $F(x)$

(1) $$F(-\infty) = 0, F(+\infty) = 1.$$ (11.3.2a)

(2)   It is a nondecreasing function of x.   (11.3.2b)

(3)   It is continuous on the right, that is,

$F(x^+) = F(x)$ (the notation will be explained below).   (11.3.2c)

Proof. The first part of (1) states that $F(-\infty) = 0$, an obvious statement since the set $(-\infty, -\infty)$ is empty. Then

$$F(+\infty) = P(X \leq +\infty) = P(\text{real line}) = P(\Omega) = 1$$

After all, this $F(x)$ still represents probabilities of sets of Ω, and must obey the same properties as the original $P(E)$.

Property (2) states that

$$F(x_2) \geq F(x_1) \qquad \text{for } x_2 > x_1$$

and this, too, is obvious, since the set $(-\infty, x_1)$ is included in the set $(-\infty, x_2)$.

Now let us look at (3). To understand it, we must explain the notation. We define

$$F(x^+) = \lim_{\epsilon \to 0} F(x + \epsilon) \tag{11.3.3}$$

$$F(x^-) = \lim_{\epsilon \to 0} F(x - \epsilon) \tag{11.3.4}$$

where $\epsilon > 0$. We will not bother to prove it but leave it as an exercise. We just needed the notation, and that is why we introduced it.

With this, we are ready for the important property of $F(x)$, the fact that it can give us (if we have it) the probability of any set in $(-\infty, +\infty)$. What do we want? The usual sets of interest are $[a, b]$, $(a, b]$, $[a, b)$, or single points $\{a_i\}$, multiple points $\{a_1, a_2, a_3\}$, and so on. This $F(x)$ can give us their probability. For example,

$$P\{(a, b]\} = F(b) - F(a) \tag{11.3.5}$$

$$P\{a_1\} = F(a_1) - F(a_1^-) \tag{11.3.6}$$

as a little thought will convince the reader. So we can say with no reservation that the probability model of a random variable *is this* $F(x)$ if you can find it and, of course, the real line as the universe, a fact that we mentioned above.

The second (and last) clever step is to take the derivative of this $F(x)$. Let us assume that it exists, and denote it as

$$p(x) = \frac{dF(x)}{dx} \tag{11.3.7}$$

This $p(x)$, called the **probability density function (p.d.f.)** of the r.v. X, is, in fact, the one that will replace $F(x)$ altogether, and from now on it will represent the probability model of a r.v. X. Let us look next at the justification for this assertion.

From calculus we know that Eq. (11.3.7) leads to

$$F(x) = \int_{-\infty}^{x} p(x)\,dx \qquad (11.3.8)$$

This means that $p(x)$—if you have it—can give you, by means of an integral, $F(x)$ and therefore the model of the r.v. We are done.

So we conclude—by these two clever steps—that the model of the r.v. is this function $p(x)$, together with the real line $(-\infty, +\infty)$ as the universe. This is quite remarkable, indeed. This $p(x)$ is a *point function*, defined on points of the universe—not a set function. Of course, this $p(x)$ *does not itself specify probabilities of sets* [$p(3)$ is not the probability of anything]. Its integral does, however. In fact, in view of (11.3.5), the probability of

$$P\{(a, b]\} = \int_{-\infty}^{b} p(x)\,dx - \int_{-\infty}^{a} p(x)\,dx = \int_{a}^{b} p(x)\,dx \qquad (11.3.9)$$

and similar expressions can give us the probability of any set that we desire. All this could not have been done without the notion of a r.v. Other things, even more important than this, are in store for us as the chapter evolves.

At this point we would like to separate our random variables into two principal types and discuss the form of their $p(x)$—*their model*—separately for each.

Continuous r.v.'s.   These are the type that take values on the entire continuum of the real line or portions of it. The $F(x)$ is continuous here and differentiable (this is not necessary, but we assume it here for simplicity), and therefore $p(x)$ exists and it is nice and smooth. Since $F(x)$ is nondecreasing,

$$p(x) \geq 0 \qquad (11.3.10)$$

and in view of $F(+\infty) = 1$,

$$\int_{-\infty}^{+\infty} p(x)\,dx = 1 \qquad (11.3.11)$$

The two properties above are the ones that identify a $p(x)$ as a p.d.f., and they give rise to problems such as Problems 11.11, 11.12, and others. We repeat that $p(x)$ does not itself specify probabilities. Its integral does. Note, for example, that the probability of a point (say a, for example) is

$$\lim_{\epsilon \to 0} \int_{a-\epsilon}^{a+\epsilon} p(x)\,dx = 0 \qquad (11.3.12)$$

which states that when the r.v. takes values in a continuum (heights of people, for example), the probability of occurrence of any one point (or finite number of points) is zero, even though it *could* occur. This is one of the philosophical oddities that we do not have time to discuss here even though we would like to.

The problem of finding the model of a continuous r.v.—the $p(x)$ or $F(x)$—remains, but it has been made somewhat simpler with this development. R.v.'s have numbers as outcomes, so electronic devices can be used to approximate their $F(x)$, by performing the experiment a great number of times and using the relative frequency interpretation of probability. The p.d.f. is then approximated as the derivative of $F(x)$. A great number of probabilistic phenomena have been studied, and their model—their $p(x)$—has been approximated. This way we have come up with a list of p.d.f.'s which represent certain phenomena—and often we know which. We have included some typical p.d.f.'s in Table 11.3.1 so that we can use them in examples and problems. Let us do one problem right away.

Example 11.3.1

The "life" of a radar device is a r.v. X with $p(x)$, the exponential type,

$$p(x) = e^{-x}u(x) \quad \text{(in hours)}$$

What is the probability that a newly purchased such device will function correctly for at least 3 hours?

Solution. To function correctly for at least 3 hours means that it must "die" after 3 hours. The probability that it will die in the interval $(3, \infty)$ is

$$\int_3^\infty e^{-x}\,dx = e^{-3} \quad \text{(not too big)}$$

The most celebrated of all p.d.f.'s is the one called Gaussian (or normal), which is shaped like a bell (see Table 11.3.1). The reasons for its fame will become apparent later. It is interesting that the most famous one is not really the easiest to use, at least from the point of view of finding probabilities. The integral of the Gaussian cannot be found in closed form, so we have included a table of its values in Appendix B. We might need this table for examples or problems.

Continuous r.v.'s are completely specified by their $p(x)$; it does not hurt to repeat it. In fact, a look at a graph of the $p(x)$ even gives you the range of values (why?), so you do not have to think of it as being the entire real line every time. You could not do anything like this for abstract probability models, could you?

Discrete r.v.'s. These are r.v.'s whose range of values (numerical outcomes) are a finite or denumerable set of points. We saw in the preceding section that we can easily treat such a case by considering its model to be $P(x_i)$, that is, the probabilities of each one of its points. The only difference here is that x_i are *always* numbers in the real line, and the notation changes to $p(x_i)$. This $p(x_i)$ for discrete r.v.'s is called the **probability mass function (p.m.f.)** of the discrete r.v. X.

In view of the above, the development involving $F(x)$ and then its derivative

TABLE 11.3.1   Some Typical p.d.f.'s and p.m.f.'s

Name of p. d. f or p. m. f.	Expression (with Mean and Variance)	Sketch of $p(x)$
Continuous uniform	$p(x) = \begin{cases} \dfrac{1}{b-a} & \text{for } a \leq x \leq b \\ 0 & \text{elsewhere} \end{cases}$ $E(X) = \dfrac{a+b}{2}, \quad \sigma_x^2 = \dfrac{(b-a)^2}{12}$	
Exponential	$p(x) = \dfrac{1}{\theta} e^{-x/\theta} u(x) \quad (\theta > 0)$ $E(X) = \theta, \quad \sigma_x^2 = \theta^2$	
Laplace	$p(x) = \dfrac{a}{2} e^{-a|x-b|} \quad (a > 0)$ $E(X) = b, \quad \sigma_x^2 = 2a^{-2}$	
Gaussian	$p(x) = \dfrac{1}{\sqrt{2\pi\sigma_x^2}} \exp\left[-\dfrac{(x-M)^2}{2\sigma_x^2}\right]$ $E(X) = M, \quad \sigma_x^2 = \sigma_x^2$	
Rayleigh	$p(x) = \dfrac{x}{a^2} e^{-x^2/2a^2} u(x)$ $E(X) = a\sqrt{\pi/2}, \quad \sigma_x^2 = \left(2 - \dfrac{\pi}{2}\right) a^2$	
Weibull	$p(x) = abx^{b-1} e^{-ax^b} u(x) \quad (a, b > 0)$	
Erlang	$p(x) = \dfrac{a^n x^{n-1} e^{-ax}}{(n-1)!} u(x)$ $(a > 0), \ (n = 1, 2, \dots)$ $E(X) = na^{-1}, \quad \sigma_x^2 = na^{-2}$	
Discrete uniform	$p(n) = \dfrac{1}{n}$	
Geometric	$p(x_i) = P(1 - P)^{x_i - 1}$ $(x_i = 1, 2, 3, \dots) \ (0 < P < 1)$ $E(X) = P^{-1}, \quad \sigma_x^2 = (1-P)P^{-2}$	

$p(x)$ is not really needed in discrete r.v.'s. Nevertheless, it still holds, so let us work it out, anyway, for the sake of uniformity.

The $F(x)$ of a discrete r.v. is of staircase form (see Fig. 11.3.1a), and this actually serves as its distinguishing mark in this development. To get the p.d.f.,

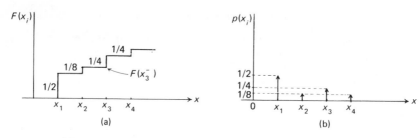

Figure 11.3.1   (a) $F(x)$ of a discrete r.v.; (b) $p(x_i)$ of a discrete r.v.

we must differentiate this $F(x)$. The result is a series of impulses (why?), each with a coefficient equal to $F(x_i) - F(x_i^-)$. Thus for a discrete r.v., the p.d.f. is

$$p(x) = \sum_i P_i \, \delta(x - x_i) \qquad (11.3.13)$$

where P_i are the jumps of the $F(x)$ at the points x_i, and in fact, the probabilities of the points x_i. [Why? The integral of $p(x)$ shows it.] This p.d.f. $p(x)$ is simply the p.m.f. $p(x_i)$, but now it poses pompously with delta functions to please our friends the mathematicians. Some examples of typical p.d.f.'s for discrete r.v.'s are given in the last rows of Table 11.3.1. To see how the table is used, look at the following example.

Example 11.3.2

The output of a quantizer is a discrete r.v. with values $-3, -2, 0, 1, 2$, and 3, and p.d.f.

$$p(x) = \sum_{n=-3}^{3} \tfrac{1}{6} \, \delta(x - n)$$

Find the probability that its next outcome will be (a) greater than 0; (b) greater than 3.

Solution.   (a) This probability is

$$\int_{0^+}^{\infty} p(x)\,dx = \tfrac{1}{6} \int_{0^+}^{3} [\delta(x - 1) + \delta(x - 2) + \delta(x - 3)]\,dx = \tfrac{1}{6} + \tfrac{1}{6} + \tfrac{1}{6} = \tfrac{1}{2}$$

(b) The probability here is zero—no impulses sitting anywhere *after* $x = 3$.

So one has to be a little careful in using the p.d.f. of a discrete r.v. Only the delta functions in the desired interval will give a value.

Continuous and discrete r.v.'s are not the only types of r.v.'s, just those we will be concerned with in this book. There are also **mixed** types, as the reader can verify by checking the Bibliography (Papoulis, 1965, for example).

Quick Summary. The probability model of a continuous r.v. is its p.d.f. $p(x)$ [or its $F(x)$]. The $p(x)$ (if we have it) can give us the probability of any subset that we want via its integral over this set. The same is true for a discrete r.v. The only difference is that in the latter, single- or multiple-point sets *have* probability.

Although the model of r.v.'s is easy to deal with, finding it to begin with remains a problem. It is actually a problem in mathematical statistics, a field concerned with estimating things about r.v.'s. We will have a thing or two to say about this problem as our discussion evolves.

11.4 MULTIPLE RANDOM VARIABLES: CONDITIONAL MODELS

Multiple r.v.'s are, of course, multiple probabilistic experiments whose outputs are numbers. We will not bother with precise definitions here, nor with the formal developments of their joint model, as we did with the single r.v.'s. We go directly to the results.

Take two r.v.'s, X and Y. Their **joint model** is given by a joint p.d.f. $p(x, y)$, where ($p(x, y) \geq 0$ and

$$\int_{-\infty}^{+\infty} \int_{-\infty}^{+\infty} p(x, y)\, dx\, dy = 1 \tag{11.4.1}$$

To find the probability that $X \in (a, b)$ and $Y \in (c, d)$, you start with,

$$P\{X \in [a, b], Y \in [c, d]\} = \int_a^b \int_c^d p(x, y)\, dx\, dy$$

To find the **marginal p.d.f.** of X or Y, we use

$$p(x) = \int_{-\infty}^{+\infty} p(x, y)\, dy \tag{11.4.2a}$$

$$p(y) = \int_{-\infty}^{+\infty} p(x, y)\, dx \tag{11.4.2b}$$

These expressions are analogous to Eqs. (11.2.14) and can easily be proved by the reader. The reader can also define the joint model of a three-dimensional r.v. X, Y, Z, and then find the expressions that give $p(x, y)$, $p(y, z)$, $p(x, z)$, $p(x)$, $p(y)$, and $p(z)$.

Stay with the two r.v.'s X and Y with joint model $p(x, y)$. We define the **conditional model** of X given that $Y = y$ as

$$p(x|y) = \frac{p(x, y)}{p(y)} \tag{11.4.3a}$$

and

$$p(y|x) = \frac{p(x, y)}{p(x)} \tag{11.4.3b}$$

and let the reader ponder why. We also let the reader reshape all the expressions above for discrete r.v.'s using their joint p.m.f.

Let us finally define **statistically independent (S.I.)** r.v.'s X and Y. They are so if

$$p(x, y) = p(x)p(y) \tag{11.4.4}$$

and if they are so,

$$p(x|y) = p(x) \tag{11.4.5a}$$
$$p(y|x) = p(y) \tag{11.4.5b}$$

easily proven expressions that actually tell the story of statistical independence.

Example 11.4.1

Consider two r.v.'s X and Y with joint p.d.f.,

$$p(x, y) = \tfrac{1}{10} e^{-x/2}e^{-y/5}u(x)u(y)$$

Find the probabilities (a) that $X \in (0, 1)$, $Y \in (3, 4)$ and (b) that $X \in (3, 4)$. Also find whether X and Y are S.I.

Solution. (a) This probability is given by

$$\tfrac{1}{10} \int_0^1 \int_3^4 e^{-x/2}e^{-y/5} \, dx \, dy = \text{(reader should supply)}$$

(b) Let us first find the marginal $p(x)$:

$$p(x) = \tfrac{1}{10} \int_0^\infty e^{-x/2}e^{-y/5}u(x) \, dy = \tfrac{1}{2} e^{-x/2}u(x)$$

Therefore,

$$P\{X \in (3, 4)\} = \tfrac{1}{2} \int_3^4 e^{-x/2} \, dx$$

We already have $p(x)$. Now

$$p(y) = \tfrac{1}{10} \int_0^\infty e^{-x/2}e^{-y/5}u(y) \, dx = \tfrac{1}{5} e^{-y/5}$$

and since $p(x, y) = p(x)p(y)$, they are S.I.

The most celebrated two-dimensional model is the one called **bivariate Gaussian**. The joint density of this model is

$$p(x, y) = \frac{1}{2\pi\sigma_x\sigma_y\sqrt{1 - \rho^2}} \exp\left\{-\left[\frac{(x - m_x)^2}{\sigma_x^2} + \frac{(y - m_y)^2}{\sigma_y^2} - 2\rho\frac{(x - m_x)(y - m_y)}{\sigma_x\sigma_y}\right]\bigg/2(1 - \rho)^2\right\} \tag{11.4.6}$$

We will see later what m_x, m_y, σ_x^2, σ_y^2, and ρ are.

If we integrate out y in the model above, we get $p(x)$ and it turns out to be Gaussian, like the one in Table 11.3.1. The same holds true for $p(y)$. Even conditional models of the form $p(x|y)$, $p(y|x)$ have a Gaussian form, as the reader can verify by using expressions (11.4.3) to find them.

Closing Comment. Multiple r.v.'s are very important for our goal—the eventual study of stochastic processes. In fact, such processes are nothing but an infinite number of r.v.'s, as we shall soon see.

11.5 TRANSFORMATION OF RANDOM VARIABLES

As we mentioned in our closing comment above, r.v.'s are eventually going to be used to represent our noises and informational signals. Both of these, as we know, go through the various devices (linear and nonlinear) of our communication and radar systems, and at the output, they are *transformed*. Keeping track of their probability model as they get so transformed is one of the most difficult problems in the area. In fact, it is an unsolved problem, in general. There exist a few methods for tracking it, but nothing that works in all cases.

Let us take a quick look at this problem for single, and then for multiple r.v.'s. Understanding its difficulty will help us appreciate the material on *averages* (coming soon).

Single r.v.'s.   The problem of the transformation of a single r.v. has the following form. Say that we have a r.v. X with known model $p_x(x)$. This X goes through a device, and its values are transformed to

$$Y = g(X) \tag{11.5.1}$$

Let us say that this Y is also a r.v. (Only mathematicians will object to this statement, and will want a proof. Can you find one?) Knowing $p_x(x)$ and $g(X)$, can we find the model of Y, $p_y(y)$?

If the r.v. X is discrete, the problem is trivial; the probabilities of the points in X transfer easily to those of Y, even if $g(X)$ is not a linear transformation. If, for example, X has values $(-1, 0, 1)$ with p.m.f. $p(x_i) = \frac{1}{3}$, then $Y = X^2$ has values $(0, 1)$ with probabilities $p(0) = \frac{1}{3}$ and $p(1) = \frac{2}{3}$. The difficulties exist when X is continuous.

One method to approach this problem involves going back and finding the cummulative distribution function of Y [denote it $F_y(y)$] and then differentiating it to get the $p_y(y)$. We must find $F_y(y)$. By definition,

$$F_y(y) = P(Y \leq y) = P(g(X) \leq y) \tag{11.5.2}$$

since $Y = g(X)$ from Eq. (11.5.1).

To continue with this method we must start putting *restrictions* on the transformation $Y = g(X)$. Let us assume that Y is a one-to-one type of transformation, such as a monotone-increasing differentiable function. This allows us to solve Eq. (11.5.1) for X,

$$X = g^{-1}(y) = h(y) \tag{11.5.3}$$

and with this, rewrite (11.5.2) as

$$F_y(y) = P(X \leq h(y)) \tag{11.5.4}$$

which tells us that

$$F_y(y) = \int_{-\infty}^{h(y)} p_x(x) \, dx \tag{11.5.5}$$

Differentiating both sides we get

$$p_y(y) = p_x(h(y))\frac{dh}{dy} \qquad (11.5.6)$$

kind of messy in terms of notation. We need an example.

Example 11.5.1

Our original r.v. X has an exponential model with $p(x) = e^{-x}u(x)$. Find the model of $Y = 3X$.

Solution. Solve $Y = 3X$ for X. It gives $X = Y/3$, so $h(y) = y/3$. Then $dh/dy = 1/3$ and

$$p_x(h(y)) = e^{-x}u(x)|_{x=h(y)=y/3} = e^{-y/3}u(y/3)$$

Finally,

$$p_y(y) = \frac{1}{3}e^{-y/3}u\left(\frac{y}{3}\right) = \frac{e^{-y/3}}{3}u(y)$$

since $u(y/3) = u(y)$. (Why?)

The whole method hinges on being able to solve $Y = g(X)$ for X. The method generalizes to the following steps, which can be given in the form of a theorem (see Papoulis, 1965, p. 126).

1. Solve $y = g(x)$ (note the lowercase letters) for x in terms of y. If for some y the equation has no real root, forget it; $p_y(y) = 0$. If all the roots $x_1, x_2, \ldots, x_n, \ldots$ are *real*, and *denumerable*, then:
2. The desired p.d.f. $p_y(y)$ is given by

$$p_y(y) = \frac{p_x(x_1)}{|g'(x_1)|} + \frac{p_x(x_2)}{|g'(x_2)|} + \cdots + \frac{p_x(x_n)}{|g'(x_n)|} + \cdots \qquad (11.5.7)$$

where $g'(x) = dg(x)/dx$.

Some examples are needed to illustrate the notation.

Example 11.5.2

Do Example 11.5.1 with the new notation of Eq. (11.5.7).

Solution. The only root is $x_1 = y/3$. So $p_x(x_1) = e^{-y/3}u(y/3)$. Then $|g'(x_1)| = 3$. The result is the same.

Example 11.5.3

Let X be of Gaussian model

$$p_x(x) = \frac{1}{\sqrt{2\pi}}e^{-x^2/2}$$

and let $Y = X^2$. Find $p_y(y)$.

Solution. First we solve $y = g(x) = x^2$ for x. There are two roots,

$$x_1 = +\sqrt{y}$$
$$x_2 = -\sqrt{y}$$

Next we note that

$$|g'(x)| = |2x|$$

and thus $|g'(x_1)| = 2\sqrt{y}$ and $|g'(x_2)| = 2\sqrt{y}$. Finally,

$$p_y(y) = \frac{1}{\sqrt{2\pi}\, 2\sqrt{y}}(e^{-y/2} + e^{-y/2}) = \frac{1}{\sqrt{2\pi y}}e^{-y/2} \qquad \text{for } y \geq 0$$

Note that $p_y(y) = 0$ for $y < 0$ since $Y = X^2$. The resultant p.d.f. is called the **chi-square p.d.f.**

Multiple r.v.'s.   We illustrate the transformations problem here with a two-dimensional model. The generalization to multiple models is immediate.

Assume that we have two r.v.'s X and Y, with known joint p.d.f. $p_{xy}(x, y)$. Find the joint p.d.f. $p_{uv}(u, v)$ if

$$U = g_1(X, Y) \tag{11.5.8a}$$

$$V = g_2(X, Y) \tag{11.5.8b}$$

The steps given in the preceding subsection generalize here as well, but now we have to solve a system of equations (the roots must be *real* and *denumerable*), and instead of the simple derivative we will have the Jacobian of the transformation.

Let the roots be $(x_1, y_1), (x_2, y_2), \ldots, (x_n, y_n), \ldots$. Then

$$p_{uv}(u, v) = \frac{p_{xy}(x_1, y_1)}{|J_{xy}(x_1, y_1)|} + \cdots + \frac{p_{xy}(x_n, y_n)}{|J_{xy}(x_n, y_n)|} + \cdots \tag{11.5.9}$$

where,

$$J_{xy} = \begin{vmatrix} \dfrac{\partial u}{\partial x} & \dfrac{\partial u}{\partial y} \\[2mm] \dfrac{\partial v}{\partial x} & \dfrac{\partial v}{\partial y} \end{vmatrix} \tag{11.5.10}$$

We will not have occasion to use the above, so we leave its application to the exercises. There is, however, an interesting case that must be mentioned because it leads to an important result.

Consider two r.v.'s X and Y which are statistically independent, with joint model $p_{xy}(x, y) = p_x(x)p_y(y)$. Then consider the r.v.

$$U = X + Y \tag{11.5.11}$$

How do we find the p.d.f. of $p_u(u)$?

To make it conform to our steps above, we add an auxiliary equation $V = X$, or $V = Y$. We then find $p_{uv}(u, v)$, and integrate out V to get $p_u(u)$. The interesting part is that if you do all that, the result is

$$p_u(u) = \int_{-\infty}^{+\infty} p_x(u - y)p_y(y)\, dy \tag{11.5.12}$$

which is the convolution of the two marginals $p_x(x)$ and $p_y(y)$. Prove this as an exercise. It is a very useful result, and can easily be extended to a sum of more than two S.I. r.v.'s.

11.6 EXPECTED VALUES OR AVERAGES

Everybody has a pretty good idea what the word *average* means—you find it in newspapers, you hear it in bars. Even kids know it—it is a big word in sports. So let us go directly to the heart of things. The only comment worth making at this point is that *averages can be defined only with random variables.* Abstract models do not have any.

11.6.1 Single R.V.

Let us assume that a group of students take an exam and the outcomes are $(6, 6, 6, 9, 3, 3, 1)$.

Then the average grade of this exam, as everybody knows, will be

$$\text{average} = \frac{6 + 6 + 6 + 9 + 3 + 3 + 1}{7}$$

or, rewriting it,

$$\text{average} = 6(\tfrac{3}{7}) + 9(\tfrac{1}{7}) + 3(\tfrac{2}{7}) + 1(\tfrac{1}{7})$$

Considering now the number 6, 9, 3, 1 as the outcomes of the r.v. "results of the exam," the terms in parentheses represent the relative frequences of these outcomes, that is, (approximately) their probabilities. Thus

$$\text{average} = 6P(6) + 9P(9) + 3P(3) + 1P(1) \tag{11.6.1}$$

The following definition will thus not be surprising.

Average, mean, or expected value of a r.v.   This is denoted by $E(X)$ and is given by

$$E(X) = \int_{-\infty}^{+\infty} xp(x)\,dx \qquad \text{for } X \text{ continuous} \tag{11.6.2}$$

$$E(X) = \int_{-\infty}^{+\infty} xp(x)\,dx = \int_{-\infty}^{+\infty} x \sum_i P_i\,\delta(x - x_i)\,dx$$
$$= \sum_i x_i P_i \qquad \text{if } X \text{ is discrete} \tag{11.6.3}$$

where $p(x)$ is the p.d.f. of the r.v. Equations (11.6.3) and (11.6.1) look very much alike, and (11.6.2) is their generalization to the continuous case.

The mean of a r.v. $E(X)$ is a very important quantity. Usually (not always), it is the value most likely to occur in a single happening of the r.v. It is often the value around which there is the interval of highest probability of the r.v.

In fact, if the p.d.f. of the r.v. is evenly symmetrical around a point $x = m$, this m is the mean. The mean (*one* number) carries a lot of meaning for a r.v. that has infinite outcomes.

Example 11.6.1

Find $E(X)$ for a r.v. with model (a) exponential; (b) Gaussian.

Solution. We use Eq. (11.6.2) and the p.d.f.'s of Table 11.3.1.
 (a) Exponential:

$$E(X) = \int_0^\infty x \frac{1}{\theta} e^{-x/\theta} \, dx = \theta \tag{11.6.4}$$

Very interesting. The parameter θ of this model turns out to be the mean. Thus, if you know that a r.v. has this model, find the mean and you have it precisely.
 (b) Gaussian: It is difficult to do, but

$$E(X) = \int_{-\infty}^{+\infty} x \frac{e^{-(x-m)^2/2\sigma_x^2}}{\sqrt{2\pi\sigma_x^2}} \, dx = m \tag{11.6.5}$$

So for this model, which looks like a bell, the parameter m is the mean, right smack in its middle at the peak. This is a good thing to know.

Now we get fancier, generalize the idea, and define what are called the **moments** of a r.v. X with p.d.f. $p(x)$.

Nth moment of a r.v. These are denoted as $E(X^n)$ $(n = 1, 2, \ldots)$ and are given by

$$E(X^n) = \begin{cases} \displaystyle\int_{-\infty}^{+\infty} x^n p(x) \, dx & \text{for } X \text{ continuous} \quad (11.6.6) \\ \displaystyle\sum_i x_i^n P_i & \text{for } X \text{ discrete} \quad (11.6.7) \end{cases}$$

Note that the notation $E(\cdot)$ is becoming standard as expressing an integral of whatever is in the argument times the p.d.f. It functions like an operator.
 Let us keep going here, define one more thing, and return to meanings later.

Nth central moments of a r.v. These are denoted as $E\{(X - M_x)^n\}$ $(n = 1, 2, \ldots)$ and are given by

$$E\{(X - M_x)^n\} = \begin{cases} \displaystyle\int_{-\infty}^{+\infty} (x - M_x)^n p(x) \, dx, & X \text{ continuous} \quad (11.6.8) \\ \displaystyle\sum_i (x_i - M_x)^n p_i, & X \text{ discrete} \quad (11.6.9) \end{cases}$$

where $M_x = E\{X\}$, the mean. These are similar to the regular moments, except that the average M_x is first subtracted from the r.v. X. If $M_x = 0$, they are exactly the same. Also, if you have one group of moments, you can find the other—there exist relationships that relate them. We shall see one of them in a second.

Aside from the mean $E(X)$, the other important moment of a r.v. for engineering applications is the second central moment ($n = 2$). It is called the *variance*, denoted by σ_x^2, and is

$$\sigma_x^2 = E\{(X - M_x)^2\} = \int_{-\infty}^{+\infty} (x - M_x)^2 p(x) \, dx \qquad (11.6.10)$$

for a continuous r.v. A similar expression exists for discrete r.v.'s, but we will let the reader find it. From now on, we will limit our discussion to continuous r.v.'s and assume that the reader can generalize to discrete r.v.'s.

If we crank out the integral of Eq. (11.6.10) and note Eq. (11.6.6) for $n = 2$, we get

$$\sigma_x^2 = E(X^2) - M_x^2 \qquad (11.6.11)$$

This is the expression that relates σ_x^2 (a central moment) to the second moment $E(X^2)$. If $M_x = 0$, $\sigma_x^2 = E(X^2)$. Expressions relating higher moments in the two groups will be omitted.

What is the meaning of the variance σ_x^2 of a r.v. X? It turns out that this number is related to the *width* of the p.d.f. *Big variance denotes a spread-out p.d.f., small variance a narrow p.d.f.*

Example 11.6.2

Find the variance of the two models in Example 11.6.1.

Solution.   (a) Exponential: We know that $M_x = \theta$:

$$\sigma_x^2 = \int_0^\infty (x - \theta)^2 \frac{1}{\theta} e^{-x/\theta} \, dx = \theta^2 \qquad (11.6.12)$$

The reader already knows that the exponential function is narrow if θ (or $\theta^2 = \sigma_x^2$) is small and wide if θ is large. The meaning checks.

(b) Gaussian: We know that $M_x = m$. It is difficult to do, but

$$\sigma_x^2 = \int_{-\infty}^{+\infty} \frac{(x - m)^2}{\sqrt{2\pi\sigma_x^2}} e^{-(x-m)^2/2\sigma_x^2} \, dx = \sigma_x^2 \qquad (11.6.13)$$

Aha! So the two parameters of the Gaussian model m, σ_x^2, are its mean and variance, respectively. So, if you know that a r.v. has a Gaussian model, and you can find these two things, you have the model precisely. This is another good thing to know.

The meaning of the variance checks here as well. The Gaussian bell shape is narrow for σ_x^2 small, and vice versa.

Now that the expressions are given, it is time for our general discussion. We need a spherical view of the situation.

The mean means what it means, and so does the variance. Together they mean even more, although everything is approximate. If you know that the r.v. X has a mean $M_x = 5$ and a variance of $\sigma_x^2 = 0.1$, you visualize a p.d.f. that is very narrow around 5. Any value you observe is surely around 5. If, on the other hand, $\sigma_x^2 = 100$, the p.d.f. is too spread out. You can observe a value, and it can be $90 \gg 5$. In other words, these two numbers give you an approximate mental image of the curve of the p.d.f. Since the p.d.f. is the **complete model** of a r.v. (the best you can do with probability theory), these two numbers (M_x, σ_x^2), which represent it approximately, can be thought of as a **partial (approximate) model** of a r.v. X. This idea can even be backed by mathematical rigor. It can be shown (Papoulis, 1965, p. 158) that if you take the p.d.f. $p(x)$, find its FT $P(\omega)$, and expand it on a Taylor series (assume that it is expandable), the coefficients of the expansion are completely determined by the moments—all of the moments. Thus the moments completely determine $P(\omega)$, or by uniqueness the p.d.f. $p(x)$. So, *two* moments determine it approximately. That two numbers determine (even approximately) the model of a r.v. is quite remarkable, indeed. Of course, such a model is not unique; there are many p.d.f.'s with the same M_x and σ_x^2. But this does not matter, as we will appreciate soon.

Now you say: "Wait a minute. To find M_x and σ_x^2, I need to know $p(x)$; that's what expressions (11.6.2) and (11.6.10) certainly state. If I have $p(x)$—the *complete* model—why bother with a *partial* one (M_x, σ_x^2)?"

A very good question. There are two very good answers to it, and we give one of them below. The other will come in the subsection that follows, and it is even more important, in our opinion.

If we know $p(x)$, finding M_x and σ_x^2 (and even all the moments) is just a trivial problem (at least mathematically), without much point *as far as we can tell right now*. But often we do not have the p.d.f. As we have already explained, this is one of the biggest problems in probability theory—how to find the p.d.f. of a virgin r.v. X. And here is the remarkable thing about the situation. You can find the M_x and σ_x^2 (approximately, of course), even if you do not have the p.d.f. How and when? Well, these are questions that are studied in mathematical statistics. There is no room to go into them here. Generally speaking, the mean can be found as we explained at the start of the subsection: Add a set of observations and divide by their number; and the second moment [from which σ_x^2 can be found from Eq. (11.6.11)] can be found by doing the same with the "squares" of the observations. If you can, then, find M_x and σ_x^2 without knowing $p(x)$, isn't this a satisfactory answer of why they are important? You will not have a *complete* model, but the partial one you *will* have (M_x, σ_x^2) is good enough for a lot of good results, as we shall see.

Conclusion. The two moments (M_x, σ_x^2) make up a *partial* model of a r.v. X. If you have $p(x)$, you can find them trivially. But you can also find them in practice, without having $p(x)$, and then the partial model is quite important.

It would not be right to close this subsection without mentioning another point. We stated above that the variance σ_x^2 (or its square root, called the **standard deviation**) is a measure of the *width* of the curve of the p.d.f. $p(x)$. But the word *width* reminds us of time duration or frequency bandwidth, concepts defined in Section 2.9. Can expression (11.6.10) be generalized to any function or signal, and then be used as the time duration of frequency bandwidth? Yes, it can, and we mentioned this possibility in Section 2.9. We will see these expressions later, in Chapter 14.

11.6.2 Functions of a Single R.V.

The problem of interest here is as follows. Let us assume that we have a r.v. X with known model $p_x(x)$. Now X is transformed to a new random variable $Y = g(X)$. What are the moments of Y, and more specifically, the two of them that interest us (M_y, σ_y^2)?

One answer is that you can find them by finding the p.d.f. $p_y(y)$ first, and then using their definition. This is okay, but finding $p_y(y)$ is not always possible, and even if it is, it is complex—something the reader must appreciate if he or she tried some of the problems associated with Section 11.5. Here is now another way to do it.

It can be shown (see Papoulis, 1964, p. 142) that the average of any function $g(X)$ of X can be given by

$$E\{g(X)\} = \int_{-\infty}^{+\infty} g(x)p_x(x)\, dx \qquad (11.6.14)$$

an expression not involving $p_y(y)$. In other words, what we are saying is that

$$M_y = E(Y) = \int_{-\infty}^{+\infty} yp_y(y)\, dy = \int_{-\infty}^{+\infty} g(x)p_x(x)\, dx \qquad (11.6.15)$$

an expression so important that no matter what we say about it, we cannot give it enough emphasis. We will not prove it because we do not have the space. Some people give it as a definition that requires no proof. Either way, its importance is immense. Why?

To begin with, you can find the mean of any function Y of X *without finding the p.d.f. of Y*. And not only the mean, but also the variance and any moment you like. The variance of Y above, for example, is

$$\sigma_y^2 = E\{(g(x) - M_y)^2\} \qquad (11.6.16)$$

and the right-hand side is the mean of a new function of X. Using (11.6.14), we get

$$\sigma_y^2 = \int_{-\infty}^{+\infty} [g(x) - M_y]^2 p_x(x)\, dx \qquad (11.6.17)$$

an expression not involving $p_y(y)$, either.

The information above means the following. If we start out with a r.v. X of known model $p(x)$, we also have its *partial* model (M_x, σ_x^2). Now if this X goes through a series of transformations (square it, take its log, etc.), we can easily find the *partial model* at each step (mean and variance) *without finding p.d.f.'s*. In other words, we can keep track of changes in the partial model quite easily, by cranking out integrals. Of course, *partial* models are not unique. But they are, often, good enough. Wait and you will be convinced. Now, do you see why partial models are so important?

Example 11.6.3

A r.v. X with $p(x) = e^{-x}u(x)$ (exponential with mean $= 1$) is subjected to a series of transformations as shown in Fig. 11.6.1. Find the partial model at each point.

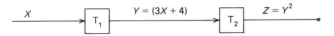

$$X \qquad \boxed{T_1} \qquad Y = (3X + 4) \qquad \boxed{T_2} \qquad Z = Y^2$$

Figure 11.6.1   Keeping track of the partial model.

Solution.   At the input we already know that $E(X) = M_x = 1$ and $\sigma_x^2 = 1$.
Now we must find the partial model of Y, M_y, and σ_y^2.

$$M_y = \int_0^\infty (3x + 4)e^{-x}\, dx$$

$$\sigma_y^2 = \int_0^\infty (3x + 4 - M_y)^2 e^{-x}\, dx$$

This is just busywork. The reader can easily write down the expression for the partial model of Z. Keeping track of partial models is an exercise in calculus.

Expression (11.6.14) leads to a number of interesting results, often called **properties of expectation**. Three of them are:

(1) $$E(C) = C \qquad \text{when } C \text{ is a constant} \qquad\qquad (11.6.18a)$$

(2) $$\sigma_c^2 = 0 \qquad \text{when } C \text{ is a constant} \qquad\qquad (11.6.18b)$$

(3) $$E\{g_1(x) + g_2(x) + \cdots + g_n(x)\} = E\{g_1(x)\} + \cdots + E\{g_n(x)\} \qquad (11.6.18c)$$

All can be proved trivially by applying Eq. (11.6.14). The first two basically tell us that a constant C which has as a complete model $p(x) = \delta(x - C)$ has as a partial model $(C, 0)$. Constants, in other words, are a trivial case of a r.v., so probability theory is a general theory that includes deterministic quantities as a subcase.

Even moments (both groups) are averages of functions of X, and as such, their defining equations can be viewed as direct results of Eq. (11.6.4). This means that averages of functions of a r.v. X have meaning—they carry information about X. This brings us to the following.

Consider a discrete r.v. X with p.d.f.

$$p(x) = \sum_i P_i\, \delta(x - x_i)$$

where the P_i, as we know, are the probabilities of the outcomes x_i. Now consider the function

$$I(X) = \log \frac{1}{P_i} \qquad\qquad (11.6.19)$$

Its average, denoted by $H(X)$, can easily be found by direct application of Eq.

(11.6.14). It is

$$H(X) = E\{I(X)\} = \int_{-\infty}^{+\infty} \log\frac{1}{P_i} \sum_i P_i \, \delta(x - x_i) \, dx$$

or

$$H(X) = \sum_i P_i \log\frac{1}{P_i} \tag{11.6.20}$$

and it is called the **(informational) entropy** of the discrete r.v. X—a mighty famous average. It can be argued convincingly that this $H(X)$ is a number that represents the average amount of *information* that we obtain by observing the outcomes of the discrete r.v. X, or the average amount of the *uncertainty* that we have about the outcomes, if we are not observing them. This simple average is, in fact, the whole foundation of modern information theory and its offshoot, coding theory. Since entropy can be defined even for discrete *abstract* probability models (ponder this carefully and try to explain why), it has crept into biology, social sciences, psychology—you name it. It has even done a "reverse play," so to speak, by turning around and helping probability theory (whose product it is) with its problem of estimating the p.d.f. of a discrete r.v. X—in the form of the **maximum entropy principle**. If there is a panacea in this world, this could be it (see Problem 11.29 for more on this).

An expression similar to Eq. (11.6.20) exists for continuous r.v.'s as well but does not have the meaning or the usefulness of the discrete case, so we ignore it.

Conclusion. Partial models (mean, variance) of functions of a r.v. X are easy to find, a lot easier than complete models (p.d.f.'s).

11.6.3 Two or More R.V.'s

Everything we said about moments for a single r.v. generalizes to multiple models as well.

Take two continuous r.v.'s X and Y with joint model $p_{xy}(x, y)$.

1. *Joint moments.* These are denoted as $E\{X^n Y^m\}$ $(n, m = 0, 1, 2, \ldots)$ and are given by

$$E(X^n Y^m) = \int_{-\infty}^{+\infty} \int_{-\infty}^{+\infty} x^n y^m p_{xy}(x, y) \, dx \, dy \tag{11.6.21a}$$

If n or $m = 0$, the expression above reduces to the moments of the marginal models, as the reader should verify.

Of all the moments above, only *one* is useful to us here, the joint moment $E(XY)$, called the **correlation** of X and Y. Keep it in mind.

2. *Joint central moments.* These are denoted and defined as

$$E\{(X - M_x)^n (Y - M_y)^m\} = \int_{-\infty}^{+\infty} \int_{-\infty}^{+\infty} (x - M_x)^n (y - M_y)^m p_{xy}(x, y) \, dx \, dy$$

$$\tag{11.6.21b}$$

with meanings similar to those above for n or $m = 0$. The only one of interest is for $m = n = 1$. It is denoted by

$$\text{cov}\,[X, Y] = E\{(X - M_x)(Y - M_y)\}$$

$$= \int_{-\infty}^{+\infty} \int_{-\infty}^{+\infty} (x - M_x)(y - M_y)p_{xy}(x, y)\, dx\, dy \quad (11.6.22)$$

and is called the **covariance** of X and Y. Factoring out the right-hand side of Eq. (11.6.22) and using property (11.6.18c), we get

$$\text{cov}\,[X, Y] = E(XY) - M_x M_y \qquad (11.6.23)$$

as the expression that relates the covariance with the correlation $E(XY)$ and the two means. If either (or both) of the means is zero, the two moments are equal.

What does the covariance mean? It turns out that it gives us a measure of how dependent X is on Y (or vice versa), in an average sense. Now we know what it means when we say that two r.v.'s are statistically independent (S.I.). But S.I. is a one-shot type of thing. If they *are* dependent, how dependent are they? The covariance has something to say about how much they depend on each other *linearly*, that is, how close we are to saying that $Y = aX + b$.

Let us expand on this a bit. We can take the covariance and divide it by its maximum possible value so that its maximum becomes unity. To find its maximum we can use the Cauchy–Schwartz inequality (where did we first see that?) on the double integral of Eq. (11.6.22). This inequality states (do it) that

$$\text{cov}\,[X, Y] \le \sigma_x \sigma_y \qquad (\sigma_x^2, \sigma_y^2 \text{ are the variances}) \qquad (11.6.24)$$

so our new quantity, the **correlation coefficient** ρ, is given by

$$\rho = \frac{\text{cov}\,[X, Y]}{\sigma_x \sigma_y} \qquad (11.6.25)$$

Let us check whether this means what we suggested above, by looking at some typical extreme cases. Let us first take the case when X and Y are S.I.—no dependence at all, linear or any other kind. In this case,

$$p_{xy}(x, y) = p_x(x)p_y(y) \qquad (11.6.26)$$

and the double integral of (11.6.22) splits up into two integrals, with the result equal to zero. When $\rho = 0$, or more officially, when

$$E(XY) = E(X)E(Y) \qquad (11.6.27)$$

(which causes cov $[X, Y]$ and ρ to equal zero), the two r.v.'s are called **uncorrelated**. We just discovered that *when X, Y are S.I., they are also uncorrelated*—but not vice versa, of course, except in some cases (see the example below).

Now take the case when $Y = AX$ (or even $Y = AX + B$). We will not actually work it out (let it be an exercise), but the result is

$$\rho = \pm 1 \qquad (11.6.28)$$

depending on whether A is positive or negative. So complete correlation appears to happen when X and Y are related linearly, as we suggested above. Now if X and Y are dependent but not linearly, the ρ takes values in $[-1, +1]$ and gives us an idea of how linear this dependence is. Before we go on with this discussion, we had best look at a brief example.

Example 11.6.4

Consider two r.v.'s X and Y whose joint model is the bivariate Gaussian given by Eq. (11.4.6). Find the correlation coefficient of X and Y.

Solution. This solution is very messy to do, so we will just give the result. Use of expression (11.6.25) in conjuction with (11.6.22) for its numerator yields

$$\rho = \rho$$

as expected. Thus the bivariate Gaussian is completely defined if the marginal means and variances and ρ are known. Note also that if $\rho = 0$, X and Y are S.I. This is one of the exceptions mentioned above.

We apologize for this interruption and return to our discussion. We have concluded so far that loosely speaking, the cov (X, Y) gives us an idea of the average linear dependence of X and Y—and thus it is useful. Not only that, but just as it happened with the moments of the single r.v., the cov (X, Y) or $E(XY)$ can be estimated in practice from observations, even when the complete model (the joint p.d.f.) is not known. So this business of partial models of r.v.'s based on moments generalizes to the case of two r.v.'s as follows.

If we have two r.v.'s X and Y, then the two marginal means (M_x, M_y), the two variances (σ_x^2, σ_y^2), and the cov $[X, Y]$ make up their **partial model**. All of these quantities can be estimated in practice, even if the joint p.d.f. is not completely known.

This two-dimensional partial model is often condensed in matrix notation as the **vector of the means**

$$\bar{M} = \begin{bmatrix} M_x \\ M_y \end{bmatrix} \tag{11.6.29}$$

and the **covariance matrix**

$$\text{cov}\,[X, Y] = \begin{bmatrix} \text{cov}\,[X, X] & \text{cov}\,[X, Y] \\ \text{cov}\,[Y, X] & \text{cov}\,[Y, Y] \end{bmatrix} \tag{11.6.30}$$

Note that the covariance matrix includes in it the variances σ_x^2 and σ_y^2 on the diagonal. Anybody can show that

$$\text{cov}\,[X, X] = \sigma_x^2, \qquad \text{cov}\,[Y, Y] = \sigma_y^2 \tag{11.6.31}$$

Note also that if $M_x = 0 = M_y$, then $\bar{M} = \bar{0}$ and the covariance matrix becomes a **correlation matrix** with diagonal elements $\sigma_x^2 = E(X^2)$ and $\sigma_y^2 = E(Y^2)$ and $E(XY)$ and $E(YX)$ in place of the covariances. The matrix is, of course, symmetric.

How do all these things generalize to n random variables $X_1, X_2, \ldots, X_n$?

We will just state the results dogmatically and let the reader ponder the issue of meanings.

There are no new moments to be defined here that have much usefulness. Just the old ones—the means and variances of each r.v., and the various covariances (or correlations) involving all combinations of pairs of r.v.'s. The **partial** n-**dimensional model** is, then, the **vector of the means**

$$\bar{M} = \begin{bmatrix} M_1 = E(X_1) \\ M_2 = E(X_2) \\ \cdot \qquad \cdot \\ \cdot \qquad \cdot \\ \cdot \qquad \cdot \\ M_n = E(X_n) \end{bmatrix} \qquad (11.6.32)$$

and the **covariance** (or **correlation** if $\bar{M} = \bar{0}$) matrix

$$\text{cov}(X_i, X_j) = \begin{bmatrix} \text{cov}(X_1, X_1) & \cdots & \text{cov}(X_1, X_j) \\ \cdot & & \cdot \\ \cdot & & \cdot \\ \cdot & & \cdot \\ \text{cov}(X_i, X_1) & \cdots & \text{cov}(X_i, X_j) \end{bmatrix} \qquad (11.6.33)$$

Do not underestimate these "partial" models; they are the key of our future endeavors.

11.6.4 Functions of Two or More R.V.'s

Let us go directly to the n-dimensional case, as it is fairly easy. We assume that we have n r.v.'s $X_1, X_2, \ldots, X_n$ with known p.d.f. $p(x_1, x_2, \ldots, x_n)$, and therefore known partial model $\bar{M}$ and $\text{cov}(X_i, X_j)$. How do we find the partial model of a function

$$y = g(X_1, X_2, \ldots, X_n) \qquad (11.6.34)$$

or of up to n such functions? We already know how (and when) we can find the p.d.f. of y (or the joint p.d.f. of many Y's), and it is no easy matter at all. The key expression in the issue is

$$E\{g(x_1, x_2, \ldots, x_n)\}$$
$$= \int_{-\infty}^{+\infty} \cdots \int_{-\infty}^{+\infty} g(x_1, x_2, \ldots, x_n) p(x_1, x_2, \ldots, x_n) \, dx_1 \, dx_2 \cdots dx_n \qquad (11.6.35)$$

a generalization of Eq. (11.6.14) to the n-dimensional case. With this you can find anything you want, as long as it is an average, and partial models are precisely that.

The mean of Y, M_y is found by direct application of expression (11.6.35); that is, it is a matter of calculus. The variance of Y, σ_y^2, is

$$\sigma_y^2 = E\{(Y - M_y)^2\} = E\{(g(x_1, \ldots, x_n) - M_y)^2\} \qquad (11.6.36)$$

another average of a function of $X_1, X_2, \ldots, X_n$, so calculus again. So the partial model of Y is easily found, as easily as we can integrate out expressions of the form (11.6.35).

When there are many Y's, the situation changes very little. The means and variances of each Y are found as before. As for the covariance matrix, it is again a new function of the old X's, and each one of its elements can be found by using Eq. (11.6.35). So partial models of transformations of r.v.'s are a lot easier to keep track of than complete models, a fact that will be of use in later discussion on stochastic processes and other topics.

Example 11.6.5

Consider the r.v.

$$Y = \frac{1}{N} \sum_{i=1}^{N} X_i \tag{11.6.37}$$

where X_i ($i = 1, \ldots, N$) are statistically independent r.v.'s each with mean M_x and variance σ_x^2 (equal means and variances). What is the partial model of Y?

Solution.   We need M_y and σ_y^2. Note that we do not even have the joint p.d.f. of the n X's, just their marginal partial models. The covariances, of course, are zero, since they are S.I. and therefore uncorrelated. Since

$$p(x_1, x_2, \ldots, x_n) = p(x_1)p(x_2) \cdots p(x_n)$$

equation (11.6.35) boils down to (do it)

$$E(Y) = \frac{1}{N}(E(X_1) + E(X_2) + \cdots + E(X_n)) = \frac{N}{N} M_x = M_x \tag{11.6.38}$$

Now σ_y^2. By definition,

$$\sigma_y^2 = E\left\{(Y - M_x)^2\right\} = E\left\{\left(\frac{1}{N} \sum X_i - M_x\right)^2\right\}$$

again an average of a function of the X_i's. Formula (11.6.35) is very useful, again, but is a bit more involved this time. The result is (do it as an exercise)

$$\sigma_y^2 = \frac{\sigma_x^2}{N} \tag{11.6.39}$$

So the partial model of Y is $(M_x, \sigma_x^2/N)$, a very useful result in statistics. Note that Y tends to become a constant [see Eqs. (11.6.18)] as $N \longrightarrow \infty$, since its variance approaches zero. The function Y of this example is used to estimate the mean of a r.v. X and is called the **sample mean**. The X_i's represent observations (outcomes of X) in this application.

The reader should note that a lot of these types of problems can be solved without the use of integrals, just by working the expectation E as an operator. The mean $E(Y)$ was found this way, and so was the variance σ_y^2 above. Of course, the integrals are behind the notation, anyway.

There is one special function of two r.v.'s whose average carries a lot of significance in the field of communications—significance nearly as great as that of entropy in the single r.v. case. This function of two r.v.'s X and Y (continuous or discrete here) with joint p.d.f. $p_{xy}(x, y)$ is

$$Z = \log \frac{p_{xy}(x, y)}{p_x(x)p_y(y)} \tag{11.6.40}$$

and its average, denoted by $I(X, Y)$, is by Eq. (11.6.35),

$$I(X, Y) = \int_{-\infty}^{+\infty} \int_{-\infty}^{+\infty} p_{xy}(x, y) \log \frac{p_{xy}(x, y)}{p_x(x)p_y(y)} \, dx \, dy \qquad (11.6.41)$$

This is called the **mutual information of X and Y**. Its meaning turns out to be "the amount of information that we get about X by observing Y, or vice versa." It, too, has a lot of applications in communication theory and other fields, particularly since it holds (with the same meaning) for both discrete and continuous r.v. models, and even discrete abstract probability models.

Incidentally, entropy can also be defined for two or more discrete r.v.'s. For more on entropy, mutual information, their relationships, and so on, see any book on information theory, such as Gallager (1968) or Ingels (1971) (see also Problem 11.29).

Conclusion. Partial models of functions of n r.v.'s can be easily found by the use of one formula, Eq. (11.6.35).

11.6.5 Conclusion for Section 11.6

Averages of r.v.'s (mean, variance, covariance) carry partial information about them and can be thought of as "partial models." They are easily estimated in practice (a lot easier than are complete models, i.e., p.d.f.'s), and are easily traced through transformations of the r.v.'s. This is one of the principal reasons why r.v.'s are so important. Abstract models have no averages. The reason that entropy and mutual information can be defined on discrete abstract models is because they are averages of the *r.v.*'s which are defined by Eq. (11.6.19) and Eq. (11.6.40) respectively. (Check this and make sure that it is clear.)

PROBLEMS, QUESTIONS, AND EXTENSIONS

11.1. Prove the following properties of P, the probability measure of an experiment.
   (a) $P(E^c) = 1 - P(E)$ where E^c is the complement of E
   (b) $P(E_1 \cup E_2) = P(E_1) + P(E_2) - P(E_1 \cap E_2)$
   (c) $P(E_1 - E_2) = P(E_1) - P(E_2)$ for $E_2 \subset E_1$

11.2. Prove Eqs. (11.2.15) and (11.2.16).

11.3. In a single probability model (Ω, P) the sets E_1 and E_2 are S.I. Show that this is also true of (a) E_1 and E_2^c and (b) E_1^c and E_2^c.

11.4. Remember the pseudorandom sequences of ± 1 in spread-spectrum systems? If they are created by flipping an honest coin in sequence, can you make a probability model for the problem? Do it for a sequence of three bits. What is the probability of $(+1, -1, -1)$?

11.5. An inventor has a machine that diagnoses cholera, or so he claims. We know that the probability of a person having cholera is $P_c = 0.01$. The inventor claims that his machine will diagnose cholera correctly, *if you have it*, with probability 0.9. Similarly, his machine will diagnose it correctly, *if you don't* *have it,* again with probability 0.9. What is the probability that you have cholera if the machine claims that you do? Interesting conclusion here if you do it right.

11.6. Consider a case of three experiments with joint model $(\Omega \times Z \times Y, p(\omega_i, z_j, y_k))$. Show that

$$p(\omega_i | z_j, y_k) = \frac{p(\omega_i, z_j | y_k)}{p(z_j | y_k)}$$

and

$$p(\omega_i, z_j | y_k) = \frac{p(\omega_i | z_j)p(y_k | \omega_i, z_j)p(z_j)}{p(y_k)}$$

11.7. Every time a PCM transmitter emits a "one" or a "zero," it does so with probabilities $P(1) = 0.6$ and $P(0) = 0.4$. The receiver makes errors due to the noise, and turns "ones" into "zeros" with probability 0.15, and "zeros" into "ones" with probability 0.1. If the receiver announces that a "one" has been received, what is the probability that this announcement is correct?

11.8. Problems 11.5 and 11.7 are actually the same. Why? Can you create a problem like that in a radar regime? It could start something like the following: "A target appears in a radar's range with probability 0.1. If it is there, the radar finds it with probability 0.9. If it is not there. . . ."

11.9. Are two mutually exclusive sets in a probability model (Ω, P) statistically independent?

11.10. Prove property (11.3.2c) for $F(x)$.

11.11. What should A be such that the function $p(x) = Ae^{-|x|}$ is a possible p.d.f. for some r.v. X?

11.12. A door-to-door salesperson is peddling p.d.f.'s. Her repertory is
(a) $p(x) = \frac{1}{\pi}\left(\frac{1}{1 + x^2}\right)$
(b) $p(x) = 3e^{-x}u(x)$
(c) $p(x) = \frac{2}{3}(x - 2)[u(x) - u(x - 3)]$
Would you buy all of them? Which are fakes? For those that are legitimate, find $F(x)$.

11.13. The joint p.d.f. of two r.v.'s X, Y is

$$p_{xy}(x, y) = \begin{cases} K(x + y), & 0 \le x \le 1, \ 0 \le y \le 2 \\ 0 & \text{elsewhere} \end{cases}$$

Find K. Then find $f_{x|y}(x | y)$.

11.14. $F(x)$ is meant to be able to give you the probabilities of any set of the real line. Give expressions [in terms of $F(x)$] for the probabilities of sets like (a) (a, b), (b) $[a, b)$, (c) $[a, b]$, and (d) (a, ∞).

11.15. The values of a r.v. X with Laplacian p.d.f. are quantized so that values in $[-1, 1]$ become zero; values in $(1, 2)$ and $(-1, -2)$, ± 2, respectively; and values in $(2, \infty)$ and $(-2, -\infty)$, ± 3, respectively. Find the p.d.f. of the output of the quantizer.

11.16. Two r.v.'s X and Y have the joint p.d.f. $p_{xy}(x, y) = xe^{-x(1+y)}u(x)u(y)$.
   (a) Are they S.I.?
   (b) Find the conditional p.d.f. $p(y|x = 3)$.
   (c) Find the *probability* that $X = 5$ if we know that $Y = 1$.

11.17. The following are problems in transformations of a single r.v., X, to a new one, $Y = g(X)$. Find the p.d.f. of Y if
   (a) (*Linear Rectification*) X is Gaussian with $m = 0$, $\sigma_x^2 = 1$, and

$$Y = \begin{cases} X & \text{for } X \geq 0 \\ 0 & \text{for } X < 0 \end{cases}$$

   (b) (*Linear Transformation*) X has any p.d.f. $p_x(x)$ and $Y = AX + B$ (A and B are constants). Find the p.d.f. of Y as a function of $p_x(x)$.
   (c) X is uniform from $[1, 2]$ and $Y = 10/x$.
   (d) X is uniform in $[-\pi, \pi]$ and $Y = \sin(X + \theta)$, where θ is a constant. Note that the equation $Y = g(X)$ has many solutions here. (How many?)

11.18. Prove Eq. (11.5.12) under the conditions stated there. Then find the p.d.f. of $Z = X_1 + X_2 + X_3 + X_4$ if all X's have a uniform p.d.f. in $[-1, +1]$ and they are S.I.

11.19. To get the flavor of multidimensional transformations, try the following problem. A particle is moving on a plane and its Cartesian coordinates X and Y are S.I. r.v.'s, with joint p.d.f. the bivariate Gaussian. Find the joint p.d.f. $p_{uv}(u, v)$, where u and v are its polar coordinates.

11.20. Find the mean and variance of $Y = AX + B$ if A and B are constants and X is a r.v. with mean M_x and variance σ_x^2.

11.21. To find the probability of sets for a Gaussian r.v. X, we use the table in Appendix B. This table assumes, however, that the p.d.f. of X is the Gaussian with $M_x = 0$, $\sigma_x^2 = 1$ (standard normal). How do you use the table for any value M_x and any value $\sigma_x^2 > 0$? (*Hint:* Use the idea of a linear transformation as in Problem 11.20.)

11.22. A r.v. X with mean M_x and variance σ_x^2 is transformed to

$$Y = (3X + B)^2 + X \cos 10^3 t$$

Find the mean and variance of Y. Note that they are functions of time.

11.23. Prove that the correlation coefficient between X and $Y = 3X + 5$ is unity.

11.24. The r.v. X is uniformly distributed (i.e., with uniform p.d.f.) in $[-1, 1]$. If $Y = X^2$, find the covariance and p of X and Y. What do you conclude?

11.25. Try Problem 11.19 again, but find only the "partial" model of u and v.

11.26. This and the next few problems pertain to the problem of finding the p.d.f. of a r.v. (i.e., its complete model). Study the literature and prepare a short paper on the "histogram" method of finding the p.d.f. of a continuous r.v. This method is based on the idea of relative frequency.

11.27. (*The central limit theorem*)   This theorem states that under certain conditions, the p.d.f. of the r.v.

$$Y = \frac{Z - E(Z)}{\sigma_z}$$

(where $Z = \sum_{i=1}^{n} X_i$, and X_i are r.v.'s) is Gaussian with $m_y = 0$ and $\sigma_y^2 = 1$. Find the conditions and prepare a paper on the proof of the theorem.

11.28. (*The central limit theorem again*)   Loosely speaking, one version of this theorem states that if $X_1, X_2, \ldots, X_n$ are S.I. r.v.'s with identical marginal p.d.f.'s, the r.v.

$$Z = \sum_{i=1}^{n} X_i$$

is approximately Gaussian. To see this, assume first that all X's are uniformly distributed in $(-1, +1)$. Then

(a) Find the p.d.f. of $Z_1 = X_1 + X_2$. (It should come out as a triangle. Convolution?)

(b) Repeat with $Z_2 = X_3 + X_4$.

(c) Find the p.d.f. of $Z = Z_1 + Z_2 = X_1 + X_2 + X_3 + X_4$. Is it beginning to look like a Gaussian curve?

11.29. It was mentioned in Section 11.6.2 that entropy can be used to estimate the p.d.f. of a discrete r.v., in the form of the maximum entropy principle. Present a paper on this issue, with some examples. A good presentation of this principle (and many examples) can be found in Tribus (1969). An extension of this idea to continuous r.v.'s can be found in Tzannes and Noonan (1973) in the form of the **mutual information principle (MIP)**.

11.30. The partial model (M_x, σ_x^2) of a r.v. X, can be estimated from measuring some of its values [i.e., by taking a *sample* $(X_1, X_2, \ldots, X_n)$ of its values]. Since this operation of *taking a sample* produces different numbers every time it is performed, the sample $(X_1, X_2, \ldots, X_n)$ *is an n-dimensional r.v.* Assuming now that $X_1, X_2, \ldots, X_n$ are identically distributed (they have the same marginal model) and statistically independent r.v.'s with means M_x and variances σ_x^2, find the mean and variances of

$$Z = \frac{1}{n} \sum_{i=1}^{n} X_i$$

$$Y = \frac{1}{N-1} \sum_{i=1}^{n} [X_i - E(Z)]^2$$

which are expressions used to estimate the mean and variance (respectively) of the original r.v. X. Note that such expressions, called **estimators** in statistics, are r.v.'s, since they are functions of n other r.v.'s.

11.31. Vector space theory (of Section 2.4.) can also be applied to random variables. Consider all the r.v.'s X_i which are defined over the same abstract space, and show that they form a vector space. Now consider the expressions

$$(X_i, X_j) = E(X_i X_j)$$

$$N(X_i) = \sqrt{E(X_i^2)}$$

$$d(X_i, X_j) = E\sqrt{\{(X_i - X_j)^2\}}$$

and show that they are legitimate inner product, norm, and distance, respectively. If the r.v.'s are complex, the expressions above are modified to account for this. (How?)

=== 12 ===
STOCHASTIC PROCESSES

12.1 INTRODUCTION

With random variables behind us, we can now tackle the theory of stochastic processes, whose usefulness was explained in the introduction to Chapter 11. In fact, that introduction is good enough for this chapter as well. Please re-read it.

12.2 STOCHASTIC PROCESSES: DEFINITIONS AND COMPLETE DESCRIPTIONS

Mathematically, a stochastic (or random) process is defined as follows. Consider a probabilistic experiment (abstract one) with outcomes ω_i and universe Ω. We recall that if to each ω_i we assign a number, the resulting collection of numbers with universe the real line is called a random variable. Now assume that instead of a number, to each outcome $\omega_i \in \Omega$ we assign (by some rule) a function of time $X(t, \omega_i)$. The resulting collection of functions (signals) $X(t, \omega_i)$ is called a **random** (or **stochastic**) **process**.

Example 12.2.1

The abstract experiment is the "flip of a coin" with $\Omega = (H, T)$. Now we assign

$$\omega_1 = H \longrightarrow X(t, H) = t$$
$$\omega_2 = T \longrightarrow X(t, T) = \cos t$$

The resultant pair of signals $(t, \cos t)$ is a random process $X(t, \omega_i)$ for $i = 1, 2$.

The definition above gives a good mental picture of what a random process is—a (finite, denumerable, or countably infinite) collection of signals, each assigned to a point $\omega_i \in \Omega$. Such a picture appears in Fig. 12.2.1a. We advise the reader to keep this picture in mind and to snap it on every time he or she hears the term "stochastic process." It will be very helpful in what is to come later. Note that the collection is assumed to have a common origin $t = 0$. The functions of the collection are called **sample functions, member functions,** or **realizations,** and the entire collection is often called an **ensemble.**

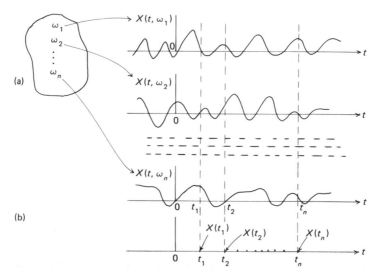

Figure 12.2.1 (a) Mental picture of a stochastic process (a collection of signals); (b) another mental picture of a stochastic process (a collection of r.v.'s).

Now we will look at the stochastic process from a slightly different point of view. The resulting notions and mental image are also helpful in what is to come.

It is obvious from what we have said so far that a stochastic process is a function of two variables, ω and t. The definition above (and the resulting mental image) come from looking at it for *each* outcome ω_i. Specify ω_i and you get a signal—that is what the definition says. But what if we were to look at it from the point of view of t? What happens if we specify t?

Well, if only t is frozen to a value t_1 (ω can be anything), the result $X(t_1, \omega)$ is a collection of numbers, *that is, a r.v.* In Example 12.2.1, if $t = 3$ say, then the first sample function has the value 3 and the second the value cos 3. The two values (3, cos 3) are numbers assigned to H and T, respectively, and therefore they make up a r.v. The same idea can also be seen in Fig. 12.2.1b. Freeze t, to say t_1, and the result is $X(t_1, \omega)$, a collection of numbers, each number being the value of each sample function at $t = t_1$. So for each value of *t*, the stochastic process $X(t, \omega)$ is a r.v. Therefore, we can say (and be just as right as before) that a stochastic process $X(t, \omega)$ is a (finite, denumerable, or uncountable) collection of r.v.'s, each defined at every value of *t*. The resulting mental image of Fig. 12.2.1b is more abstract, but quite necessary for complete understanding of random processes. That, too, has to be "snapped on" every time the term "random process" appears.

Symbolically, a random process will be designated from now on as $X(t)$, that is, the dependence on ω will be kept silent. This symbol $X(t)$ is, of course, a little vague. It could mean the entire process (all the sample functions, or all the r.v.'s), or it could mean the r.v. at $t = t$. The symbol $X(t_1)$ will mean the r.v. at $t = t_1$—no problem there. The symbol $x(t_1)$ will mean a specific value x of the r.v. at $t = t_1$. The symbol $X_r(t)$ will mean some sample function of the process—*r* for realization. Usually, the context of the discussion will make the notation obvious. If it does not, we will spell it out.

Example 12.2.2

Consider the stochastic process (S.P.) defined by

$$X(t) = At + 3 \qquad (12.2.1)$$

where A is a r.v. with exponential p.d.f. (a) Find one of its realizations. (b) Find the r.v. of the process at $t = 3$.

Solution.   First, let us make sure that $X(t)$ is a S.P. Do we have a collection of signals each assigned to an outcome of an abstract probability phenomenon? Well, we certainly have a collection of functions, each assigned to the values of the r.v. A. Now A is not abstract; it is a r.v., but that is all right. R.v.'s are special cases of abstract probability experiments, so everything is okay.

(a) Since A takes values in $(0, \infty)$, a typical realization of $X(t)$ (the one when $A = 20$) will be $X(t) = 20t + 3$—a straight line.

(b) Let us freeze *t* to the value $t = 15$. The result will be $X(15) = 15A + 3$. Is this a r.v.? Of course, it is. It is a function of the r.v. A. In fact, we could even find its p.d.f. if we wanted to, since the p.d.f. of A is given. (How?)

Example 12.2.3

Consider the S.P.'s defined by

$$X(t) = A \cos(\omega_c t + \theta) \qquad (12.2.2)$$

Find a realization and a r.v. of the process defined if (a) A and ω_c are constants and θ is a r.v. uniformly distributed in $(-\pi, +\pi)$; and (b) if ω_c is constant and A and θ are r.v.'s of unknown joint p.d.f. but with range of values $A \geq 0$ and $\theta \in (-\pi, +\pi)$.

Solution.   (a) This case is often called the "random phase generator." A typical realization is $X_r(t) = A \cos(\omega_c t + \pi/6)$. A typical r.v. is $X(3) = A \cos(\omega_c 3 + \theta)$ a function of the r.v. θ. The random phase generator S.P. can be used as a model of the local oscillator at the receiver of a DSB system if there is no synchronism with the incoming AM signal. Not only there—any system that has a sinusoidal signal whose phase is not specified can be modeled as a random phase generator.

(b) Now both A and θ are r.v.'s. In view of their range of values, a typical

$X_r(t) = 10 \cos(\omega_c t + \pi/4)$, and a r.v. is $X(5) = A \cos(\omega_c 5 + \theta)$. Note that the r.v. $X(5)$ is a function of two r.v.'s A and θ. To find its p.d.f. you would need to have the joint p.d.f. of A and θ. Note also that each realization of $X(t)$ in this case is a function of time assigned to the *joint* outcomes (A, θ). The definition of the S.P. *implies* that the probability model behind the process could be two-dimensional or even n-dimensional. Why not?

Let us now concentrate on the second "definition" of a S.P., the one that talks about the collection of r.v.'s (snap mental image of Fig. 12.2.1b). We should be fairly comfortable with this definition, as it represents an extension of the n-dimensional random variables of Chapter 11.

Depending on the nature of these r.v.'s and the points t on which they are defined, we can distinguish four types of S.P.'s. A **continuous-continuous** S.P. process is one whose r.v.'s are continuous, *and* they are defined for all values of t. A **continuous-discrete** S.P. will have continuous r.v.'s, but defined only on a finite (or denumerable set) of t's. **Discrete-continuous** and **discrete-discrete** would be processes whose r.v.'s are discrete, and they are defined on all t's or finite (or denumerable) set of t's, respectively. Note that when the t points are finite, in essence we have an n-dimensional r.v. model $(X(t_1), X(t_2), \ldots, X(t_n))$, like the ones we discussed in Chapter 11.

With all this as background, we are now ready to discuss the complete probabilistic model of a S.P. We will do so assuming that the process is continuous-continuous, and let the reader modify the model for the other three cases.

Since the S.P. is a collection of r.v.'s, its complete probabilistic description is obviously the joint p.d.f. of all the r.v.'s. Of course, the number of r.v.'s is uncountably infinite in the most general case, and we could not even write this p.d.f. symbolically. To go around this problem, we state that a S.P. $X(t)$ is completely modeled if the n-dimensional p.d.f.

$$p\{x(t_1), x(t_2), \ldots, x(t_n)\} \tag{12.2.3}$$

is known, for all values of n and any group of values $(t_1, t_2, \ldots, t_n)$. If the reader will study this expression, he or she will read it for what it really is—a joint p.d.f. of infinitely many r.v.'s—a real mess, in other words. It is bad enough trying to find an n-dimensional p.d.f., but imagine an infinite number of them (for any n) and an infinite number of combinations $(t_1, t_2, \ldots, t_n)$. It is hopeless.

Even so, there is one stochastic process that can be defined this way, the Gaussian one, and for this (and other reasons) it enjoys immense popularity in engineering and other fields.

The joint n-dimensional p.d.f. of a Gaussian S.P. is given by

$$p\{x(t_1), x(t_2), \ldots, x(t_n)\}$$

$$= \frac{1}{\sqrt{|C|(2\pi)^n}} \exp\left\{-\frac{1}{2} \sum_{i=1}^{n} \sum_{j=1}^{n} [x(t_i) - \mu(t_i))d_{ij}(x(t_j) - \mu(t_j)] \right. \tag{12.2.4}$$

for any integer $n \geq 1$ and $t_1, t_2, \ldots, t_n$ in $(-\infty, +\infty)$, where $\mu_j = E\{X(t_j)\}$, $C_{ij} = E\{(X(t_i) - \mu(t_i))(X(t_j) - \mu(t_j))\}$, C = matrix $[C_{ij}]$, and d_{ij} = elements of matrix C^{-1}.

The expression looks somewhat formidable, but it really is not. We will not have occasion to use it in this book, so we will not discuss it further. It should be noted, however, that the expression is completely specified by knowing the means of the r.v.'s and their covariance matrix, denoted here as C.

Now let us go back to the first definition of a S.P., the one that identifies it as a family (collection) of signals $X(t)$, and think about the possibility of describing it in the frequency domain. Let us assume that each sample function $X_r(t)$ (realization) of $X(t)$ is reasonably well behaved, so that it has a FT, which we denote as $X_r(\omega)$ (here ω is frequency). This being the case, we end up with a family of FTs, one for each realization, and that, too, is a stochastic process, although not of time but of ω. Thus we can say that the FT of a S.P. $X(t)$ is another S.P. $X(\omega)$. Zeroing in now on $X(\omega)$, we can apply to it the second definition of a S.P. and think of it as a collection of r.v.'s, each defined at a value of $\omega \in (-\infty, +\infty)$. A complete probabilistic model of $X(\omega)$ would then be a joint p.d.f. of these r.v.'s, something like expression (12.2.3) extended to r.v.'s $X(\omega_1)$, $X(\omega_2)$, and $X(\omega_n)$. We will not concern ourselves with this any more, but it is essential to an understanding of the importance and usefulness of "partial" models of stochastic processes, which we take up in the next section. A little more on complete models will be found in Problem 12.1.

Conclusion. S.P.'s are either (1) a collection of signals or (2) a collection of r.v.'s, identified at time (or frequency or any other parameter) points $t \in (-\infty, +\infty)$. Their complete probability model is the joint p.d.f. of all the r.v.'s, almost impossible to obtain in practice.

At this point, the reader should ponder the overall definition of a S.P. and try to answer questions such as the following. (1) How can noise from a heated resistor be thought of as an example of a S.P.? (2) How about atmospheric noise in a communication or radar system? (3) In radar we often get return echoes from objects (mountains, trees, sea) which are located around the radar sight, and these returns make up a noise called **clutter**. Does it fit the model of S.P.? (4) What about human speech itself? Is it also a S.P.?

12.3 STOCHASTIC PROCESSES: PARTIAL MODELS

Lucky for us (and the world), there is another way to model stochastic processes, an approximate but useful one, which we shall call here **partial**. As the reader has already guessed, it involves averages of the r.v.'s of the process. In fact, it is nothing but a generalization of the partial model of a n-dimensional r.v. discussed in Section 11.6.3, especially expressions (11.6.32) and (11.6.33). The generalization is necessary because in a S.P. the r.v.'s could be (uncountably) infinite. Vectors and matrices become one- and two-variable functions in this type of generalization.

The **partial model** of a S.P. is defined as the **mean of the process**

$$E\{X(t)\} = M_x(t) \tag{12.3.1}$$

and the **covariance of the process**

$$\text{cov}\,[t_1, t_2] = \cos\,[X(t_1), X(t_2)] = E\{(X(t_1) - M(t_1))(X(t_2) - M(t_2))\} \tag{12.3.2}$$

Let us take them one at a time and see what they represent.

The mean of the process $M_x(t)$ is, of course, the generalization of the vector of the means of an n-dimensional model [expression (11.6.32)]. For each value of t (say t_1) there is a r.v. $X(t_1)$, and its mean is, say, $M(t_1)$. Since t in the general case takes values in $(-\infty, +\infty)$, and there is a r.v. at each one of these values, the end result is a function—the mean of the process $M_x(t)$. What more can you say about it?

The covariance of the process is, of course, a generalization of expression (11.6.33). It represents the covariance between the r.v. at t_1 and the r.v. at t_2, where both t_1 and t_2 can take values in $(-\infty, +\infty)$. It is a function of t_1 and t_2, two time variables. If, for example, a S.P. has $\text{cov}\,[t_1, t_2] = e^{-|t_1|}e^{-|t_2|}$, its value at $t_1 = 3$, $t_2 = 5$, $\text{cov}\,[3, 5] = e^{-3}e^{-5} = e^{-8}$ represents the covariance of the r.v.'s $X(3)$ and $X(5)$. Note also that

$$\text{cov}\,(t_1, t_2) = E\{X(t_1)X(t_2)\} - M_x(t_1)M_x(t_2) \tag{12.3.3}$$

an expression obtained by factoring out the right-hand side of Eq. (12.3.2) and using the properties of expectation.

We now denote the first term of Eq. (12.3.3) as

$$R_x(t_1, t_2) = E\{X(t_1)X(t_2)\} \tag{12.3.4}$$

and call it the **statistical correlation function** of the process $X(t)$. The partial model of $X(t)$, can, in fact, be thought of as $M_x(t)$ and $R_x(t_1, t_2)$, since knowledge of these two provide us with knowledge of $\text{cov}\,[t_1, t_2]$, via Eq. (12.3.3).

The point we have reached is quite remarkable, or it will prove to be later when we discover how useful partial models of S.P.'s are. Think about it for a minute. The partial model of a S.P., which is an infinite collection of signals or r.v.'s, turns out to be made up of two simple (deterministic) signals $M_x(t)$ and $R_x(t_1, t_2)$. It is a gigantic simplification. Add to it the fact that both of these functions can be estimated by statistical methods (which we will not discuss), and you end up with something practical as well. The proof of the pudding will, of course, come a little later, when the usefulness of this model is ascertained.

Example 12.3.1

A random process $X(t)$ is given by

$$X(t) = At^2 + 3t$$

where A is a r.v. with Laplacian p.d.f., $p(A) = \frac{1}{2}\,e^{-|A|}$. Find $M_x(t)$ and $R_x(1, 2)$.

Solution. We have that

$$M_x(t) = E\{X(t)\} = E\{At^2 + 3t\}$$

where the expectation is with respect to the r.v. A, the only one present. Continuing,

$$M_x(t) = t^2 E(A) + 3t$$

since the t terms behave like constants with regard to the expectation (they are not functions of A). In view of the fact that $E(A) = 0$ for the Laplacian model, the result is

$$M_x(t) = 3t$$

Let us now find $R_x(t_1, t_2)$, and it will give us $R_x(1, 2)$ immediately.

$$\begin{aligned}
R_x(t_1, t_2) &= E\{X(t_1)X(t_2)\} = E\{(At_1^2 + 3t_1)(At_2^2 + 3t_2)\} \\
&= t_1^2 t_2^2 E(A^2) + 3t_1^2 t_2 E(A) + 3t_1 t_2^2 E(A) + 9t_1 t_2 \\
&= t_1^2 t_2^2 E(A^2) + 9t_1 t_2
\end{aligned}$$

But $E(A^2) = \sigma_A^2$ since $E(A) = 0$, and therefore $E(A^2) = 2$ for a Laplacian model, as the reader can easily verify. Thus

$$R_x(t_1, t_2) = 2t_1^2 t_2^2 + 9t_1 t_2$$

and $R_x(1, 2)$ is immediate.

What about a partial model of the Fourier transform of a process $X(t)$, [the process $X(\omega)$] whose realizations are the FT's of the realizations of $X(t)$? That, too, is possible. It would be made up of the mean of the process $X(\omega)$, and the (statistical) correlation function of the process, call it $\Gamma(\omega_1, \omega_2)$. Now the process $X(\omega)$ can be written down as

$$X(\omega) = \int_{-\infty}^{+\infty} X(t)e^{-j\omega t}\, dt \qquad (12.3.5)$$

a symbolic notation, of course, meaning that each realization in $X(\omega)$ is the FT of each realization of $X(t)$. In terms of r.v.'s, expression (12.3.5) means that each r.v. of $X(\omega)$ [say $X(\omega_1)$] is an integral transformation of all the r.v.'s of the process $X(\omega)$. To find $E\{X(\omega)\}$, which we can denote as $M(\omega)$, we have

$$M(\omega) = E\{X(\omega)\} = E\left[\int_{-\infty}^{+\infty} X(t)e^{-j\omega t}\, dt\right] \qquad (12.3.6)$$

and assuming that the operator E (it is an integral) and the integral can be interchanged, we end up with

$$M(\omega) = \int_{-\infty}^{+\infty} M_x(t)e^{-j\omega t}\, dt \qquad (12.3.7)$$

that is, with the fact that $M(\omega)$ and $M_x(t)$ are FT pairs. Good!

Now let us see about $\Gamma(\omega_1, \omega_2)$, the correlation function of the FT process $X(\omega)$. Since $X(\omega)$ may have complex r.v.'s at the various values of ω, we must redefine the notion of the correlation function a bit. In such cases,

$$\Gamma(\omega_1, \omega_2) = E\{X(\omega_1)X^*(\omega_2)\} \qquad (12.3.8)$$

where $X^*(\omega)$ is the complex conjugate of $X(\omega)$ (see also Problem 11.31 in this respect). Now

$$X(\omega_1) = \int_{-\infty}^{+\infty} X(t_1) e^{-j\omega_1 t_1} \, dt_1 \qquad (12.3.9)$$

$$X^*(\omega_2) = \int_{-\infty}^{+\infty} X(t_2) e^{+j\omega_2 t_2} \, dt_2 \qquad (12.3.10)$$

and thus (ignoring mathematical rigor)

$$\Gamma(\omega_1, \omega_2) = E\left\{ \int_{-\infty}^{+\infty} \int_{-\infty}^{+\infty} X(t_1) X(t_2) e^{-j(\omega_1 t_1 - \omega_2 t_2)} \, dt_1 \, dt_2 \right\} \qquad (12.3.11)$$

Bringing the expectation inside and recognizing that $E\{X(t_1)X(t_2)\} = R_x(t_1, t_2)$ (the rest are not r.v.'s), we have

$$\Gamma(\omega_1, \omega_2) = \int_{-0}^{+0} \int_{-\infty}^{+\infty} R_x(t_1, t_2) e^{-j(\omega_1 t_1 - \omega_2 t_2)} \, dt_1 \, dt_2 \qquad (12.3.12)$$

which will be recognized by those "in the know" as the two-dimensional FT of $R_x(t_1, t_2)$ (see Papoulis, 1965, p. 466).

So $M(\omega)$ and $\Gamma(\omega_1, \omega_2)$ [the partial model of $X(t)$ in the frequency domain] are the FTs of $M_x(t)$ and $R_x(t_1, t_2)$, respectively. Quite interesting, but not altogether surprising.

Conclusion. The mean of a process $M_x(t)$ and its correlation function $R_x(t_1, t_2)$ can be thought of as a partial (approximate) probabilistic model of a S.P. $X(t)$, in the time domain, of course. In the frequency domain the partial model is $M(\omega)$ and $\Gamma(\omega_1, \omega_2)$, the FTs of $M_x(t)$ and $R_x(\tau)$, respectively. It all makes good sense, doesn't it?

12.4 WIDE-SENSE STATIONARY PROCESSES

Surprising as it may sound, we can actually make the partial model of a S.P. even simpler, by restricting ourselves to the study of a subgroup of processes called wide-sense stationary processes. We will not loose much by this restriction, as most noises (or useful processes) in communications and radar appear to belong in this group.

A **wide-sense stationary** process (or simply **stationary**[1] in this book) is one whose partial model $M_x(t)$ and $R_x(t_1, t_2)$ obeys the two conditions

(1) $M_x(t) = C$ (a constant)   (12.4.1a)

(2) $R_x(t_1, t_2) = R_x(t_1 + t, t_2 + t) = R_x(|t_1 - t_2|)$

                                = $R_x(\tau)$ $(\tau = |t_2 - t_1|)$ (12.4.1b)

---

[1] There is a group of processes called stationary in the **strict-sense**, with conditions on their complete model, the p.d.f.'s. We shall not be concerned with this group here, so the word "stationary" will mean the wide-sense type only.

The first condition is quite obvious—it just demands that all r.v.'s of the process have the same mean C, called the dc component of the process. The second condition needs a little more elaboration, and it is given immediately below. The notation is the biggest buggaboo here, and we must be quite careful with it.

Recall that $R_x(t_1, t_2)$ is given by

$$R_x(t_1, t_2) = E\{X(t_1)X(t_2)\} \qquad (12.4.2)$$

and that it is the correlation between the r.v.'s at t_1 and t_2. The notation $R_x(t_1, t_2)$ is meant to indicate that this correlation depends on the values t_1 and t_2 since these values determine the r.v.'s we are correlating.

In a stationary process (wide sense, of course), the correlation between the two r.v.'s depends on their time distance $|t_2 - t_1| = \tau$, not on the specific times t_1 and t_2 on which they are defined. Thus the r.v.'s at $(t_1 = 0, t_2 = 1)$ have the same correlation as the r.v.'s at $(t_1 = 1, t_2 = 2)$, and so on. This property is called *stationarity*, because the partial model $\{M_x, R_x(t_1, t_2)\}$ is invariant under time shifts; it does not depend on where you put the origin $t = 0$.

It is obvious that the partial model of stationary processes is quite a lot simpler. The mean is a constant, and what is even more important, the $R_x(t_1, t_2)$ is no longer a function of two variables, but of one, τ. Often we write

$$R_x(\tau) = E\{X(t)X(t + \tau)\} \qquad (12.4.3)$$

since the distance of the two r.v.'s is $t + \tau - t = \tau$, a fact that we wish to emphasize. The notation here is tricky, as we have already pointed out.

So stationary processes have a relatively simple partial model, but how are we to know whether a given process $X(t)$ is stationary or not? Well, if you have an expression for the process, then all you have to do is check the two conditions above.

Example 12.4.1

Consider the "random phase generator" process

$$X(t) = A \cos(\omega_c t + \theta) \qquad (12.4.4)$$

where A and ω_c are constants and θ is a r.v. with uniform model in $(0, 2\pi)$. Is it wide-sense stationary?

Solution. We must check first whether the mean $M_x(t)$ is a constant. Now for any t, the r.v. $X(t)$ has the mean

$$M_x(t) = AE\{\cos(\omega_c t + \theta)\} \qquad (12.4.5)$$

The expectation is on the r.v. $\cos(\omega_c t + \theta)$, which is a function of the r.v. θ. Using expression (11.6.14) (we told you how useful it is), we have

$$M_x(t) = \frac{A}{2\pi} \int_0^{2\pi} \cos(\omega_c t + \theta)\, d\theta \qquad (12.4.6)$$

since $p(\theta) = 1/2\pi$ for $0 \le \theta \le 2\pi$, and zero elsewhere. The result of the integration is zero, and it holds for all r.v.'s $X(t)$, so the first condition is satisfied.

Now we will check whether $R_x(t_1, t_2)$ can boil down to a function of $(t_2 - t_1)$ only. By definition,

$$R_x(t_1, t_2) = E\{X(t_1)X(t_2)\}$$

$$= E\{A^2 \cos(\omega_c t_1 + \theta) \cos(\omega_c t_2 + \theta)\}$$

$$= \frac{A^2}{2} E\{\cos \omega_c(t_2 - t_1) + \cos(\omega_c t_1 + \omega_c t_2 + 2\theta)\}$$

$$= \frac{A^2}{2} \cos \omega_c(t_2 - t_1) + E\{\cos(\omega_c t_1 + \omega_c t_2 + 2\theta)\}$$

since the first term inside the brackets is not a r.v. Now the second term has an expected value of zero (check it). The final result is

$$R_x(t_1, t_2) = \frac{A^2}{2} \cos \omega_c(t_2 - t_1) = \frac{A^2}{2} \cos \omega_c \tau \qquad (12.4.7)$$

and thus the process is stationary.

Of course, in practice, we seldom have an expression such as (12.4.4) for the process; we usually have just a few of its realizations. In this case, there exist some tests that one can perform to check for stationarity, but we shall not discuss them here. The interested reader is referred to Bendat and Piersol (1971) for further reading on this issue.

We have reached a point in our discussion that merits a little repetition and emphasis. From now on, we will deal primarily with stationary processes $X(t)$ whose partial model is $(M_x, R_x(\tau))$. In fact, since M_x is a constant (and can be subtracted in a practical problem), we can assume it zero, and ignore it. Thus our partial model of a stationary process is only this $R_x(\tau)$. It is an approximate model, of course, a "partial" one, as we keep emphasizing. It does not represent the process uniquely from a probabilistic point of view—only the joint p.d.f.'s can do that. Still it is plenty good enough, as we will start to appreciate following this chapter.

This $R_x(\tau)$ is, of course, a partial model of a stationary process in the time domain. What about a representation of the process in the frequency domain ω? We recall from the preceding section that in ω, the partial model of a process $X(t)$ is the mean $E\{X(\omega)\} = M(\omega)$ and the correlation function $\Gamma(\omega_1, \omega_2)$ of its Fourier transform process $X(\omega)$. But we showed that [see Eq. (12.3.7)] $M(\omega)$ and $M_x(t)$ are FT pairs. Thus, for $M_x(t) = C$, a constant,

$$M(\omega) = C\pi\delta(\omega) \qquad (12.4.8)$$

and if $M_x(t) = 0$, as we will usually assume from now on, $M(\omega) = 0$ also.

The second member of the frequency-domain partial model will be taken to be the FT of $R_x(\tau)$, a logical consequence of the fact that $R_x(t_1, t_2)$ and $\Gamma(\omega_1, \omega_2)$ are two-dimensional FT pairs [see Eq. (12.3.12)]. In fact, this FT, denoted by $S_x(\omega)$,

$$S_x(\omega) = \int_{-\infty}^{+\infty} R_x(\tau)e^{-j\omega\tau}\, d\tau \qquad (12.4.9)$$

is called the **power spectral density** of the process $X(t)$ and plays the same role in stationary processes as the energy (or power) density spectrum does in deterministic signals. Being the FT of an average [$R_x(\tau)$ is an average], its total area in $(-\infty, +\infty)$ is denoted by P_{ave}, given by

$$P_{ave} = \frac{1}{2\pi} \int_{-\infty}^{+\infty} S_x(\omega)\, d\omega \tag{12.4.10}$$

and represents the average power of the process $X(t)$. The power in frequency bands $(\omega_1 \leq \omega \leq \omega_2)$ is found by integrating over both the ranges, $(-\omega_1, -\omega_2)$ and (ω_1, ω_2), as usual.

We have now arrived at an important point in our development of random processes. By sticking with w.s. (wide-sense) stationary ones, and assuming their constant mean equal to zero, we have as their partial models, $R_x(\tau)$ in time and its FT $S_x(\omega)$ in frequency. Just two functions, that's all—representing an infinite collection of realization both in time and frequency domains. If this is not remarkable, we don't know what is.

Let us now go on and list some important properties of $R_x(\tau)$ which make our life easier.

(1) $R_x(\tau) = R_x(-\tau)$ (12.4.11)

(2) $R_x(0) \geq R_x(\tau)$ for all $\tau \neq 0$ (12.4.12)

(3) $R_x(0) = P_{ave}$ (12.4.13)

The proofs of (1) and (3) are quite immediate.

(1) This is true by definition, since it depends only on the distance $|t_2 - t_1|$. It is also kind of nice to have as a property (even symmetry), because it means that $S_x(\omega)$ (its FT) is always *real* and *even* (prove it).

(3) This is proven as follows. Since $R_x(\tau)$ and $S_x(\omega)$ are FT pairs,

$$R_x(\tau) = \frac{1}{2\pi} \int_{-\infty}^{+\infty} S_x(\omega) e^{j\omega\tau}\, d\omega \tag{12.4.14}$$

and thus

$$R_x(0) = \frac{1}{2\pi} \int_{-\infty}^{+\infty} S_x(\omega)\, d\omega \tag{12.4.15}$$

which is P_{ave}, by Eq. (12.4.10).

We will leave property (2) as an exercise for the reader. It makes sense, of course, since it says that the correlation between a r.v. $X(t)$ and itself $[R_x(0) = E\{X(t)X(t)\}]$ should not be less than that between $X(t)$ and some other r.v. $X(t + \tau)$. In fact, we should be rather worried if it did not hold.

With that, we close our *basic* theory of random processes. The principal point to remember is that from now on we will assume them wide-sense stationary and usually with zero mean. This being the case, their partial model is $R_x(\tau)$ or $S_x(\omega)$ in time and frequency, respectively.

We will close this section with some examples. Read them carefully. We are at a point where we can begin to correlate random process theory with material that we have covered in the past, as some of the examples will show.

Example 12.4.2

A stationary process $X(t)$ has $M_x(t) = 0$ and $R_x(\tau) = 10e^{-|\tau|}$. Find its average power P_{ave}. Find its power spectral density (PSD) and sketch it.

Solution. $P_{ave} = R_x(0) = 10$ W. That was quick. Now $S_x(\omega)$ is (by Table 2.6.1)

$$S_x(\omega) = \frac{20}{1 + \omega^2}$$

and its sketch is given in Table 2.6.1.

Example 12.4.3

The PSD of a stationary random process with $M_x(t) = 0$ is $S_x(\omega) = 2\pi$ rect $(\omega/100)$. (a) Find its P_{ave}. (b) Find two r.v.'s of the process which are uncorrelated with each other.

Solution. (a) P_{ave} can be found either by integrating $S_x(\omega)$ or by finding $R_x(0)$. Let us do it by finding $R_x(0)$, since we need $R_x(\tau)$ for the second part.

From Table 2.6.1 we have that

$$R_x(\tau) = 100 \text{ Sa } (50\tau) = 100\frac{\sin 50\tau}{50\tau}$$

and thus $R_x(0) = 100$ W.

(b) Now we know that two r.v.'s $X(t_1)$ and $X(t_2)$ are uncorrelated if $R(t_1, t_2) = E\{X(t_1)X(t_2)\} = E\{X(t_1)\}E\{X(t_2)\}$. Since $E\{X(t_1)\} = 0$ for all t_1, two r.v.'s of this process are uncorrelated if $R_x(t_1, t_2) = 0$. Now we note that $R_x(\tau) = 0$ at $50\tau = \pi$ (and other places) or at $\tau = \pi/50$. Thus any two r.v.'s that are located $\pi/50$ time units apart are uncorrelated. Two such r.v.'s (not the only ones) are $X(0)$ and $X(\pi/50)$. The whole idea is that $R_x(\tau)$ gives the correlation between r.v.'s τ units apart from each other.

Example 12.4.4

What is the P_{ave} of the random phase generator [Eq. (12.4.4)]?

Solution. Since $R_x(\tau)$ is given by Eq. (12.4.7), then

$$P_{ave} = R_x(0) = \frac{A^2}{2}$$

which is the same as the power of a deterministic cosine $A \cos \omega_c t$.

Example 12.4.5

Consider the process

$$Y(t) = X(t)A \cos (\omega_c t + \theta) \qquad (12.4.16)$$

where $X(t)$ is a stationary process with $M_x = 0$ and the rest a random phase oscillator. It is also assumed that $X(t)$ (all its r.v.'s) and θ are statistically independent. Is the process $Y(t)$ stationary? If so (and it is so), find its $R_y(\tau)$ and $S_y(\omega)$ in terms of the presumed known partial model of $X(t)$, $R_x(\tau)$, or $S_x(\omega)$.

Solution. Here is an example that ties this theory in with our previous material. $X(t)$ could easily represent a message (speech) or a radar signal (both stochastic), and $Y(t)$ the output of an AM modulator. The carrier is assumed of constant amplitude and frequency (A, ω_c), but of a random phase angle of known p.d.f. It is a more general model of an AM modulated carrier than what we took it to be before.

Let us see if $E(Y(t))$ is a constant.

$$E\{Y(t)\} = AE\{X(t)\cos(\omega_c t + \theta)\}$$

Now, since $X(t)$ and θ are independent, it can be shown that so is $X(t)$ with a function of θ, in this case $\cos(\omega_c t + \theta)$. We will assume it proven. Therefore,

$$M_Y(t) = AE\{X(t)\}E\{\cos(\omega_c t + \theta)\} = 0$$

So far, so good. Now let's find $R_y(t_1, t_2)$ to see if it boils down to some $R_y(\tau)$.

$$R_y(t_1, t_2) = E\{Y(t_1)Y(t_2)\}$$

$$= E\left\{\frac{A^2}{2}X(t_1)X(t_2)[\cos\omega_c(t_2 - t_1) + \cos(\omega_c t_1 + \omega_c t_2 + 2\theta)]\right\}$$

which due to independence reduces to

$$R_y(t_1, t_2) = \frac{A^2}{2}R_x(t)\cos\omega_c\tau \tag{12.4.17}$$

Therefore, $Y(t)$ is (w.s.) stationary. Its PSD $S_y(\omega)$ is obviously

$$S_y(\omega) = \frac{A^2}{4}[S_x(\omega - \omega_c) + S_x(\omega + \omega_c)] \tag{12.4.18}$$

So even though the mathematics are sophisticated, the result is pretty much the same (as what?).

In this business of tying things together, see also Problem 12.6. It establishes the partial model of the random ± 1 sequences used in spread-spectrum systems and elsewhere. (Where?)

12.5 MODELS OF TWO OR MORE PROCESSES

In a communication or radar system there are often two or more random processes and we must learn how to deal with them in a *joint manner*. We will generalize our previous results to two processes and let the reader do the same for three or more.

Consider two random processes $X(t)$ and $Y(t)$, both defined over the same abstract probability model (Ω, P).

A **complete joint description** of $X(t)$ and $Y(t)$ is knowledge of the joint p.d.f.

$$p\{x(t_1), x(t_2), \ldots, x(t_n), y(t_1'), y(t_2'), \ldots, y(t_m')\} \tag{12.5.1}$$

for all values of n, m, and all combinations $t_1, t_2, \ldots, t_n$ and $t'_1, t'_2, \ldots, t'_m$. We forget about this immediately.

A **partial joint model** (in the time domain) of the two processes consists of

1. The two means $M_x(t)$ and $M_y(t)$
2. The two correlation functions $R_x(t_1, t_2)$ and $R_y(t_1, t_2)$
3. The cross-correlation function[2]

$$R_{xy}(t_1, t_2) = E\{X(t_1)Y(t_2)\} \tag{12.5.2a}$$

The new element is, of course, (3), a function that provides us with the correlation between the r.v.'s of the two processes in pairs.

If this $R_{xy}(\tau)$ reduces to

$$R_{xy}(t_1, t_2) = E\{X(t_1)\}E\{Y(t_2)\} \tag{12.5.2b}$$

the processes are called **uncorrelated**.

We shall deal here mainly with processes that are called **cross-stationary in the wide sense**. They are so called if they are individually stationary, and, in addition,

$$R_{xy}(t_1, t_2) = R_{xy}(\tau) \tag{12.5.3}$$

If the two processes are **cross-stationary** and **uncorrelated**, then

$$R_{xy}(\tau) = E\{X(t)Y(t + \tau)\} = E\{X(t)\}E\{Y(t + \tau)\} = M_x M_y \tag{12.5.4}$$

and if one or both of the means are zero (as we will usually assume),

$$R_{xy}(\tau) = 0 \tag{12.5.5}$$

Now let us go to the frequency domain, where these two processes have also a joint partial model. Assuming that their constant means are zero, this model consists of

1. The two PSDs $S_x(\omega)$ and $S_y(\omega)$
2. The so-called **cross-power spectral density**,

$$S_{xy}(\omega) = \int_{-\infty}^{+\infty} R_{xy}(\tau)e^{-j\omega\tau}\, d\tau \tag{12.5.6}$$

that is, the FT of $R_{xy}(\tau)$. It all follows the same reasoning, doesn't it?

Let us do an example involving the sum of two processes, a common case in communications and radar systems. It gives us an idea of how we find models of processes that are functions of other processes.

Example 12.5.1

We are given two cross-stationary random processes $X(t)$ and $Y(t)$ of known joint partial model. Find the partial model of the process

$$Z(t) = X(t) + Y(t) \tag{12.5.7}$$

[2]If their r.v.'s are complex, $R_{xy}(t_1, t_2) = E\{X(t_1)Y^*(t_2)\}$.

Solution. We must find the mean $M_z(t)$ and correlation $R_z(t_1, t_2)$. While doing it, we will discover whether $Z(t)$ is w.s. stationary.

$$E\{Z(t)\} = E\{X(t) + Y(t)\} = M_x + M_y \qquad \text{(a constant)}$$

The first condition for stationarity of $Z(t)$ holds. Now

$$R_z(t_1, t_2) = E\{Z(t_1)Z(t_2)\} = E\{(X(t_1) + Y(t_1))(X(t_2) + Y(t_2))\}$$

Factoring out the product and taking the expectations of each term while keeping in mind the cross-stationarity of $X(t)$ and $Y(t)$, we have

$$R_z(t_1, t_2) = R_x(\tau) + R_{xy}(\tau) + R_{yx}(\tau) + R_y(\tau) = R_z(\tau) \qquad (12.5.8)$$

all known since the joint partial model is known. $[R_{xy}(\tau) = R_{yx}(-\tau)$. Show it.] The sum process is clearly stationary.

Incidentally, if the processes $X(t)$, $Y(t)$ are uncorrelated with zero means,

$$R_z(\tau) = R_x(\tau) + R_y(\tau) \qquad (12.5.9)$$

which is a good relationship to keep in mind.

12.6 ERGODIC STATIONARY PROCESSES

We have established the nature of the "partial model" of one or more stochastic processes, but how is one to measure it (estimate it) in practice? The only answer to this question that we have implied so far is that mathematical statistics provides us with formulas that can estimate its parts, and that any further discussion of this issue is beyond the scope of the book.

Well, there is a class of w.s. stationary stochastic processes for which discussion of the estimation of the partial model is within our scope, at least up to a point. This class of processes is called **ergodic**. Their definition and how their partial model is estimated is taken up immediately.

Generally speaking, the property of ergodicity implies that some member of the partial model of a process equals some time average of the realizations of the process. Since the partial model members M and $R_y(\tau)$ are statistical averages, we often think of ergodicity as: *Time averages equal statistical averages.* So much for generalities. Now we become specific.

1. *Ergodicity in the mean.* Consider a w.s. stationary process $X(t)$ with partial model M_x and $R_x(\tau)$. This process is called **ergodic in the mean** if all its realizations $X_r(t)$ have the *same* time mean,

$$\lim_{T \to \infty} \frac{1}{T} \int_{-T/2}^{T/2} X_r(t)\, dt = \eta \qquad (12.6.1)$$

and furthermore,

$$\eta = M_x \qquad (12.6.2)$$

It is quite obvious that if a process is ergodic in the mean, we can find its M_x by using only one of its realizations. That is pretty nice.

2. *Ergodicity in correlation.* Here each realization of the stationary process has the same time autocorrelation (or ambiguity) function $\phi_{xx}(\tau)$, where

$$\phi_{xx}(\tau) = \lim_{T \to \infty} \frac{1}{T} \int_{-T/2}^{T/2} X_r(t)X_r(t + \tau)\, dt \tag{12.6.3}$$

and furthermore, this $\phi_{xx}(\tau)$ equals $R_x(\tau)$, the statistical correlation function of the process, that is,

$$R_x(\tau) = \phi_{xx}(\tau) \tag{12.6.4}$$

It is again obvious that if this property holds, one realization is sufficient for estimating $R_x(\tau)$. If both properties hold, one realization enables us to find the partial statistical model of the process $X(t)$. Not bad.

3. *Ergodicity in cross-correlation.* This property involves two cross-stationary processes, $X(t)$ and $Y(t)$. If it holds, all pairs of realizations $X_r(t)$ and $Y_r(t)$ have the same time cross-correlation $\phi_{xy}(\tau)$:

$$\phi_{xy}(\tau) = \lim_{T \to \infty} \frac{1}{T} \int_{-T/2}^{T/2} X_r(t)Y_r(t + \tau)\, dt \tag{12.6.5}$$

and the statistical cross-correlation function $R_{xy}(\tau)$ is

$$R_{xy}(\tau) = \phi_{xy}(\tau) \tag{12.6.6}$$

So if you have two stationary processes and wish to estimate their joint partial model, one realization from each is sufficient if all ergodic properties are in effect. Quite a good deal!

When do these ergodic properties hold? That is a tough question to answer, at least in a way that it can be used in practical problems. Theoretically, it can be answered, but the conditions involved have little practical use. Even in the simplest case (ergodicity in the mean), the condition for its validation involves knowledge of $R_x(\tau)$, and if we knew that, we would not need ergodicity in the first place.

Even so, ergodicity seems to hold in practice, for most processes. One usually assumes it until something happens to contradict the assumption. Not only that. Ergodicity helps us also to understand the power spectral density a little bit better. How? Witness the following.

Let us go back to a deterministic energy signal, say $f(t)$, with FT $F(\omega)$. Recall that the energy density spectrum was defined as $S(\omega) = |F(\omega)|^2$ and that it is equal to the FT of the time autocorrelation function of $f(t)$ (the Wiener–Khintchine theorem). Well, if a process is ergodic, we have pretty much the same story for any of its realizations. So, defining the PSD of a process as the FT of $R_x(\tau)$ makes perfect sense for ergodic processes, and therefore good sense for nonergodic ones.

In practice, one usually assumes it, as we noted above, and goes after estimating the PSD of the process, which is, of course, equivalent to estimating $R_x(\tau)$. This area is so important that a tremendous amount of work has been pouring in about it, uninterruptedly. Even entropy and mutual information (see Sections 11.6.2 and 11.6.4) and other averages have been used in this respect, as can be ascertained by checking Childers (1978), Newman (1977), Avgeris et al. (1980), Tzannes and Avgeris (1981), or Tzannes et al. (1983).

From now on we will assume that the processes are ergodic (in everything) unless stated otherwise. We do not really need this property for our future analysis. What we do need is that processes always be stationary, so that their partial model is $(M, R_x(\tau))$.

12.7 RANDOM PROCESSES THROUGH LINEAR SYSTEMS

We have now reached the point where the partial model of a stationary process is well understood (by the writer, anyway—hopefully by the reader, too). The time is ripe to tackle the problem of how this model changes as a stochastic process goes through devices—a case most common in the analysis of communication and radar systems.

This section considers the problem of model transformations as a stochastic process goes through *linear systems*. Of course, communications and radar systems contain nonlinear systems as well; an FM modulator is a highly nonlinear device, as we know. Still, this topic will not be covered here—it is too complicated and beyond the scope[3] of the book. If in any part of our future analyses we come up against a nonlinear system, we will see what we can do on the spot. The interested reader can pursue the matter by checking any book on stochastic processes, notably the one by Deutsch (1965), which is devoted entirely to this subject.

So linear systems is our topic, and let's get with it immediately. First, let us quickly review what we know about input/output relations in a linear system if the signals are deterministic. Let us assume that the input is $f(t)$, the impulse response of a system is $h(t)$, and the output $o(t)$. Then we know that in the time domain,

$$o(t) = f(t) * h(t) = \int_{-\infty}^{+\infty} f(\tau)h(t - \tau)\,d\tau \tag{12.7.1}$$

whereas in the frequency domain,

$$O(\omega) = F(\omega)H(\omega) \tag{12.7.2}$$

where the capital letters designate the FTs of our signals. Note also that Eq. (12.7.2) leads to

$$|O(\omega)|^2 = |F(\omega)|^2 |H(\omega)|^2 \tag{12.7.3}$$

and if we denote $|O(\omega)|^2$ and $|F(\omega)|^2$ by $S_o(\omega)$ and $S_f(\omega)$, respectively (the spectral densities), then

$$S_o(\omega) = |H(\omega)|^2 S_f(\omega) \tag{12.7.4}$$

---

[3] A terrific phrase to do away with undesirable material.

That is all we need from the past. Now we take up stochastic processes.

Consider a linear system with impulse response $h(t)$, whose input is a stationary process $X(t)$ of known partial model M_x, $R_x(\tau)$. For a *specific* realization $X_r(t)$, the techniques above are sufficient. What we are interested in, however, is what happens if the input is *any* realization of $X(t)$, that is, when the input is the process $X(t)$.

Let us assume that the output is denoted by $Y(t)$. We can write

$$Y(t) = \int_{-\infty}^{+\infty} X(\tau)h(t - \tau)\, d\tau = \int_{-\infty}^{+\infty} h(\tau)X(t - \tau)\, d\tau \qquad (12.7.5)$$

which shows that $Y(t)$ is also a stochastic process—for each realization of $X(t)$ that you put in, you get one out for $Y(t)$.

Having established that $Y(t)$ is a process, our next concern is whether it is also stationary, and if so, how its partial model can be found using the model of $X(t)$ and the known $h(t)$. For stationarity, we first look at the mean of the process $Y(t)$, $M_y(t)$. In view of Eq. (12.7.5) and our known method of finding means of functions of r.v.'s,

$$M_y(t) = E\{Y(t)\} = E\left\{\int_{-\infty}^{+\infty} h(\tau)X(t - \tau)\, d\tau\right\}$$

$$= \int_{-\infty}^{+\infty} h(\tau)E\{X(t - \tau)\}\, d\tau = M_x \int_{-\infty}^{+\infty} h(\tau)\, d\tau = \text{constant} \qquad (12.7.6)$$

since $E\{X(t - \tau)\} = M_x$ and the integral of $h(\tau)$ is a number. So the first condition for stationarity of $Y(t)$ holds, and, in fact, Eq. (12.7.6) tells us how to find the mean of $Y(t)$ from known quantities.

Now we must go after $R_y(t_1, t_2)$, and see if it boils down to some $R_y(\tau)$. By definition,

$$R_y(t_1, t_2) = E\{Y(t_1)Y(t_2)\} \qquad (12.7.7)$$

and in view of Eq. (12.7.5),

$$Y(t_1) = \int_{-\infty}^{+\infty} h(\lambda)X(t_1 - \lambda)\, d\lambda \qquad (12.7.8)$$

$$Y(t_2) = \int_{-\infty}^{+\infty} h(v)X(t_2 - v)\, dv \qquad (12.7.9)$$

where $Y(t_1)$ and $Y(t_2)$ are r.v.'s—functions of the r.v.'s $X(t_1)$ and $X(t_2)$, respectively. With these, Eq. (12.7.7) becomes (check it)

$$R_y(t_1, t_2) = \int_{-\infty}^{+\infty} h(\lambda)\, d\lambda \int_{-\infty}^{+\infty} h(v)E\{(X(t_1 - \lambda)X(t_2 - v)\}\, dv \qquad (12.7.10)$$

Now $X(t)$ is stationary, which means that

$$E\{X(t_1 - \lambda)X(t_2 - v)\} = R_x(t_1 - \lambda - t_2 + v) = R_x(\tau + v - \lambda) \qquad (12.7.11)$$

where $\tau = t_1 - t_2$. Thus Eq. (12.7.10) becomes

$$R_y(t_1, t_2) = \int_{-\infty}^{+\infty} h(\lambda)\, d\lambda \int_{-\infty}^{+\infty} h(v)R_x(\tau + v - \lambda)\, dv \qquad (12.7.12)$$

which shows that $R_y(t_1, t_2) = R_y(\tau)$, and therefore $Y(t)$ is stationary. Not only that, but Equation (12.7.12) also provides us with an expression for finding $R_y(\tau)$, and in conjuction with Eq. (12.7.6), the partial model M_y, $R_y(\tau)$ of the output process $Y(t)$. Granted, Eq. (12.7.12) is messy. But do not despair yet. All that was in the time domain. We still have the frequency domain to worry about, and that will be much easier to handle.

We can forget about the mean of $Y(t)$ as regards the frequency domain; it is simply the FT of a constant, and therefore an impulse at $\omega = 0$. Let us find $S_y(\omega)$, the FT of $R_y(\tau)$. By definition,

$$S_y(\omega) = \int_{-\infty}^{+\infty} R_y(\tau)e^{-j\omega\tau}\, d\tau = \int_{-\infty}^{+\infty} h(\lambda)\, d\lambda \int_{-\infty}^{+\infty} h(v)\, dv \int_{-\infty}^{+\infty} R_x(\tau + v - \lambda)e^{-j\omega\tau}\, d\tau$$

$$(12.7.13)$$

It looks formidable now, but let us go on. We set $\gamma = \tau + v - \lambda$, which means that $\tau = \lambda + \gamma - v$. The limits of the integrals stay the same. With this change of variable,

$$S_y(\omega) = \int_{-\infty}^{+\infty} h(\lambda)\, d\lambda \int_{-\infty}^{+\infty} h(v)\, dv \int_{-\infty}^{+\infty} R_x(\gamma)e^{-j\omega(\lambda+\gamma-v)}\, d\gamma$$

$$= \int_{-\infty}^{+\infty} h(\lambda)e^{-j\omega\lambda}\, d\lambda \int_{-\infty}^{+\infty} h(v)e^{+j\omega v}\, dv \int_{-\infty}^{+\infty} R_x(\gamma)e^{-j\omega\gamma}\, d\gamma$$

or finally,

$$S_y(\omega) = H(\omega)H(-\omega)S_x(\omega) = |H(\omega)|^2 S_x(\omega) \qquad (12.7.14)$$

Very nifty, indeed. In fact, exactly the same relationship that holds for deterministic signals if we compare it with Eq. (12.7.4).

With the results above, we can easily solve problems that demand the tracing of the partial model of a process as it goes through linear systems, like the example below.

Example 12.7.1

A process $X(t)$ with $M_x = 0$ and

$$R_x(\tau) = 10 \cos(10^2\tau)$$

is processed as shown in Fig. 12.7.1. The first linear filter has impulse response $h_1(t) = \delta(t - 10)$ and the second one

$$h_1(t) = 2 \times 10^3 \frac{\sin 10^3(t - 3)}{10^3(t - 3)}$$

while the third one is a differentiator. Find the partial models of $Y(t)$, $Z(t)$, and $O(t)$ and the P_{ave} of each.

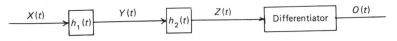

Figure 12.7.1 Processing of a process.

Solution. In view of $M_x = 0$, all means are zero and we ignore them. We must find $R_y(\tau)$, $R_z(\tau)$, and $R_o(\tau)$, or equivalently $S_y(\omega)$, $S_z(\omega)$, and $S_o(\omega)$. With these, P_{ave} is easy—$R(0)$ or the integral of $S(\omega)$.

Although Eq. (12.7.12) can be used for finding $R_y(\tau)$ [it is not bad, since $h_1(t)$ is an impulse], we will use Eq. (12.7.14), which is even easier. Since

$$h_1(t) = \delta(t - 10)$$

then

$$|H(\omega)|^2 = 1$$

and

$$S_y(\omega) = S_x(\omega), \qquad \text{or} \qquad R_y(\tau) = R_x(\tau)$$

so its P_{ave} is

$$R_y(0) = R_x(0) = 10 \text{ W}$$

Now, in view of $h_2(t)$, from Tables 2.6.1 and 2.7.1 we have that

$$H_2(\omega) = 2\pi \text{ rect}\left(\frac{\omega}{2 \times 10^3}\right) e^{-j\omega 3}$$

and

$$|H_2(\omega)|^2 = 4\pi^2 \text{ rect}\left(\frac{\omega}{2 \times 10^3}\right)$$

Thus

$$S_z(\omega) = 40\pi^2 \text{ rect}\left(\frac{\omega}{2 \times 10^3}\right) \pi[\delta(\omega - 10^2) + \delta(\omega + 10^2)]$$

$$= 40\pi^3[\delta(\omega - 10^2) + \delta(\omega + 10^2)]$$

in view of the known property $f(t)\delta(t) = f(0)\delta(t)$, or more correctly here, $f(t)\delta(t - t_0) = f(t_0)\delta(t - t_0)$. Inverting now this last FT we obtain

$$R_z(\tau) = 40\pi^2 \cos 10^2\tau$$

Naturally, its $P_{ave} = 40\pi^2$ W.

The differentiator has $H(\omega) = j\omega$, so

$$S_o(\omega) = \omega^2 \times 40\pi^3 [\delta(\omega - 10^2) + \delta(\omega + 10^2)]$$

$$= 10^2 \times 40 \times \pi^3 [\delta(\omega - 10^2) + \delta(\omega + 10^2)]$$

Finally,

$$R_o(\tau) = 4 \times 10^3\pi^2 \cos 10^2\tau$$

and P_{ave} is obvious.

The reader should now try some of the problems at the end of the chapter. Processing random processes seems to give trouble to beginners. We think that this is because with processes we work on *two* "levels." The first level is $X(t)$ and what happens to it as it goes through the devices. The second level is M and $R_x(\tau)$, its partial model. A linear system acts on the first level with expressions that are mainly symbolic—a convolution, for example. It makes sense, but it is not the partial model of the output process; that has to be found with other expressions. Ponder this issue for a while, as it is necessary that it be cleared up.

One more side topic, and we will be done with this section. We established at the start that if the input to $h(t)$ is a stationary process, so is the output. In fact, these two $X(t)$ and $Y(t)$ can be viewed simultaneously, and their joint partial model can be evaluated. Their joint model is, of course, both the marginal ones and $R_{xy}(\tau)$, or $R_{yx}(\tau)$ if the processes are cross-stationary.

Well, *they are*. Work similar to what we did above (we leave it as an exercise) yields

$$R_{xy}(\tau) = \int_{-\infty}^{+\infty} R_x(\tau + \lambda)h(\lambda)\, d\lambda \qquad (12.7.15)$$

$$R_{yx}(\tau) = \int_{-\infty}^{+\infty} R_x(\tau - \lambda)h(\lambda)\, d\lambda \qquad (12.7.16)$$

and cross-spectral densities

$$S_{xy}(\omega) = S_x(\omega)H(-\omega) \qquad (12.7.17)$$

$$S_{yx}(\omega) = S_x(\omega)H(\omega) \qquad (12.7.18)$$

We will not have much occasion to use them, but they are useful in other studies.

Conclusion. Keeping track of the changes of the partial model of a stationary process as it goes through linear systems is easy using Eq. (12.7.14). Complete models (which we did not discuss) are pretty hopeless unless the input process is Gaussian, a case also not discussed.

12.8 SOME TYPICAL NOISE MODELS

We will now take up and study a few typical noise processes from the point of view of their partial model $R_x(\tau)$ or $S_x(\omega)$ (their M_x will be assumed zero). These models are, of course, mostly theoretical, but practical noises appear to approximate them well enough to legitimate their use in the analysis of communication and radar systems.

We start with the most general one, which is also the most theoretical—white noise.

12.8.1 White Noise

This is a stationary process whose partial model $R_x(\tau)$ is given by

$$R(\tau) = \frac{N_0}{2}\delta(\tau) \qquad (12.8.1)$$

and consequently, its PSD (or simply *spectrum*) is

$$S(\omega) = \frac{N_0}{2} \qquad (12.8.2)$$

where N_0 is a constant (the $N_0/2$ is traditional).

This noise takes its name from its constant spectrum—it resembles the spectrum of white light. Of course, such a noise is largely a figment of our imagination. Equation (12.8.2) implies that its P_{ave} is infinite [the integral of $S(\omega)$], and no practical noise could have that. In the time domain, $R(\tau)$ is an impulse (another figment of someone's imagination—Mr. Heaviside's, we presume). Since the impulse is supposed to be nonzero at $\tau = 0$ and zero everywhere else (the reader will allow us this charlatanism as to the nature of the impulse), this means that the r.v.'s of the noise process are all uncorrelated with each other. Even $X(1)$ and $X(1.000001)$ are completely uncorrelated, and $X(1)$ could have the value 3, whereas the other is 10^8. A realization of such a process would be hard to sketch with this kind of property.

Even so, there are practical processes that are almost white. Thermal noise from a resistor is such a case. Naturally, any practical thermal noise has finite P_{ave}, and this is reflected in $S(\omega)$—it is constant up to a point ($\omega = 2\pi \times 10^{13}$), and then it drops off gradually to zero. In fact, thermal noise can be shown to be approximately a Gaussian process, so theoretically, at least, we could even come up with its complete probabilistic model, involving joint p.d.f.'s. For our purposes, if the noise spectrum is flat and its bandwith much wider than that of the useful signals it is interfering with, we can think of it as white.

Let us take a look now at the case when white $N(t)$ enters an ideal LPF with

$$H(\omega) = \text{rect}\left(\frac{\omega}{2\omega_c}\right) \tag{12.8.3}$$

What is the output's $S_y(\omega)$ and $R_y(\tau)$? Well, all we have to do to find $S_y(\omega)$ is to use the known relation

$$S_y(\omega) = |H(\omega)|^2 S_n(\omega) = |H(\omega)|^2 \frac{N_0}{2} \tag{12.8.4}$$

and in view of the nature of $H(\omega)$,

$$S_y(\omega) = \frac{N_0}{2} \text{rect}\left(\frac{\omega}{2\omega_c}\right) \tag{12.8.5}$$

and inverting it,

$$R_y(\tau) = \frac{\omega_c N_0}{2\pi} \frac{\sin \omega_c t}{\omega_c t} \tag{12.8.6}$$

with $P_{ave} = R_y(0) = \omega_c N_0/2\pi$. This type of noise is often called **ideal low-pass filtered white noise**. What else could it be called?

Now let us try the same idea, with an ideal BPF. If the input to it is white noise, the output's spectrum will have the same shape as the $H(\omega)$ of the BPF, and such noise is called (you guessed it) **ideal bandpass filtered white noise**.

This does it for partial model of processes whose spectra are flat in some region of ω. They can be thought of as outputs of ideal filters, whose input is white noise, and that's that.

What if their spectra are not flat?

12.8.2 Low-Frequency Noise

If the noise is real and has a spectrum around $\omega = \omega_0$, all you can say is that its $S_n(\omega)$ is whatever it is, and that $R_n(\tau)$ is its inverse FT. If $S_n(\omega) = 0$ for $|\omega| \geq \omega_m$, we can call the noise **bandlimited up to ω_m**, in analogy with such a name for deterministic signals. In fact, a sampling theorem exists here as well, but we shall leave it for the reader to find and study it in the literature.

12.8.3 Narrowband Noise

Another very useful model of noise for the analysis of communication and radar systems is the one whose partial model in the frequency domain $S_n(\omega)$ is located somewhere symmetrically around $\omega = \omega_c$, but it is not flat as it would be if it were bandpass filtered white noise.

Let us assume that a noise has a $S_n(\omega)$ defined for $-\omega_c - \omega_m \leq |\omega| \leq \omega_c + \omega_m$, with a symmetrical shape such as the one shown in Fig. 12.8.1. Recalling our Example 12.4.5, this noise can be expressed as

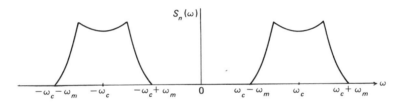

Figure 12.8.1 $S_n(\omega)$ of narrowband noise.

$$N(t) = X(t) \cos (\omega_c t + \theta) \tag{12.8.7}$$

where $X(t)$ is a random process with zero mean and $S_x(\omega)$ around $\omega = 0$, and θ is a r.v. with uniform p.d.f. in $[0, 2\pi]$, and $X(t)$, θ statistically independent. We have already discovered in Example 12.4.5 that this noise's correlation function $R_n(\tau)$ is given by

$$R_n(\tau) = \frac{R_x(\tau)}{2} \cos \omega_c \tau \tag{12.8.8}$$

and its $S_n(\omega)$ by

$$S_n(\omega) = \tfrac{1}{4}[S_x(\omega - \omega_c) + S_x(\omega + \omega_c)] \tag{12.8.9}$$

so the shape of $S_n(\omega)$ around $\omega = \omega_c$ is the same as that of $S_x(\omega)$ around $\omega = 0$, and everything checks. Why are we looking at this again? Be patient.

Let us expand out the cosine in Eq. (12.8.7). This yields

$$N(t) = X(t) \cos \theta \cos \omega_c t - X(t) \sin \theta \sin \omega_c t \tag{12.8.10}$$

and setting

$$n_c(t) = X(t) \cos \theta \tag{12.8.11a}$$

$$n_s(t) = X(t) \sin \theta \tag{12.8.11b}$$

we obtain

$$N(t) = n_c(t) \cos \dot{\omega}_c t - n_s(t) \sin \omega_c t$$

as another expression for our narrowband noise.

The two processes (they are processes, aren't they?) $n_c(t)$ and $n_s(t)$ are called the **quadrature components** of $N(t)$. We want to study these two components briefly, as they will prove quite useful in our eventual analysis of communication and radar systems in the presence of additive noise. Specifically, we wish to show that these two processes have the following two properties:

1. They are both with spectrum around $\omega = 0$.
2. The P_{ave} (average power) of $n_c(t)$, $n_s(t)$, and $N(t)$ are all equal.

To begin with, we know that P_{ave} of $N(t)$, call it P_n, is from Eq. (12.8.8),

$$P_n = R_n(0) = \frac{R_x(0)}{2} \tag{12.8.12}$$

Now let us find $R_c(\tau)$ the correlation function of $n_c(t)$, call it $R_c(t_1, t_2)$. By definition,

$$R_c(t_1, t_2) = E\{n_c(t_1)n_c(t_2)\} = E\{X(t_1)X(t_2) \cos^2 \theta\} \tag{12.8.13}$$

and in view of the above-mentioned independence (really?),

$$R_c(\tau) = E\{X(t_1)X(t_2)\}E\{\cos^2 \theta\} = R_x(\tau)E\left\{\frac{1}{2} + \frac{\cos 2\theta}{2}\right\} = \frac{R_x(\tau)}{2} \tag{12.8.14}$$

Similar mathematical work leads to the correlation function of $n_s(t)$, call it $R_s(\tau)$,

$$R_s(\tau) = \frac{R_x(\tau)}{2} \tag{12.8.15}$$

and with this we have shown both (1) and (2) above. Clearly, the average power of both $n_c(t)$ and $n_s(t)$ is equal to that of $N(t)$ [i.e., $R_x(0)/2$]. And certainly the spectrum of both of these processes is around $\omega = 0$, since they are nothing but $S_x(\omega)/2$. Good.

Now if the spectrum of $N(t)$ is not symmetrical around $\omega = \omega_c$, $N(t)$ can be written as

$$N(t) = X(t) \cos (\omega_c t + \varphi(t)) \tag{12.8.16}$$

where $X(t)$ and $\varphi(t)$ are both processes. Each realization of such a noise will have the form

$$N_r(t) = X_r(t) \cos (\omega_c t + \varphi_r(t)) \tag{12.8.17}$$

that is, it could look like an AM and FM wave, much like the one we encountered in CW-FM radar. The noise of Eq. (12.8.16) is the most general expression for narrowband noise. It can also be written in terms of quadrature components, that is,

$$N(t) = n_c(t) \cos \omega_c t - n_s(t) \sin \omega_c t \tag{12.8.18}$$

where now

$$n_c(t) = X(t) \cos \varphi(t) \qquad\qquad (12.8.19a)$$

$$n_s(t) = X(t) \sin \varphi(t) \qquad\qquad (12.8.19b)$$

and one can come up with various things about them depending on the assumptions put on $X(t)$ and $\varphi(t)$. We will not pursue this matter further here. Anytime we have narrowband noise in a system, we will assume that its spectrum is symmetrical around some $\omega = \omega_c$ (an approximation) and use the results we uncovered above.

In closing, we emphasize that Eqs. (12.8.7), (12.8.11a) and (12.8.11b) are *expressions* for the processes of interest, not their partial probability models. The latter are given by Eqs. (12.8.8), (12.8.14) and (12.8.15), respectively.

PROBLEMS, QUESTIONS, AND EXTENSIONS

12.1. We did not develop the subject of complete models of random processes to the degree that it deserves, and this may give the impression that it is completely useless. We will try to correct that somewhat here. A complete model of a process $X(t)$ is defined by Eq. (12.2.3). Often we do not have it (the Gaussian process is an exception) for any n, but we can find it for $n = 1$ or even $n = 2$. The p.d.f.'s $p\{x(t)\}$ ($n = 1$) and $p\{x(t_1), x(t_2)\}$ ($n = 2$) are called **first-** and **second-order p.d.f.'s** of the process $X(t)$, respectively. Let us try a problem that involves finding them. (We shall return to this topic in Problems 12.15 and 12.16.) A process is given by

$$X(t) = at + bt^2 + ct^3$$

(a) Find its first-order p.d.f. $p\{x(t)\}$ if a and b are constants and c is a r.v. with exponential p.d.f. of mean unity.

(b) Find its first- and second-order p.d.f.'s if a is a constant and b and c are r.v.'s with joint p.d.f.,

$$p(b, c) = 5e^{-c}e^{-5b}u(c)u(b)$$

(*Hint:* Use the theorems on model transformations of r.v.'s.)

12.2. If $X(t) = A \cos(\omega_c t + \theta)$, with ω_c and θ constants and A a r.v. with exponential p.d.f., determine if $X(t)$ is a wide-sense stationary process.

12.3. If $X(t) = A \cos(\omega_c t + \theta)$, with A and ω_c constants and θ a r.v. with $p(\theta)$ in $(-\infty, +\infty)$ but unknown, show that $X(t)$ is stationary if

$$P(1) = P(2) = 0$$

where $P(\omega)$ is the FT of $p(\theta)$.

12.4. We have a stochastic process $X(t)$ and the expression $\int_a^b X(t)\,dt$, where a and b are constants. What is this expression—a constant, a r.v., a process, or what?

12.5. Consider the two processes

$$X(t) = A_1 \cos(\omega_1 t + \theta_1)$$

$$Y(t) = A_2 \cos(\omega_2 t + \theta_2)$$

where A_1, A_2, ω_1, and ω_2 are constants and θ_1 and θ_2 are r.v.'s. Now θ_1 is uniformly distributed in $[0, 2\pi]$, and $\theta_2 = 2\theta$. Find $R_{xy}(t_1, t_2)$. What do you conclude, and why?

12.6. (*The Telegraph Signal or Full-Random Binary Waveform*)   A stochastic process $X(t)$ has realizations of the form shown in Fig. P12.6, that is, they resemble our

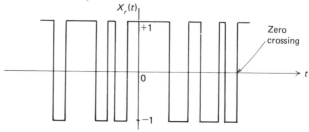

Figure P12.6   Sample function of the process in Problem 12.6.

random ± 1 codes discussed for spread-spectrum systems or pulse compression radar. Now assume that the average number of "zero crossings" per unit time is λ. Assume also that the probability of getting k crossings in time τ seconds is a r.v. with Poisson p.m.f. (a p.d.f. without the impulses), that is,

$$p(k) = e^{-\lambda \tau} \frac{(\lambda \tau)^k}{k!}$$

Show that the $R_x(\tau)$ of this process is

$$R_x(\tau) = e^{-2|\lambda|\tau}$$

12.7. There are some other random processes with realizations like Fig. P12.6, going by such names as **semirandom binary sequences and linear maximal sequences**. Prepare a paper on such processes, tying the result with SS communication systems. (You will find help in Papoulis, 1965, p. 288, or Haykin, 1978, p. 184, etc.)

12.8. Consider a process $X(t)$ with zero mean and partial model $R_x(\tau)$.
   (a) Find the correlation function and spectrum of (1) $aX(t)$ (a constant), (2) $dX(t)/dt$, (3) $d^n X(t)/dt^n$, and (4) $X(t)e^{\pm j\omega_c t}$. Make a table of the results for easy reference.
   (b) Find the cross-correlation function between $X(t)$ and $dX(t)/dt$, and then the cross-correlation function between $d^n X(t)/dt^n$ and $d^m X(t)/dt^m$. [You may find Papoulis (1965) helpful in answering both (a) and (b).]

12.9. White noise $N(t)$, with $S_n(\omega) = N_0/2$, enters the filters of Fig. P12.9a and b. Find $R_o(\tau)$, $S_o(\omega)$, and P_{ave} for $O(t)$ in both cases.

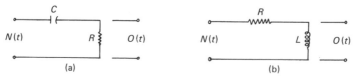

Figure P12.9   Filters for Problem 12.9.

12.10. Repeat Problem 12.9 if

$$R_n(\tau) = 30e^{-10|\tau|}$$

12.11. White noise with $R_n(\tau) = (N_0/2)\delta(\tau)$ enters a filter with

$$H(\omega) = K \exp\left[\frac{-(\omega - \omega_0)^2}{2\sigma^2}\right]$$

where $\omega_0 \gg \sigma$. Find the partial model of the output of the filter as well as its P_{ave}.

12.12. Consider the circuit shown in Fig. P12.12. The process $X(t)$ has $M_x = 1$ and $R_x(\tau) = 1 + e^{-4|\tau|}$. Find $M_o(t)$ and $S_o(\omega)$ of the process $O(t)$, $S_i(\omega)$ of the process $I(t)$, and $S_{oi}(\omega)$, the cross-spectral density of $O(t)$ and $I(t)$.

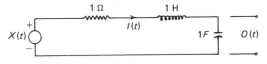

Figure P12.12   Circuit of Problem 12.12.

12.13. A stationary process $X(t)$ with partial model $M_x = 0$, $R_x(\tau) = (\sin 10\tau/10\tau)^2$ enters an ideal LPF filter with

$$|H(\omega)| = u(\omega + \omega_0) - u(\omega - \omega_0)$$

If the output process has P_{ave} which is half of that of the input process, find ω_0.

12.14. (*Matched Filter in the Presence of Additive White Noise*)   Consider a causal filter with impulse response $h(t)$ whose input is

$$X(t) = s(t) + N(t)$$

where $s(t)$ is a known signal (a pulse, a triangle, etc.), and $N(t)$ is a white noise with $S_n(\omega) = N_0/2$. Now at the output of the filter we will have a signal term $s_o(t) = s(t)*h(t)$ and a noise term $N_o(t) = N(t)*h(t)$ (* indicates convolution). Next we define an output signal-to-noise ratio (S/N) as

$$\frac{S}{N} = \frac{\text{instantaneous power of } s_o(t)}{P_{ave} \text{ of } N_o(t)}$$

Find the S/N and $h(t)$ so that the S/N is maximum for all values of t. The result should be

$$h(\xi) = s(t - \xi)$$

and if t is taken as the end of the signal $s(t)$, this filter is called a **matched filter**. Compare this analysis with that of Section 2.8, where the matched filter was first discussed. (*Hint:* In finding the maximum of S/N, use the Cauchy–Schwartz inequality.)

12.15. Recall that the general narrowband noise is given by

$$N(t) = X(t) \cos(\omega_c t + \varphi(t))$$

or

$$N(t) = n_c(t) \cos \omega_c t - n_s(t) \sin \omega_c t$$

(a) Show that if $N(t)$ is a zero mean, σ^2 variance, Gaussian white noise (i.e., all its r.v.'s are uncorrelated and therefore independent—an exception), so are $n_c(t)$ and $n_s(t)$ with the same mean and variance (i.e., the same P_{ave}). In fact, they are statistically independent.

(b) Now do the problem backwards. Assume that $n_c(t)$ and $n_s(t)$ are statistically independent, zero mean, σ^2 variance Gaussian processes. Then find the first-order p.d.f. of

$$X(t) = \sqrt{n_c^2(t) + n_s^2(t)}$$

the first-order p.d.f. of

$$\varphi(t) = \tan^{-1}\frac{n_s(t)}{n_c(t)}$$

and their joint p.d.f. $p\{X(t), \varphi(t)\}$. If you need some help, search the literature. It is usually done as part of the theory.

12.16. (*Sine-Wave plus Narrowband Noise*)   Consider the process

$$X(t) = A\cos\omega_c t + N(t)$$

where A and ω_c are constants and $N(t)$ is a zero mean, σ^2 variance, white Gaussian process. Now write $X(t)$ as

$$X(t) = [A + n_c(t)]\cos\omega_c t - n_s(t)\sin\omega_c t$$
$$= n_c'(t)\cos\omega_c t - n_s(t)\sin\omega_c t$$

(a) Show that $n_c'(t)$ and $n_c(t)$ are Gaussian and statistically independent.

(b) Show that the mean of $n_c'(t)$ is A, the mean of $n_s(t) = 0$, and the variances of both are σ^2.

(c) Define

$$R(t) = \sqrt{[n_c'(t)]^2 + n_s^2(t)}$$
$$\varphi(t) = \tan^{-1}\frac{n_s(t)}{n_c'(t)}$$

and find their joint p.d.f. as well as their marginals.

12.17. State and prove a "sampling theorem" for stationary stochastic processes. Note that the points $X(n\pi/\omega_m)$ (where ω_m is the last frequency of the process) are r.v.'s.

===*13*===

NOISE
IN COMMUNICATION
SYSTEMS

13.1 INTRODUCTION

Noise enters the signal path in a communication system in many places—in the channel, the antennas, even the devices that process the signal. Some noises can be avoided, but most are there and cannot be removed—you just have to live with them and do your best to minimize their detrimental effects. These effects are generally to contaminate the information somewhat, and in extreme cases to bury it completely within itself.

We are now in a position to analyze communication systems, taking noise into account. We have all the background we need in stochastic processes that model noise mathematically. We will, of course, consider only additive noise, as any other type is hard to handle. Needless to say, we cannot possibly cover all the communication systems that we analyzed with the added presence of noise; there is not enough room for that. Our purpose here is to show that such an analysis is possible, and how it is done. We have picked only a few communication systems, with an eye to good pedagogy. The rest, as well as comparisons between them all, can be found in the literature.

One more comment. It will be seen that noises as well as useful signals can be described adequately by their partial models (as we promised in Chapter 12). Only in a very few cases is a complete model (a p.d.f.) needed, and even then the n-dimensional model can be simplified to a single-dimensional one. That is a good thing to notice.

13.2 METHOD OF ANALYSIS: THE SIGNAL-TO-NOISE RATIO

Before we take up the analysis of some communication systems in the presence of additive noise, it is well to consider the *method of analysis* itself. In fact, the method may be more important than the results. We are not going to cover all communication systems here, just a few, for the sake of illustrating the method. The reader will have to do the rest, so a good understanding of the methodology is essential.

To evaluate system performance and then compare systems with each other, we need a **criterion of performance**. When we were comparing systems on the basis of their bandwidth (or economics), the bandwidth itself was the criterion of performance—narrow bandwidth was good, a wide bandwidth was bad. Now noise is being allowed in the system and the criterion of performance will have to change. With it, of course, the ranking of the systems may also change, demonstrating once again that there is no panacea in any field of endeavor.

Criteria of performance constitute a field in themselves, with deep philosophical and mathematical connotations. Since fundamentally we wish to know how well the output of the receiver of a system approximates the signal transmitted originally, *distance* functions of vector space theory play a principal role. Of course, a detailed discussion of performance criteria is beyond the scope of this book (there is that phrase again, so we will just throw out the criterion and leave the philosophical questions for the reader to ponder.

Our criterion of performance in analog systems will be the **output signal-to-noise ratio**, which we shall denote as S_o/N_o. What is it, and how does it indicate performance in the presence of additive noise? First, we shall concentrate our noise analysis at the *receiver* of a system; after all, it is there that the output will emerge. At the input to the receiver we will assume that the signal has the form $s_i(t) + N_i(t)$ (see Fig. 13.2.1), where $s_i(t)$ is the modulated wave (the useful part of our received signal) and $N_i(t)$ the total of all noise accumulated up to that point from the channel, antenna, devices, and the like. The receiver will operate on the incoming signals and produce at its output two terms, $s_o(t)$ (the useful part) and $N_o(t)$ (the noise part). At this point, then, the output, we will calculate the *power* of the signal term S_o and the *power* of the noise term N_o. Their ratio, S_o/N_o, is a number that will indicate the performance of the system. If it is large, we will accept that the system performs well—the signal predominates. If it is small, the system is performing poorly—the noise predominates. Systems can be compared to each other by looking at their S_o/N_o under the same conditions and ascertaining which of the two has the advantage.

Figure 13.2.1 Illustration of the S_o/N_o criterion of performance.

How good is this criterion? It is hard to say. Practical results seem to agree with theoretical results, at least most of the time, and those that do not may be due to the many approximations involved in the analysis. The criterion is good from other points of view—tractable analysis, relative ease of handling with a limited background in stochastic processes, and the like—and all these things contribute to its high popularity. Anyway, this is what we can handle at this point, so this is what we will use. A brief introduction to another criterion, more appropriate for PCM or mixed systems, will also be presented later, toward the end of the chapter.

Incidentally, how do we justify looking only at the receiver of a system? After all, the input at the receiver also depends on the transmitter. Why are we ignoring this aspect of the overall communication system?

The answer to that (not completely satisfactory) is that although we look only at the receiver, we do not completely ignore the transmitter. The nature of the useful signal at the input to the receiver is due to the transmitter. Even the receiver structure is dictated by the incoming modulated wave and therefore the transmitter. Thus the work of the transmitter is there all right—but behind the scenes. Not very convincing, but that is all we have.

In most cases, the S_o/N_o of a system's receiver can be found as a function of the input signal-to-noise ratio, S_i/N_i. In such cases we can also calculate their ratio, a quantity called **demodulation gain** (D.G.):

$$\text{D.G.} = \frac{S_o/N_o}{S_i/N_i} \qquad (13.2.1)$$

which represents another way of looking at the same criterion. D.G. specifies a little more clearly how the receiver improves (if it does) the incoming situation.

13.3 THE DSB SYSTEM

Our first analysis involves the DSB system, whose receiver is shown in Fig. 3.2.1 and reproduced in Fig. 13.3.1 for easy reference. Note also that we have kept only the essential part of this receiver—the multiplicator and LPF [i.e., we picked it up after the amplifier(s)]. The input to this receiver is

$$m(t) + N_i(t) = f(t) \cos \omega_c t + N_i(t) \qquad (13.3.1)$$

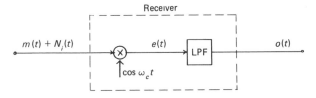

Figure 13.3.1 Product detector in DSB systems.

where $f(t)$ is the information signal transmitted originally, $\cos \omega_c t$ the carrier, and $N_i(t)$ the noise—an accumulation of all noise effects up to this point of the system.

We are interested in obtaining S_o/N_o. Before we do that, however, let us first discuss the nature of $m(t)$ and $N_i(t)$ and find S_i/N_i as well. It will be useful at the end in calculating our D.G. We will take $f(t)$ to be a deterministic signal, even though, with slight modifications, we could assume it to be a random process as well. The results will be the same, anyway.

Let us assume that the original power of $f(t)$ is S, that is,

$$\lim_{T \to \infty} \frac{1}{2T} \int_{-T}^{+T} f^2(t)\, dt = S \tag{13.3.2}$$

The power of the useful signal $m(t)$ at the input of the homodyne detector will be

$$S_i = \lim_{T \to \infty} \frac{1}{2T} \int_{-T}^{+T} f^2(t) \cos^2 \omega_c t\, dt = \frac{S}{2} \tag{13.3.3}$$

as the reader can easily verify by breaking down the cosine to $\frac{1}{2} + \cos 2\omega_c$ and arguing that the second term is zero (check the result in the frequency domain also). So much for the power of the useful signal $m(t)$. Now we look at the noise $N_i(t)$.

It is quite reasonable to assume that this noise will be of the narrowband type, with a spectrum around ω_c. Most noises are of wider spectrum to begin with, but after going through all the narrowband amplifier stages, and so on, they will end up narrowband, anyway.

This being the case, and from our background of Section 12.8, we can write

$$N_i(t) = n_c(t) \cos \omega_c t - n_s(t) \sin \omega_c t \tag{13.3.4}$$

as we have already established with Eq. (12.8.18). Let us assume that this noise $N_i(t)$ has power equal to P, that is,

$$N_i = R_{ni}(0) = P \tag{13.3.5}$$

Equations (13.3.3) and (13.3.5) combine to give us the input S_i/N_i,

$$\left(\frac{S_i}{N_i}\right)_{\text{DSB}} = \frac{S}{2P} \tag{13.3.6}$$

Now we go after the S_o/N_o. If we multiply the input signal by the locally generated $\cos \omega_c t$ and then keep only the low-frequency terms (the action of the LPF), the result is

$$o(t) = \frac{f(t)}{2} + \frac{n_c(t)}{2} \tag{13.3.7}$$

since the $n_s(t)$ term of $N_i(t)$ will disappear. The first term is our useful signal and its power is

$$S_o = \frac{S}{4} \tag{13.3.8}$$

The second term is our noise term. Recalling that the power of $n_c(t)$ is equal to that of $N_i(t)$ (i.e., P), the power of this term is

$$N_o = \frac{P}{4} \tag{13.3.9}$$

Thus

$$\left(\frac{S_o}{N_o}\right)_{\text{DSB}} = \frac{S}{P} = 2\left(\frac{S_i}{N_i}\right)_{\text{DSB}} \tag{13.3.10}$$

and,

$$\text{D.G.} = \left(\frac{S_o/N_o}{S_i/N_i}\right)_{\text{DSB}} = 2 \tag{13.3.11}$$

These last two expressions give us an idea of the behavior of the DSB system with a homodyne detector in the presence of additive noise. The signal-to-noise ratio has improved at the output by a factor of 2, since one of the two quadrature components of the $N_i(t)$ [the $n_s(t)$] has been rejected. At this point we cannot be sure whether that is good or bad, but all this will become clearer as the discussion evolves and other systems are analyzed for comparison.

Anyway, the analysis in the presence of additive noise was not too difficult after all the background we have accumulated in the Chapters 11 and 12.

13.4 THE SSB SYSTEM

Everything remains just about the same here as it was in the DSB system, except that the useful signal at the input of the homodyne receiver has a form similar to Eq. (3.3.16), that is,

$$m'_{\pm}(t) = \tfrac{1}{2}[f(t)\cos \omega_c t \mp \hat{f}(t) \sin \omega_c t] \tag{13.4.1}$$

where $\hat{f}(t)$ is the Hilbert transform of $f(t)$. We already know that the FT of $m'_{\pm}(t)$ has only one of the two sidebands. For reasons of comparison with the DSB case, we can assume that the transmitter power of $m'_{\pm}(t)$ is also $S_i = S/2$ [i.e., the power of $f(t)$ is $2S$], now packed in one sideband rather than two. We can keep the same expression for the noise, but we will assume that its power is $N_i = P/2$, since it now has half the bandwidth it had in the DSB case. (Why?) Thus

$$\left(\frac{S_i}{N_i}\right)_{\text{SSB}} = \frac{S}{P} \tag{13.4.2}$$

After the multiplication and the LPF operation, the result will be (check it)

$$o(t) = \tfrac{1}{4} f(t) + \tfrac{1}{2} n_c(t) \tag{13.4.3}$$

so

$$\left(\frac{S_o}{N_o}\right)_{SSB} = \frac{S/8}{P/8} \tag{13.4.4}$$

and

$$\text{D.G.} = 1 \tag{13.4.5}$$

half that of the *DSB* case, *assuming that both transmit the same power* and *that the noise power at the input of the SSB detector is half that of the DSB one.*

We should point out that comparisons between systems are a bit messy, because in practice conditions are not the same in all systems as we try to assume in theory. In any event, the output signal-to-noise ratio is not worse than the input one in any of our two systems so far. No degradation takes place due to the action of the receiver in DSB and SSB systems.

13.5 THE AM SYSTEM

The transmitted signal in this case has the form

$$m(t) = [A + f(t)] \cos \omega_c t \tag{13.5.1}$$

and this is the form that we will assume the signal has at the input of the receiver, which, in this case, is an envelope detector. Note, however, that the useful part of this signal is only $f(t) \cos \omega_c t$, since the carrier carries no information.

We will assume that the noise $N_i(t)$ is again narrowband around ω_c, with power P for comparison purposes. So

$$N_i = P \tag{13.5.2}$$

What about the power in the useful signal? We must be a bit careful here if we are to compare the result with the previous two systems.

Let us assume that the useful transmitted power is again $S/2$; that is, the term $f(t) \cos \omega_c t$ has this power. In view of the above,

$$\left(\frac{S_i}{N_i}\right)_{AM} = \frac{S}{2P} \tag{13.5.3}$$

To find the output of the envelope detector, we must write its input as a single cosine.

$$m(t) + N_i(t) = [A + f(t)] \cos \omega_c t + n_c(t) \cos \omega_c t - n_s(t) \sin \omega_c t$$
$$= c(t) \cos (\omega_c t + \psi(t)) \tag{13.5.4}$$

where the envelope is

$$c(t) = [(A + f(t) + n_c(t))^2 + n_s^2(t)]^{1/2} \tag{13.5.5a}$$

and the phase angle is

$$\psi(t) = \tan^{-1}\frac{n_s(t)}{A + f(t) + n_c(t)} \tag{13.5.5b}$$

Clearly, the output of the envelope detector is approximately $c(t)$, but to continue we must hit the approximations road.

1. *Large-signal case.* Let us assume that the signal magnitude is always (or nearly always) much larger than that of the noise, or mathematically that

$$|A + f(t)| \gg \sqrt{n_c^2(t) + n_s^2(t)} \tag{13.5.6}$$

With this, it can easily be argued that

$$o(t) \approx c(t) \approx A + f(t) + n_c(t) \tag{13.5.7}$$

an approximation, of course, as we have already mentioned.

Now we recognize as the useful signal the term $f(t)$, whose power is S, since the power of $f(t)\cos\omega_c t$ was assumed to be $S/2$. The power of the noise is P, as we recall that $N_i(t)$ and $n_c(t)$ have the same power. Thus

$$\left(\frac{S_o}{N_o}\right)_{\text{AM}} = \frac{S}{P} \tag{13.5.8}$$

and

$$\text{D.G.} = 2 \tag{13.5.9}$$

which is the result obtained in the previous two systems. This result is, however, misleading. The large-carrier AM, we should recall, wastes a lot of its transmitted power in the carrier. In our analysis above we assumed that the useful term $f(t)\cos\omega_c t$ has power $S/2$. But to have this much power in this AM system one may need three times more power in $m(t)$, as we established in Section 3.4 [see Eq. (3.4.16)]. Thus this system, even with the approximation above, is much worse in terms of performance in the presence of noise than the previous two, a price one pays (another one) for the cheap receiver.

2. *Small-signal case.* Now if

$$|A + f(t)| \ll \sqrt{n_c^2(t) + n_s^2(t)} \tag{13.5.10}$$

the output is approximately (why?)

$$o(t) \approx \sqrt{n_c^2(t) + n_s^2(t)} + \frac{n_c(t)}{\sqrt{n_c^2(t) + n_s^2(t)}}[A + f(t)] \tag{13.5.11}$$

a disaster. The useful signal $f(t)$ is actually lost here, since in the only term in which it appears it is multiplied by noise and gets completely annihilated. No point going on any further here.

The cases above are, of course, extremes. Between the two there is also the case when signal and noise have about the same power. Somewhere around that point the signal gets lost, and this type of transition is usually called a **threshold effect**. We will encounter this type of effect again in the next subsection in the analysis of FM systems.

So far, then, we have established that DSB and SSB perform pretty similarly in the presence of additive noise, and that AM with the peak detector is worse. We could try other types of AM (vestigial, carrier reinsertion, etc.), but the method of analysis is the same, and we leave them as further study for the reader (see Problem 13.2).

13.6 THE FM SYSTEM

We have often remarked that in an FM system one tolerates large bandwidth for the sake of good performance in the presence of noise. The time has come to prove this assertion with mathematics. It will not be easy, nothing is with FM—of this we are keenly aware. Approximations will always play a key role in anything having to do with FM analysis.

We will not bother with narrowband FM (the reader can check it in the literature) but plunge directly into the wideband FM case, the most interesting one. Phase modulation systems will also be omitted, for reasons mentioned in Chapter 5.

Let us start out by putting down a diagram of an ideal FM receiver, the one shown in Fig. 13.6.1. The signal arriving at the receiver is assumed to be $m_f(t) + N(t)$, where $m_f(t)$ has the form

$$m_f(t) = A \cos(\omega_c t + k_f g(t)) \tag{13.6.1}$$

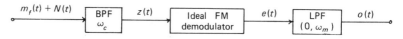

Figure 13.6.1   FM receiver.

with

$$g(t) = \int s(t)\, dt \tag{13.6.2}$$

and $s(t)$ the original message signal with spectrum in $(0, \omega_m)$. The constant k_f, we recall, is a device constant indicating the amount of FM modulation. Recall also that the bandwidth of $m_f(t)$ is, by Carson's rule,

$$B \approx 2(\Delta\omega + \omega_m) \approx 2\,\Delta\omega \tag{13.6.3}$$

where $\Delta\omega$ is the maximum frequency deviation of the carrier's frequency, from ω_c. It is probably best for the reader to restudy Section 5.4 so that everything about FM without noise is fresh in his or her mind. Note that B above is approximated further simply by $2\,\Delta\omega$, to make our analysis a bit easier to handle.

In Fig. 13.6.1, the noise $N(t)$ is assumed to be originally white, with power $N_o/2$. Right at the start of the receiver system we have placed an ideal BPF filter which represents the various amplifiers and other devices and whose bandwidth around ω_c is just enough to pass the signal $m_f(t)$. There is no need to make it wider—it will just allow more noise to go through. The output of this filter is denoted as $z(t)$, and it is this signal that we will assume as the input to our FM receiver. At this point, then, we will find the S_i/N_i.

To begin with, the power of the useful signal $m_f(t)$ is obviously

$$S_i = \frac{A^2}{2} \qquad (13.6.4)$$

as it is just a cosine and its phase angle does not change its power.

Now let us look at the noise in $z(t)$, call it $N_i(t)$. Well, $N(t)$ at the input of the BPF filter was assumed white with power $N_o/2$ (i.e., with a flat spectrum of $N_o/2$ height). At the output of the filter the noise will be bandpass filtered white noise, and its power will be the area under its spectral density, or

$$N_i = \frac{1}{\pi}\frac{N_o}{2}(2\,\Delta\omega) = \frac{N_o\,\Delta\omega}{\pi} \qquad (13.6.5)$$

where $2\,\Delta\omega$ is the bandwidth of the filter [equal to the approximate bandwidth of $m_f(t)$].

Having found both S_i and N_i, we have

$$\left(\frac{S_i}{N_i}\right)_{\mathrm{FM}} = \frac{\pi A^2}{2N_o\,\Delta\omega} \qquad (13.6.6)$$

So far, so good. Now we must find the powers S_o and N_o at the output of the receiver—not an easy task, since even an ideal demodulator is a nonlinear device. What do we do? Make an approximation, of course.

There are convincing theoretical arguments (see Schwartz et al., 1966) which indicate that for large input S_i/N_i [so let Eq. (13.6.6) be large], the powers of the signal and noise at the output of an FM receiver can be found "independently." By this it is meant that we can proceed to find the output signal power S_o by assuming that $N_i(t)$ is zero, and the output noise power N_o, by assuming that the *modulating* signal $s(t)$ is zero. We naturally accept these arguments and go on with the analysis.

The ideal FM demodulator will give as an output the derivative of the angle of $m_f(t)$, that is, $K_D(\omega_c + k_f s(t))$, where K_D is a device constant. The useful signal is $K_D k_f s(t)$, which will go through the ideal LPF in fine shape and appear at $o(t)$. Its power is, of course,

$$S_o = K_D^2 k_f^2 S \qquad (13.6.7)$$

where S the power of the modulating signal $s(t)$.

Next, we assume that $s(t) = 0$, that is,

$$z(t) = A\cos\omega_c t + N_i(t) \qquad (13.6.8)$$

Flat spectrum or not, the noise $N_i(t)$ is of the narrowband type and can be written in terms of its components $n_c(t)$ and $n_s(t)$ as in Eq. (13.3.4). So

$$z(t) = [A + n_c(t)] \cos \omega_c t - n_s(t) \sin \omega_c t \qquad (13.6.9)$$

or

$$z(t) = c(t) \cos (\omega_c t + \psi(t)) \qquad (13.6.10)$$

where $c(t)$ is as usual, and $\psi(t)$, which is of interest to us, is

$$\psi(t) = \tan^{-1} \frac{n_s(t)}{A + n_c(t)} \qquad (13.6.11)$$

We have already assumed that the signal power is larger than the noise power, so the approximation

$$\psi(t) \approx \tan^{-1} \frac{n_s(t)}{A} \approx \frac{n_s(t)}{A} \qquad (13.6.12)$$

is not going to surprise anyone, we hope.

The ideal FM demodulator (a discriminator say) will give as an output

$$K_D \omega_c + K_D \frac{d}{dt} \frac{n_s(t)}{A} \qquad (13.6.13)$$

the second term being the one that interests us.

Now we know that the power of $n_s(t)$ is equal to the power of $N_i(t)$, which we found in Eq. (13.6.5). How do we find the power of the noise term in Eq. (13.6.13)? It is not that difficult if we do our work in the frequency domain, where a differentiator has a system function equal to $j\omega$. We recall that the spectrum of $n_s(t)$ has the same flat shape as that of $N_i(t)$, but it is sitting around $\omega = 0$ [check Eqs. (12.8.9) and (12.8.15)]. Since $n_s(t)$ has the same average power and half the bandwidth of $N_i(t)$, the magnitude of its spectrum must be N_o. Using our known input/output spectral relationship in linear systems, Eq. (12.7.14), we get the noise spectrum as

$$\left(\frac{K_D}{A}\right)^2 N_o \, |j\omega|^2 = \left(\frac{K_D}{A}\right)^2 N_o \omega^2 \qquad (13.6.14)$$

This is at the point just before the LPF. The LPF is ideal, and the above will be cut off at $\omega = \pm \omega_m$, so the output noise power will be

$$N_o = \frac{1}{\pi} \int_0^{\omega_m} \frac{K_D^2 N_o}{A^2} \omega^2 \, d\omega = \frac{K_D^2 N_o \omega_m^3}{3\pi A^2} \qquad (13.6.15)$$

and finally,

$$\left(\frac{S_o}{N_o}\right)_{FM} = \frac{3\pi A^2 k_f^2 S}{N_o \omega_m^3} \qquad (13.6.16)$$

with

$$\text{D.G.} = \frac{6 S k_f^2 \, \Delta\omega}{\omega_m^3} \qquad (13.6.17)$$

not just a number, as it was with the AM systems.

This last expression is mighty important, as it more or less tells the story of our ability to exchange bandwidth for performance in the presence of noise—something we have been harping on during the entire book. To see this more clearly, let us take our simple case where $s(t) = a \cos \omega_m t$, whose power is $S = a^2/2$, and $\Delta \omega = ak_f$. Then

$$\text{D.G.} = 3\left(\frac{ak_f}{\omega_m}\right)^3 = 3\beta^3 \qquad (13.6.18)$$

by recalling that $\beta = ak_f/\omega_m$ [Eq. (5.4.5)]. So the D.G. varies proportionally to the cube of the modulation index; that is, theoretically, at least, we can make it as large as we would like. Of course, high β increases the bandwidth of the FM wave, which we recall is

$$B = 2\omega_m(\beta + 1) \qquad (13.6.19)$$

and that is how the exchange we mentioned takes place in practice.

In FM broadcasts, the maximum frequency present is $\omega_m = 15$ kHz and $ak_f = 75$ kHz. Inserting these values into Eq. (13.6.18), we obtain

$$\text{D.G.} = 375$$

a lot bigger than 2 (or less) that we were getting in AM systems. We finally did get something to pay off for the large bandwidth of wideband FM. The results even seem to justify the complexity that we encountered in the analysis of FM systems. It was about time.

Of course, this whole analysis was based on many approximations and conditions, the most important of which is that the signal-to-noise ratio (SNR) at the input to the receiver is large. There is a point at which this ratio is not large enough for the analysis to hold, and this point is called the **FM threshold**. This point is usually around 10 to 12 dB. If the input SNR falls below this threshold, it can be shown theoretically (and verified practically) that the D.G. is smaller than that predicted by our analysis above; it can even get smaller than the one in AM systems. The threshold problem in FM is quite important, and a lot of work has gone into lowering it by studying it and using special circuits. It is a good topic for further independent study by the reader, as it must be omitted here. Also check Problem 13.7, which extends this section further.

In closing, let us remark that our analyses have shown FM to be best on the basis of the D.G. criterion (above threshold, anyway), while AM appears to be worst. Of course, what is best here may be worst elsewhere, a point that will be clarified in Chapter 15.

13.7 THE PCM SYSTEM

PCM systems are rapidly gaining in preference in modern communications, so we will look at them with some care. In fact, we will take this opportunity to fill in a gap that remains (by necessity) in their analysis up to now, a gap that is nicely filled with knowledge of stochastic processes.

Take the way the pulses are received in a baseband PCM (not mixed) system. We have often remarked that this is done with a *matched* filter. We have even given a justification for this in Section 2.8, the best we could do then with the background we had at that point (see also Problem 12.14). Now we will show that the matched filter is *optimum* even in the presence of noise—optimum with respect to the signal-to-noise ratio (SNR) criterion we have been talking about. More specifically, we will show that the output SNR of a linear filter is maximum when this filter is matched to the input signal which enters the filter with additive white noise.

Let us assume, then, that the input to a linear filter with impulse response $h(t)$ is

$$\text{input} = s(t) + N(t) \tag{13.7.1}$$

where $s(t)$ is a useful signal (a pulse in PCM) and $N(t)$ is zero mean white noise with partial model

$$R_n(\tau) = \frac{N_o}{2} \delta(\tau) \tag{13.7.2}$$

or

$$S_n(\omega) = \frac{N_o}{2} \tag{13.7.3}$$

The output of the filter $o(t)$ will be

$$o(t) = s_o(t) + N_o(t) \tag{13.7.4}$$

Assuming that the filter is causal, we have

$$s_o(t) = \int_0^\infty h(\xi)s(t - \xi) \, d\xi \tag{13.7.5}$$

and the instantaneous power of this output deterministic signal will be the square of the expression above.

Now let us look at the output noise $N_o(t)$. Using our knowledge of linear systems with random processes as inputs, we can easily show that the output power $R_{no}(0)$ will be

$$R_{no}(0) = \frac{N_o}{2} \int_0^\infty h^2(v) \, dv \tag{13.7.6}$$

With this and Eq. (13.7.5), we can define the output SNR as

$$\frac{S_o}{N_o} = \frac{2}{N_o} \frac{\left[\int_0^\infty h(\xi)s(t - \xi) \, d\xi \right]^2}{\int_0^\infty h^2(v) \, dv} \tag{13.7.7}$$

where the numerator is the *instantaneous* power of the signal. This SNR is a little bit different from our previous ones (those had total average power of the signal), but still an SNR.

Now we seek $h(t)$ which will maximize expression (13.7.7). Recalling the

Cauchy–Schwartz inequality and using it on the numerator, we can write

$$\left[\int_0^\infty h(\xi)s(t-\xi)\,d\xi\right]^2 \le \int_0^\infty h^2(\xi)\,d\xi \int_0^\infty s^2(t-\xi)\,d\xi \qquad (13.7.8a)$$

with equality iff

$$h(\xi) = As(t-\xi) \qquad (13.7.8b)$$

If Eq. (13.7.8) is inserted in Eq. (13.7.7), we end up with

$$\frac{S_o}{N_o} \le \frac{2}{N_o}\int_0^\infty s^2(t-\xi)\,d\xi \qquad (13.7.9)$$

This shows that the maximum value of S_o/N_o is the left-hand side, and it occurs when $h(\xi)$ is the one dictated by Eq. (13.7.8b). Note, however, that our S_o/N_o is a function of t. What is the value of t that makes the right-hand side of Eq. (13.7.9) a maximum?

Well, since $s(t)$ is a pulse (square, triangle, etc.), it is not hard to see that this value is the end point of the pulse (prove it). Thus, finally the maximum value of S_o/N_o takes place when

$$h(\xi) = As(\xi - T) \qquad (13.7.10)$$

where T is the end of the pulse, and it occurs at the value $t = T$. Figure 13.7.1a and b shows an example, with $s(t)$ a rectangle. The filter $h(\xi - T)$ turns out to be exactly of the same shape—that is why it is called a **matched filter**.

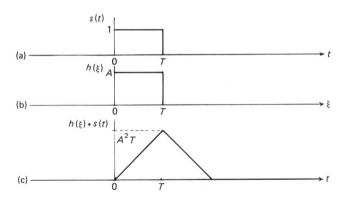

Figure 13.7.1   (a) $s(t)$; (b) its matched filter; (c) their convolution.

Our conclusion so far is that the best way (in the SNR sense) to receive a PCM pulse (or any other signal, actually) in the presence of additive white noise is by means of a matched filter. That is exactly what we do in PCM and most other systems (pulse radar, for instance) when the noise can be considered white.

Let us assume now that the incoming signal is a rectangular pulse, and that it is processed by a matched filter. The output—the convolution of the two—

is known to be a triangle, as shown in Fig. 13.7.1c. The maximum value of the output occurs at $t = T$, as expected from our discussion above. It all checks out. Of course, the output no longer looks like the original $s(t)$—now it looks like a triangle. So when we try to decide whether a "+1" has arrived, we must look for triangles, not rectangles. In the presence of noise, the triangles are expected to distort, and the decision as to whether a "one" is present may not be so easy. We will have more to say later about this decision. For the time being, we will return to our problem of analyzing a general PCM system in the presence of additive noise. The gap has now been filled.

Let us now shift the scene to a PCM receiver, to see what we can do about an analysis in the presence of noise. We ignore its actual form (it will have a matched filter, a decoder, and an LPF smoothing filter) and go directly to its output. If we assume that the PCM is baseband, we can also assume that enough repeaters have been placed along the channel so that no errors have taken place in reconstructing $f_c(t)$, the coded form of the original message $f(t)$. The output of the PCM receiver can be written as

$$o(t) = f_q(t) + N(t) \tag{13.7.11}$$

where $f_q(t)$ is the quantized form of $f(t)$ and $N(t)$ is the receiver noise, usually thermal. Next we recall that

$$f_q(t) = f(t) + \epsilon(t) \tag{13.7.12}$$

where $\epsilon(t)$ is the quantization noise, a noise always present in such systems. Thus

$$o(t) = f(t) + \epsilon(t) + N(t)$$

and we seek the S_o/N_o for this output. The useful signal is our original $f(t)$, and the rest two are noise terms that can be assumed uncorrelated (with zero means) since they are produced by entirely different mechanisms. If we denote the partial model of $\epsilon(t)$ and $N(t)$ by $R_\epsilon(\tau)$ and $R_n(\tau)$, respectively, we have that

$$\left(\frac{S_o}{N_o}\right)_{PCM} = \frac{S}{R_\epsilon(0) + R_n(0)} \tag{13.7.13}$$

where S is the power of the original message $f(t)$. Now $R_\epsilon(0)$ is constant once the quantization mechanism has been decided, so it is the term $R_n(0)$ that could give rise to a threshold effect in PCM systems.

In most baseband PCM systems $R_n(0)$ is small compared to $R_\epsilon(0)$, so

$$\left(\frac{S_o}{N_o}\right)_{PCM} = \frac{S}{R_\epsilon(0)} \tag{13.7.14}$$

an expression valid when both channel noise and repeater noise are negligible. In mixed systems this cannot be assumed, but in regular baseband PCM, this assumption is pretty close to reality.

At this point we are going to derive another form for the S_o/N_o of Eq. (13.7.14), a form that relates it to the bandwidth of the emitted signal $f_c(t)$. It is a useful result, and it will enable us to compare the PCM with the systems analyzed previously. The comparison will be a little bit unfair, of course, since all noises except quantization will be ignored, but it will be better than nothing.

Since we are dealing only with the quantization noise, we can switch our attention to the transmitter of the PCM system, where the quantization mechanism is performed. Such a transmitter was first shown in Fig. 8.4.1, and we have redrawn it here in Fig. 13.7.2 for easy reference. Note that we have changed the order of sampling and quantization, a change that has no effect on the system, as a little thinking will reveal.

Figure 13.7.2   Typical PCM transmitter.

We will start by assuming that $f(t)$ is a zero mean stationary process with known partial model $R_f(\tau)$ [or $S_f(\omega)$] and therefore with power $R_f(0) = S$. There is no problem with this assumption; we have often stated that useful signals (say speech) can also be modeled as such processes.

At the input to the quantizer, then, the signal power is S and the noise power is zero. The S_i/N_i is infinite, but we are not interested in that in this analysis. Our concern here is to find $R_\epsilon(0)$, the quantization noise power, so that we can use it in expression (13.7.14). To find it we proceed as follows.

We assume that the r.v.'s of the message process $f(t)$ at each t have a *uniform* marginal p.d.f. (it is called the first order p.d.f. of the process; see Problem 12.1) in a region of values $[-M, +M]$. The quantizer is assumed to have n steps, all equal to the value a. This means (prove it) that the r.v.'s of the process $\epsilon(t)$ have a range of values $[-a/2, +a/2]$, and a uniform p.d.f. in this range as well. Since the mean of the process $\epsilon(t)$ is zero (why?), the variance of its r.v.'s is equal to $R_\epsilon(0)$, which is also the power of $\epsilon(t)$. Now the variance of the r.v.'s of $\epsilon(t)$ is given by

$$R_\epsilon(0) = E\{\epsilon^2(t)\} = \frac{1}{a} \int_{-a/2}^{a/2} e^2(t)\, de(t) = \frac{a^2}{12} \tag{13.7.15}$$

and this is also the value of the power we have been seeking. Therefore,

$$\left(\frac{S_o}{N_o}\right)_{\text{PCM}} = \frac{12S}{a^2} \tag{13.7.16}$$

but we will have to work on it a bit before we can begin to discuss it. Let it be noted, though, that the whole derivation is based on the assumption that the p.d.f.'s of the r.v.'s of $f(t)$ are uniform. If not, Eq. (13.7.16) will be different, but not enough to change our eventual conclusions.

Now since the range of values of the continuous $f(t)$ is $2M$ (from $-M$ to

$+M$) and there are n steps, the step size a is

$$a = \frac{2M}{n} \qquad (13.7.17)$$

which makes

$$\left(\frac{S_o}{N_o}\right)_{PCM} = \frac{3n^2 S}{M^2} \qquad (13.7.18)$$

To relate this expression to the bandwidth of the transmitted signal so that we can compare this system with the previous ones, we proceed as follows.

First, let's do away with this sampling business, which as we already remarked can take place before or after quantization. If the original continuous signal is bandlimited up to ω_m, the sampling is done every $T_s = \pi/\omega_m$ seconds. The fact that $f(t)$ is now a process does not change anything here, since the sampling theorem still holds, as indicated in Problem 12.17.

The quantized-sampled values of the r.v.'s are then coded. Let us assume that the code is of the binary arithmetic type. This means that if the quantization levels are n, the length of the code words are $\log_2 n$, and this number of bits must be squeezed in T_s if we are not multiplexing. Assuming no guard bands between pulses, each pulse width will be

$$\frac{\pi}{\omega_m \log_2 n} \qquad (13.7.19)$$

and the approximate bandwidth of the emitted signal will be (why?)

$$B = 2\omega_m \log_2 n \qquad (13.7.20)$$

from which

$$n^2 = 2^{B/\omega_m} \qquad (13.7.21)$$

and therefore,

$$\left(\frac{S_o}{N_o}\right)_{PCM} = \frac{3S}{M^2} 2^{B/\omega_m} \qquad (13.7.22)$$

Here is the expression that we have been aiming for. The performance of the PCM can be exchanged with bandwidth—in an exponential way. Of course, the performance here is in the presence of quantization noise, so it is not altogether fair to compare the result with those for the previous systems. Basically, Eq. (13.7.22) tells us that if you want to decrease the quantization noise in a PCM, quantize it as little as possible (i.e., make n very large). If you do that, $\log_2 n$ gets large and so does B, and therefore S_o/N_o. But we knew that, anyway.

We close by reminding the reader that our analysis was based on ignoring channel noise completely. Only in deriving the matched filter was it taken into account. After that we assumed it negligible. If it is not, there will be errors in the detection of the coded signal, and such errors, as well as their effects in performance, are best handled with another criterion—the probability of error. We will see this criterion and how it comes about in the last section of this chapter.

13.8 SIGNAL-TO-NOISE RATIO COMPARISONS

Comparing things is part of the human condition; it enables us to classify them as good, bad, and the like. It must always be done, of course, on the basis of a specific criterion, and the results are valid for this criterion alone. The chapter so far has been devoted to the signal-to-noise ratio type of criterion, and on the basis of this we can now make some comparisons, at least for the systems we have analyzed.

To be able to include the PCM in our comparisons, we stick to the (S_o/N_o) and ignore D.G. Let us summarize most of our results up to now.

1. *For DSB, AM*

$$\left(\frac{S_o}{N_o}\right) = 2\left(\frac{S_i}{N_i}\right) \tag{13.8.1}$$

2. *For FM*

$$\frac{S_o}{N_o} = \frac{3\pi A^2 S}{N_o \omega_m^3}(k_f)^2 \tag{13.8.2}$$

3. *For PCM*

$$\frac{S_o}{N_o} = \frac{3S}{M^2} 2^{B/\omega_m} \tag{13.8.3}$$

The picture is quite clear, at least as a generality. The PCM seems to be best, followed by the FM. In both of these, the S_o/N_o can be increased at will (by suffering an increase in bandwidth), but in PCM the increase is an exponential function of bandwidth, whereas in FM it is a function of the square of the bandwidth (k_f^2). All AM systems lack the ability to exchange S_o/N_o with bandwidth.

Of course, sweeping generalizations are worth nothing—and this one is no exception. The conclusions above are not only contigent on the criterion, but on the way in which the criterion was used, the conditions of derivations of the expression, and other factors. For AM and FM, the expressions were approximations based on large S_i/N_i. For PCM, the analysis was based only on quantization noise. If such conditions are removed, the picture could change— and does. Keep all that in mind, and avoid being absolute in making statements.

Even if we accept the generalizations, they are only as good as the criterion —let us state that one more time. If your criterion is economics, you might pick AM as best. If your criterion is "lowest bandwidth," you will probably pick SSB as best. Things are never as they first seem.

This business of exchanging S/N with bandwidth is the most important conclusion of the use of the SNR criterion and leads to the following important question. What is the "ideal" way of performing this exchange, or the ideal system for accomplishing it? In other words, is the PCM (exponential type of exchange) as good as we can do, or should we keep looking for new systems that can do better?

There is an answer to this question, provided by the field of information theory, whose fundamental notions (entropy and mutual information) were presented briefly in Chapter 11. The answer is based on the capacity[4] of a channel in the presence of white additive Gaussian noise, given by an expression usually called the *Shannon–Hartley theorem*. This expression is then used to define an ideal receiver as a channel that loses no information, and with this, it can be shown that such a receiver obeys

$$\left(\frac{S_o}{N_o}\right)_{\text{ideal}} \approx \left(\frac{S_i}{N_i}\right)^{B/f_m} \tag{13.8.4}$$

an exponential type of exchange. The PCM has such an exchange but in a questionable way, since only quantization noise was taken into account, and the conditions of the Shannon–Hartley theorem are not necessarily met.

We will not discuss this further, but it makes an excellent topic for further reading. The fundamental Shannon–Hartley theorem can be found in Gallager (1968).

13.9 MIXED SYSTEMS: THE PROBABILITY OF ERROR CRITERION

Mixed systems (ASK, FSK, PSK, SS, etc.) take the coded signal $f_c(t)$ and put it on a carrier so that it can cross the BPF channel. Channel noise may not be negligible now, and the output of the matched filter will be distorted and difficult to recognize. Errors in bits will take place—"ones" will be called "zeros," and vice versa, and this will cause errors in the reconstruction of $f_q(t)$. The SNR criterion is a bit weak in handling such situations and a new one must be developed, one that is also useful in handling detection of echoes in radar signals. Let us briefly discuss the origin and the definition of this criterion. A detailed, rigorous presentation and actual application of the criterion are beyond the scope of this book. It requires knowledge of mathematical statistics (hypothesis testing) and complete descriptions of stochastic processes, which we have deemphasized herein.

One way to simplify the situation is to stick to regular baseband PCM, assume that the channel noise is high, and look at it as follows. All right, the output of the matched filter is distorted, and a triangle (say) no longer looks like anything recognizable. Still, we know from the preceding section that in the absence of noise, the output attains its maximum value at $t = T$, the end of the rectangular pulse. Why don't we just look at that point, and if the value of the output is "large" call it a "one," otherwise call it a "zero" (or a "minus one"). Good idea, but how large is "large"?

Obviously, we have got to decide on a value, say C, and if the output of

---

[4]Capacity of a channel is defined as the maximum mutual information between input and output r.v.'s of the processes, maximum with respect to the input probabilities.

the filter at $t = T$ exceeds it, we must declare to the world that we have received a "one," otherwise a "zero" [if the incoming $f_c(t)$ is bipolar, i.e., ± 1, the situation is similar]. But how do we pick this value C, called the **threshold**, in an optimum way?

To answer this question we need the theory of hypothesis testing, from mathematical statistics. This theory suggests the following, although with much more rigor than we are about to describe. No matter where you set this threshold C, you are going to make two types of errors; (1) you will decide on a "one" when there is actually a "zero" (**false alarm**), and (2) you will decide on a "zero" when there is actually a "one" (**a miss**). The "optimum" way to set this C is at a value where the *probability* of making both of these errors is minimized. It turns out, however, that the C cannot be set so that *both* probabilities of error are minimized, as the errors are interrelated; if you decrease one, you raise the other. So what we decide to do is to set one of them at a very small acceptable value (say 10^{-4}), and then pick C so that the probability of the other is minimized. In most PCM (and radar) systems the case is even simpler; setting one probability automatically sets the other, and thus setting C turns out to be not so difficult, after all.

Time for some mathematics. The word "probability" implies complete models, of course, but luckily we have simplified the problem enough so that only the first-order p.d.f.'s (the marginal ones) of the processes are needed. The key to this is that we have drawn our attention only to the point $t = T$, and at single time points processes are random variables, as we well know.

We now concentrate our attention at the output of the matched filter at $t = T$. If there is a "zero" present in $f_c(t)$, this output will be $N(T)$ (i.e., only noise), and specifically, the r.v. of the noise defined at that point. If we assume that the noise is Gaussian in terms of the complete model, this r.v. is univariate Gaussian (i.e., with our known Gaussian p.d.f.). We further assume that its mean is zero and its variance σ^2, and this variance is also $R_n(0)$ (i.e., the output noise power).

If there is a "one" present, the output will be $N(T) + f_o(T)$, where $f_o(t)$ is the output of the useful signal (the triangle) at $t = T$, which has a value of, say A^2T (see Fig. 13.7.1c), a constant.

Summarizing, the output of the matched filter at $t = T$ is

$$O_1(T) = N(T) + A^2T \quad \text{(a "one" present)} \quad (13.9.1)$$

or

$$O_0(T) = N(T) \quad \text{(a "zero" present)} \quad (13.9.2)$$

We have already established that $O_0(T)$ is a r.v. with Gaussian p.d.f., and so on. So, of course, is $O_1(t)$, except for the fact that its mean is not zero but A^2T (check it), since adding a constant to a r.v. changes only its mean, not its p.d.f. The upshot of the above is that every time we look at the value of the output at $t = T$, we must decide whether it came from a Gaussian r.v. with zero mean (a "zero") or from a Gaussian r.v. with mean A^2T. In other words, we have reduced the problem to looking at a value and deciding which of the two p.d.f.'s it came from, p.d.f.'s that are illustrated in Fig. 13.9.1.

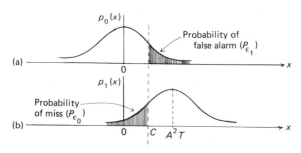

Figure 13.9.1 P.d.f.'s for the binary decision problem: (a) $p_0(x)$, a "zero" is present; (b) $p_1(x)$, a "one" is present.

Now let us assume that the threshold is set at some value C. The probability of declaring a "one" incorrectly is

$$P_{\epsilon_1} = \int_C^\infty p_0(x)\, dx \qquad (13.9.3)$$

and of declaring a "zero" incorrectly is

$$P_{\epsilon_0} = \int_{-\infty}^C p_1(x)\, dx \qquad (13.9.4)$$

where both $p_0(x)$ and $p_1(x)$ are known.

As we have already ascertained, C will have to be picked so that the sum of the two errors P_ϵ,

$$P_\epsilon = P_{\epsilon_1} + P_{\epsilon_0} \qquad (13.9.5)$$

is minimum. This last P_ϵ is a simplified form of what we call the **probability-of-error criterion**. We can find it for various systems and then compare them with each other to decide which does better. It is the new criterion that we promised at the start of the present section.

A first look at Fig. 13.9.1 reveals that a good place to set C is at $A^2T/2$, or $A^2/2$ if $T = 1$ s. That is the value of the threshold for which $P_{\epsilon_1} = P_{\epsilon_0}$, a case of interest in PCM. In radar the value of P_{ϵ_1} may be more important than that of P_{ϵ_0}, and the C may be set elsewhere.

A second look at the figure also reveals that both errors can be decreased (and therefore P_ϵ) if A is large, or if σ^2 (the variance of the p.d.f.) is small. But A represents the area under the pulse (i.e., the square root of the average power of the pulse). And σ^2 is the power of the noise, as we have already remarked. We are thus led to conclude that our ubiquitous friend the SNR is with us again. If the SNR is large (A large, or σ^2 small), then P_ϵ is small.

We will not pursue the matter much further. We will leave it for the reader to show that

$$P_\epsilon = \text{erfc}\left(\sqrt{S/2N}\right) = \int_{\sqrt{S/2N}}^{\infty} \frac{1}{\sqrt{2\pi}} e^{-x^2/2}\, dx \qquad (13.9.6)$$

for our "one" or "zero" PCM, where S is the average power of the signal and N the power of the noise (i.e., σ^2). If the reader wants to do even more work, he or she can try showing that under similar conditions (which?), P_ϵ for the ± 1 PCM turns out to be

$$P_\epsilon = \text{erfc}\left(\sqrt{S/N}\right) \qquad (13.9.7)$$

which means that the latter is better—the power of the signal at the former must be twice as much to achieve the same probability of error as the ± 1 type of PCM.

Example 13.9.1
What should the SNR be in a bipolar PCM (± 1 type) so that $P_\epsilon = 10^{-7}$?

Solution.   This is a straight plug-in type of an example. We go directly to Eq. (13.9.7) and set $P_\epsilon = 10^{-7}$. Then we use Appendix B to find the value of $\sqrt{S/N}$ which makes the erfc $\left(\sqrt{S/N}\right)$ equal to this P_ϵ. Its square is the SNR desired. Good luck.

The probability-of-error criterion is used in all mixed systems, always in relation to the SNR. The reader may wish to try to apply it by doing Problems 13.10 and 13.11. Most of the present material will be reexamined in Section 14.2 when we take up the problem of detection in radar systems.

PROBLEMS, QUESTIONS, AND EXTENSIONS

13.1. Repeat the analyses of AM systems in the presence of additive noise assuming that the noise is *bandpass filtered white noise* with a spectrum around ω_c and a bandwidth equal to that of the message. Find the D.G. in all three cases, DSB, SSB, and AM (large-signal case).

13.2. Prepare a paper with noise analysis of other AM systems (vestigial, carrier reinsertion, etc.).

13.3. We mentioned at the start of Section 13.2.2 that the analysis of analog systems in the presence of additive noise can be performed by assuming that the message $f(t)$ is also a random process. If we did that, we would have to find $R(\tau)$ of the useful signal at the input and the output of the receiver, so that we could calculate $R(0)$ the power. What assumptions would you need about $f(t)$ (as a process), and the carrier $\cos \omega_c t$, to be able to carry through this analysis? Find the assumptions and do the analysis. (*Hint:* Be careful with the phase angle of the carrier, and of the locally generated cosine at the receiver. Should they be random oscillators? If so, how do we justify the necessary synchronization?)

13.4. How do you argue successfully that condition (13.5.6) leads to (13.5.7)? [*Hint:* Use the known approximation $(1 \pm x)^2 \approx 1 + qx$ for $|x| \ll 1$.]

13.5. Find the D.G. in an FM system if $k_f = 1000$ and the original information signal is:

(a) $\dfrac{\sin 10^3 t}{10^3 t}$

(b) $\dfrac{\sin 10^3(t - 10^4)}{10^3(t - 10^4)}$

(c) $\left(\dfrac{\sin 10^3 t}{10^3 t}\right)^2$

(d) The product of the signals in parts (a) and (c)

13.6. Try to invent three problems that need Eq. (13.6.17) for their solution.

13.7. It was noted in Section 13.6 that the noise spectrum at the output of the FM receiver, just before the LPF, was of parabolic shape [see Eq. (13.6.14)]; that is, most of the power of the noise is away from $\omega = 0$. On the other hand, speech and music have a lot of power in the low frequencies (i.e., where the noise is low). Determine how this is exploited by the **preemphasis** and **deemphasis** techniques, by searching the literature.

13.8. Prepare a paper on **threshold** in FM and methods for limiting its effect on performance.

13.9. Derive the S_o/N_o of a PCM (with only quantization noise present) if the code is assumed trinary (0, 1, 2, say).

13.10. Let us see how the probability-of-error criterion is applied on ASK systems. The signal received has the form, say,

$$m_{\text{ASK}}(t) = \begin{cases} A \sin \omega_c t & \text{for } 0 < t < T \\ 0 & \text{otherwise} \end{cases}$$

Find the matched filter. Sketch its output (it should look like the product of a triangle with the foregoing sine wave). Now look at $t = T$ and proceed to derive that

$$P_\epsilon = \text{erfc}\left(\frac{S}{2N}\right)^{1/2}$$

where $S = \frac{1}{2}(A^2/2)$ and N is the power of the noise within the bandwidth of $m_{\text{ASK}}(t)$.

13.11. Using P_ϵ for FSK and PSK is a bit more difficult, but not much. Prepare a paper on both cases by searching the literature [try Stremler (1977), for example].

13.12. Once the P_ϵ has been found for a system—in terms of S/N—a plot can be drawn of its values versus S/N, for easy reference. Try sketching such a plot for regular PCM (on–off type), with the S/N in decibels. Discuss its usefulness.

=== 14 ===

NOISE IN RADAR SYSTEMS

14.1 INTRODUCTION

This chapter is the apotheosis of time and frequency correlation functions and their generalization, the **radar ambiquity function**. We have repeatedly argued in the past that their role is critical in the analysis of radar systems, and this time we are about to prove it.

We will not bother here to take typical radar systems and analyze them in the presence of additive noise. That is too cumbersome and, in fact, boring. Instead of that, we will take a more general point of view and examine how *any* radar can best perform its functions in the presence of noise, functions starting from detection, running through range and velocity estimation, and ending with resolution. Quite obviously, the emphasis will be on radar signals and not antennas, and thus angle estimation and tracking will be ignored. All our results will be general, easily specialized to specific radar systems—tasks left to the reader as additional projects.

Noise comes into radar systems from the same sources as in communication systems, plus some additional ones. If a radar is pointing low, for example, the radar signal hits terrain objects, such as mountains or trees, and the end result is a sum of many unwanted echoes called **clutter** noise. Changes in the target aspect with respect to the radar produce noise called **angle scintillation** or **glint**. We will not bother discussing these or other noises, but it should be pointed out that they exist and cause plenty of trouble in radar performance.

We started out this chapter by stating that it will be mainly a demonstration of the "power of analysis" of correlation functions. Let us repeat it one more time. In fact, there is no radar property that cannot be analyzed with the use of such functions. Detection may seem so at first, but it isn't really. The optimum processor turns out to be a "matched" filter, and it, too, is related to correlation functions, as we know very well by now.

14.2 DETECTION IN THE PRESENCE OF NOISE

The first job of any radar system is to detect the target, and only after detection has been ensured can it start measuring some of the target parameters of interest: range, velocity, angles, and the like. So detection will have to be our first concern, with additive noise finally accepted as part of the return echo.

In reality we do not have to say anything new regarding this issue; the whole thing is pretty much the same as in carrier PCM systems, which we covered in Section 13.9. Pulse radars are used mainly in detection, and detection of pulses is the same problem as in a unipolar ("one" or "zero") type of PCM. A matched filter (or correlation receiver) processes the envelope of the return echo, and if the output of the matched filter exceeds the predecided threshold, we declare *an echo present*; otherwise, we keep quiet.

Let us review this whole problem briefly and note the differences (there are but a few) between radar and PCM.

The radar receiver accepts the return signal $r(t)$ and a decision must be made as to whether

$$r(t) = N(t) \qquad \text{(no echo present)} \qquad (14.2.1)$$

or

$$r(t) = N(t) + f(t) \qquad \text{(echo present)} \qquad (14.2.2)$$

We will assume that $N(t)$ is a Gaussian white noise and $f(t)$ a known signal of short duration (usually a pulse). Note that we are concentrating on only one pulse (i.e., ignoring illumination time, etc.). In actuality, the decision is based on more than one pulse, and this decreases the probability of making errors (but adds range ambiguities). Note also that $f(t)$ is assumed without carrier modulation (i.e., the pure original signal of short duration), with frequency content around $\omega = 0$.

We have established—on the basis of maximum output SNR—that the "optimum"[1] way to process this signal is by a matched filter (this type of data processing is called a *statistic* in hypothesis testing, but we do not care about that here), and we concentrate our attention at time $t = T$ of the output of this

---

[1]The matched filter can be shown to be optimum for processing our return signal in the presence of additive Gaussian white noise by other more advanced criteria as well (Neyman–Pearson, Bayes, etc.).

filter, where T is the duration of $f(t)$. Denoting the output of the matched filter by $O(t)$, then

$$O(T) = \int_0^T r(\tau)h(T - \tau)\, d\tau \tag{14.2.3}$$

is this point of interest. In view of the fact that the matched filter is given by

$$h(\xi) = Kf(T - \xi) \qquad (\text{let } K = 1) \tag{14.2.4}$$

this point can also be written as

$$O(T) = \int_0^T r(t)f(t)\, dt \tag{14.2.5}$$

which is also the time cross-correlation $R_{rf}(\tau)$ or $r(t)$ and $f(t)$ at $\tau = 0$. If the fact that $r(t)$ is a process in this last statement of cross-correlation bothers the reader, it shouldn't. We can assume all needed ergodicity and forget it. (Why?) All of this we know, of course, from previous discussions.

This point $O(T)$ is, of course, a random variable and its p.d.f. turns out to be Gaussian with mean zero if $f(t)$ *is not* present. If $f(t)$ *is* present, its mean is the value of $O(T)$ without the noise (prove it), which is

$$\int_0^T f^2(t)\, dt = S \tag{14.2.6}$$

the energy of the signal $f(t)$ (or its average power if $T = 1$). Even the variance of $O(T)$ can be found with the background we have. It is simply the power of the output noise, and we know enough about linear systems with processes for input/outputs to find eventually,

$$\text{var}\,(O(T)) = \frac{N_o}{2} \int_0^T f^2(t)\, dt = \frac{N_o S}{2} \tag{14.2.7}$$

where $N_o/2$ is the PSD of the input noise. We also note that this variance is the same whether or not $f(t)$ is present in $O(T)$.

The upshot of the discussion above is that our problem really boils down to observing $O(T)$ (a value of this r.v.) and deciding whether its value came from a Gaussian p.d.f. $p_0(x)$ with mean zero, or from a Gaussian p.d.f. $p_1(x)$ with mean S. If we decide that it came from a mean zero, we have decided that there is no echo, otherwise that there is.

This decision (illustrated also in Fig. 13.9.1) is made by setting a threshold C and if

$$O(T) > C \tag{14.2.8}$$

we decide "yes, there is an echo," or if

$$O(T) < C \tag{14.2.9}$$

we decide "no echo." If $O(T) = C$, a rare case, we cannot make a decision and might just as well flip a coin.

We also know that no matter how we set this C, we are going to make errors—of two types—with probabilities P_{ϵ_0} and P_{ϵ_1},[2] and that these errors are interrelated, so we cannot minimize them both. We have already seen what is done in PCM about C, so let us now discuss the radar case; it is actually a little easier.

The probabilities of errors P_{ϵ_1} and P_{ϵ_0} are not of equal significance here, as it is judged that a "miss" is more important than a false alarm (missing a missile headed for a city is more important than deciding there is one when there isn't). In fact, what we are really interested in in radar is the probability *that we do not miss*, that is, the probability of *detecting a target if it is there*. We denote this by P_D (**probability of detection**) and note in Fig. 13.9.1 that for a given C,

$$P_D = \int_C^\infty p_1(x)\, dx \qquad (14.2.10)$$

or that

$$P_D = 1 - P_{\epsilon_0} = 1 - \int_C^\infty \frac{1}{\sqrt{2\pi\sigma^2}} e^{-(x-S)^2/2\sigma^2}\, dx \qquad (14.2.11)$$

where $\sigma^2 = N_o S/2$.

With a change of variable, P_D can also be written as (prove it)

$$P_D = \operatorname{erfc}\left(C\sqrt{\frac{2}{N_o S}} - \sqrt{\frac{2S}{N_o}}\right) \qquad (14.2.12)$$

where $\sqrt{2S/N_o}$ is a kind of signal-to-noise (SNR) ratio. (Why?) It is quite interesting that P_D turns out to be a function of a SNR, but it is not surprising in view of our results in Section 13.9.

So in radar, one way to proceed is as follows. First, we can decide on the value of P_D we want, say $P_D = 0.99$. This can be inserted into Eq. (14.2.11), and the threshold C can be found by the use of Gaussian p.d.f. tables (see Appendix B). This C is the key to the decision. If we now want to find P_{ϵ_1}, there is no problem; we just use Eq. (13.9.3) and evaluate it. The value of P_{ϵ_0} is 1 $- P_D$. Of course, this is not the only way to proceed; it all depends on your interests. You may wish to start with P_{ϵ_1} or P_{ϵ_0}. You may be stuck with a $\sqrt{2S/N_o}$ and wish to start with that. But in any event, solutions of radar detection are facilitated a great deal with various plots of all these quantities, typical examples of which are given in Fig. 14.2.1. Of those, Fig. 14.2.1b even has its own name; it reflects **receiver operating characteristics**. Both plots are, of course, only indicative of the types of curves one comes up with in such problems.

The final conclusions of this discussion are immediate. The key to good radar detection in the presence of Gaussian white noise is the matched filter and the SNR. No wonder, then, that every analysis of radar systems always includes a discussion of those two things: how to construct the matched filter and how to pack plenty of power (or energy) in the signal for large SNR. The actual shape of the pulse is not important here, but it is important in reducing

---

[2]The names of these errors, "false alarm" and "miss," came from radar analysis. In hypothesis testing they are called type I and type II errors.

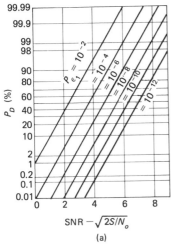

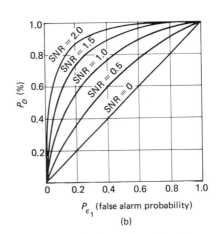

Figure 14.2.1 (a) Probability of detection versus SNR for various P_{ϵ_1}; (b) receiver operating characteristics.

range ambiguities and in the other functions of radar, as we are about to witness in the sections that follow.

Incidentally, the conclusions above hold, with a small theoretical modification, even in the presence of Gaussian noise that is not white. The matched filter is not matched to the signal $f(t)$ but to another signal related to $f(t)$ and the partial model of the noise $R_N(\tau)$, via an integral equation. This is, of course, much harder to find and implement in practice. Interested readers can pursue this matter further in the literature, but they will need much more mathematical background to follow the developments.

In closing we should point out that in radar (unlike in PCM) the point $t = T$ is not known a priori, but this causes no practical difficulties. After the emitted signal leaves the transmitter, we simply observe the output of the receiver until we locate a point where the threshold has been exceeded. Such a point hopefully will designate the presence of an echo and the radar detection of a target. When detection has been assured by additional checking, the radar (or other radars) is alerted to proceed with range estimation, velocity estimation, and so on, depending on the particular application.

14.3 THE RADAR EQUATION AGAIN

What we have learned in the preceding section can now be incorporated into some of the old forms of the radar equation, producing new and "improved" forms in the process. Consider, for example, the radar equation (4.4.6), which was

$$R_{\max}^4 = \frac{KP_T}{S_{\min}} \qquad (14.3.1)$$

This $S_{\min}$ can be related to the radar detection problem as follows.

Let us assume that the detection is based on one pulse (not a group of them as is really the case in practice). Let us also assume that for some P_D, we desire a SNR ≥ 10. Recalling that the SNR was

$$\text{SNR} = \sqrt{\frac{2S}{N_o}} \tag{14.3.2}$$

the condition SNR ≥ 10 gives

$$S \geq 100 N_o \tag{14.3.3}$$

as the $S_{\min}$ that can be detected by the radar with the P_D we picked. Thus

$$R_{\max}^4 \leq \frac{KP_T}{100 N_o} \tag{14.3.4}$$

is yet another form of the **radar equation**, this time including the effects of additive Gaussian white noise. The story can go on, but we will end it right here, as we have more important things to discuss and we are rapidly running out of space.

14.4 RANGE ESTIMATION OF A SINGLE TARGET

We already know from our noiseless analysis of radar that to estimate the range (radial distance) of a target, one needs to measure the time delay (t_o, say) between the transmitted and received pulses (see Fig. 14.4.1a). If there is no noise (Fig. 14.4.1a), the problem is straightforward. We zero in on a reference point, and measure away. It is a good idea to use a well-defined point as a reference (a maximum say), because time constants, channel bandwidths, and the like can round off our shapes and cause difficulties with starting or ending points. Even

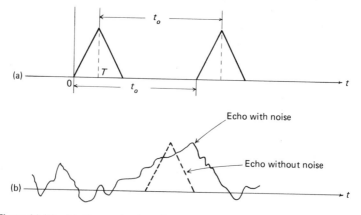

Figure 14.4.1 (a) Transmitted and received envelopes after matched filters, without noise; (b) an echo with noise.

without noise a nice sharp maximum is desirable, particularly if there is more than one adjacent target, as we will see later.

In practice, the noiseless case does not exist. The time has come to discuss range estimation in the presence of additive noise, whose effect on the return echo is to shift the reference point as illustrated in Fig. 14.2.1b. Estimating *anything* in the presence of noise is a highly developed field called estimation theory, another subfield of mathematical statistics. We would need an entire book to do it justice (there are, in fact, plenty of books on the market). We summarize the essential ideas below, enough to enable us to continue with this and the next couple of sections. Our summary will be given with reference to the problem of estimating the delay time t_o, which represents the target range of our interest.

Most estimation problems typically have the following common characteristics.

1. There are some data (r.v.'s or a random process), also called **the sample**, which contain the parameter that we wish to estimate. In our problems above, the data include the return echo embedded in noise, and it is a random process. We know that the echo is there, of course, because we are assuming that the existence of the target has been ascertained (detection has been confirmed).

2. The complete model of the data is known a priori. In our problem the noise is assumed to be Gaussian, and the samples' complete model can be derived.

With these assumptions, the next order of business in estimation theory is the following: to find a way to process the data (to find a function of the data, that is) which gives us the "optimum" estimate of the parameter of interest. This function of the data is, of course, a *random variable* (being a function of r.v.'s), and it is called an **estimator** of the parameter. Every time you get data and use them, you get one value called an **estimate** of the parameter. In our problem our estimator is already known; it is the distance between the two peaks, say in Fig. 14.4.1a. There is no doubt that it is a r.v., since the noise shifts the peak of the return echo, and (assuming that the target is stationary) every time you use the estimator, you get a different value (an estimate).

As we said above, estimation theory *attempts to find optimum* estimators (never mind that in our problem we already have one). We have seen enough of the word "optimum" in the past to know that it implies a criterion. This is true here, as well. Our next task, then, is to decide on a *criterion of optimality* and then use it to find the estimator that satisfies it. After that, you are done, or almost done, depending on the criterion.

What about the criterion of optimality? Well, the *criteria* is more like it, as there are quite a few, most popular of which are **maximum likelihood** and **Bayes'**, but a thorough discussion of any of them is beyond the scope of this book. In any event, the reader can easily guess that they are probably some kind of "distance" function that has to be minimized. In fact, there is one criterion (a subcase of the Bayesian) whose goal is to find the estimator that minimizes the distance between it and the parameter of interest, and the distance is our usual expression, the mean-squared type. In mathematical terms, if the estimator is denoted by $G(X)$, where X represents the data, and t_o is, say, the parameter of interest, this criterion claims that we should pick as $G(X)$ the one that minimizes

$$d(G(X), t_o) = \sqrt{E(G(X) - t_o)^2} \qquad (14.4.1)$$

Let us stick to this criterion, then, for it is going to lead us to the results that we want in this and the next few sections.

Aside from the criterion above, we often add a desirable property to the sought-after estimator, whose usefulness is obvious. We demand that its mean (it is a r.v., after all, and it has a mean) be equal to the parameter we wish to estimate. Such estimators are called **unbiased**. The property is desirable because if you try the estimation operation a few times and take the average (as you do in radar, since you have more than one echo from a target), you get closer to the true value of the parameter. If, now, we stick to unbiased estimators, that is, such that

$$E\{G(X)\} = t_o \qquad (14.4.2)$$

in the notation above, Eq. (14.4.1) can be written as

$$d(G(X), t_o) = \sqrt{E[G(X) - E\{G(X)\}]^2} \qquad (14.4.3)$$

which is nothing but the square root of the variance of $G(X)$. It is no wonder, then, that unbiased estimators that minimize Eq. (14.4.1) are often called **minimum variance unbiased estimators** (MVUEs).

All of this mathematics is simply camouflaged common sense, of course. If your $G(X)$ has as mean the value t_o, and variance zero (you cannot get smaller than that), who can argue that you didn't hit the jackpot? A r.v. with a mean t_o and variance (width of p.d.f.) zero is like a constant t_o, so your $G(X)$ always gives you what you want. In the absence of variance zero, the next best thing is "minimum variance," and that is what this criterion is all about.

All this is fine in theory, but how do we apply it in practice—and more specifically to our own problem here, of the estimation of t_o? Well, if you do not have an estimator to begin with, you have to work with your data and minimize Eq. (14.4.3) to find one, a rather messy business that takes many pages. We will not bother with such problems here.

In most practical problems the estimator is obvious, and so is our case here and in the next few sections (they can be found by mathematical criteria as well, but we will not do that here). So we take what we got and try to see how good it is—guided, of course, by the theory above. This means that we check it first to see if it is unbiased. If it is, fine. If it is not, it can be changed to be, by subtracting from it the difference from the desired mean. After that, we find its variance. If it is zero (hardly expected), we jump with joy. If it is not, we look at it and see what effects it, and how we can make it small. This last step is necessary even if you use the theory to find your $G(X)$, because even if its variance is minimum, it may still be large and might need improving. To do these last two steps, you usually have to have the p.d.f. of $G(X)$, and that is where you need a complete model of the data, assumption (2) above. Hopefully, this model can lead you, by known methods, to the p.d.f. of $G(X)$, and then its mean and variance will be found easily. Sometimes the joint p.d.f. of the data is not necessary, and the mean and variance of $G(X)$ (they are a partial model) can be found from the partial model of the data by other known methods.

With this background, we now come to our range-estimation problem and can polish it off in a hurry. Some complex mathematical steps will be omitted, as they are not judged illustrative enough to include here. The reader can pursue the details in the literature (e.g., Burdic, 1968; Hellstrom, 1968).

All right, then, let us assume that the pulse emitted has the general form

$$m_e(t) = a(t) \cos (\omega_c t + \varphi(t))$$

which covers every possible radar. $a(t)$ is, of course, the envelope function, $\varphi(t)$ the phase function, and ω_c the carrier frequency. It is assumed that $a(t)$ and $\varphi(t)$ are low-frequency signals (i.e., with spectra around $\omega = 0$) and that ω_c is very large, much larger than the maximum frequency of $a(t)$ or $\varphi(t)$.

We can rewrite the above as

$$m_e(t) = \text{Re}\,[a(t)e^{j\varphi(t)}e^{j\omega_c t}] = \text{Re}\,[A(t)e^{j\omega_c t}] \qquad (14.4.4)$$

where $A(t)$ is the complex envelope of the signal.

We next assume that the target is stationary ($\omega_d = 0$) or of known radial velocity (ω_d — known). The single return echo from such a target will be

$$m_r(t) = K\,\text{Re}\,[A(t - t_o)e^{j\omega_c(t - t_o)}] \qquad (14.4.5)$$

where K is the loss of power due to the travel. Here we will set it equal to unity, as it does not affect our theoretical conclusions. The signals are, of course, given before they enter the matched filters, even though they may be measured after the filters.

We have already stated that the estimator is the distance of the peaks (or some other reference point) between emitted and received pulses, or the outputs of their matched filters. This estimator can be shown to be unbiased. The next step is to find its variance, which gives you an idea of how good the estimator is. This variance turns out to be

$$\sigma_{t_o}^2 = \frac{1}{\beta_o^2 2S/N_o} \qquad (14.4.6)$$

where $N_o/2$ is the average power of the noise (assumed white and Gaussian), and S the energy of the received signal. Obviously, $S/2N_o$ is the SNR. What about β_o^2?

In the complex mathematics that we omitted, it is shown that

$$\beta_o^2 = \frac{\int_{-\infty}^{+\infty} \omega^2 \, |A(\omega)|^2 \, d\omega}{\int_{-\infty}^{+\infty} |A(\omega)|^2 \, d\omega} \qquad (14.4.7)$$

where $A(\omega)$ is the FT of $A(t)$ [or $A(t - t_o)$], the complex envelope of the *received* waveform. Naturally, $|A(\omega)|^2$ is the energy density spectrum of the received complex waveform $A(t)$, and the denominator represents the total energy of $A(t)$, except for the missing factor of $1/2\pi$.

This β_o^2, we claim, is a new type of bandwidth of $A(t)$, one that we had warned the reader to expect way back in Section 2.9. It is often called **effective bandwidth** or **rms bandwidth**. Is it legitimately called a bandwidth? Yes, indeed. $|A(\omega)|^2$ is everywhere positive, and if you divide it by its integral, you normalize it, and you have a function that behaves just like a p.d.f. (positive which integrates to unity). This being the case, expression (14.4.7) represents the second moment of a r.v. with such a p.d.f., and if its mean is zero [$|A(\omega)|^2$ is symmetric around $\omega = 0$], then β_o^2 is a variance, and therefore represents the width of $|A(\omega)|^2$. That is why we stated earlier that this type of bandwidth needs probability theory to be understood.

Now we are finally at the point of drawing our conclusions. For good estimates of t_o and therefore range, $\sigma_{t_o}^2$ must be made small. This can obviously be done by high SNR, for if the noise is low, it does not affect the measurement much. But it can also be done by increasing the rms bandwidth of $A(t)$, the width of $|A(\omega)|^2$. All along in our analyses of various radar systems we have been saying that we need a good narrow signal (an impulse?) for accurate range measurements. In fact, we have been saying that the time ambiguity of $A(t)$ should be narrow. Now, we have finally proved it, indirectly, of course. Narrow time ambiguity implies wide $|A(\omega)|^2$, which is its FT. Why does it imply it? Basic Fourier transform property—that of time scaling—that's why. So all is fine and dandy, and these conclusions are valid for all radar systems. Can you put this section and the last section together and argue that chirp radar (or any other pulse compression) is good? Can you argue why a CW radar is bad? Good or bad for range estimation of stationary targets, of course.

Example 14.4.1

Find the error in the range estimate if the return signal is

$$m_r(t) = \text{rect}\left(\frac{t}{T}\right) \cos \omega_c t$$

and the SNR is known.

Solution. Well

$$A(t) = \text{rect}\left(\frac{t}{T}\right)$$

and it is real. Now we calculate β_o^2. It turns out that (do it)

$$\beta_o^2 = \infty$$

so the error (the variance) of the range estimate is zero—well, not really. First, the mathematics that leads to Eq. (14.4.6) has some approximations in it which we did not mention, and which do not hold for the $A(t)$ above. Then again in practice there is no $A(t)$ like the above, with sharp edges, and so on, so the result is only for armchair scientists. Still a pulse is a good signal for measuring range. Of this there can be no serious doubt in anybody's mind.

In closing this section we should like to remind the reader that in estimating range there is always the problem of range ambiguities, caused of course by the fact that the radar emits more than one pulse. It could also be caused by the use of a signal without a single sharp, clear maximum in its time ambiguity function. We will not take up this problem again, as we feel that it has been discussed sufficiently in the past.

14.5 VELOCITY ESTIMATION OF A SINGLE TARGET

The scene now changes to the frequency-domain representation of the signals, where the target velocity manifests itself as a (Doppler) shift ω_d in the spectrum of the return signal. Detection is again assumed certain, and the range known, or not of interest. Our fundamentals of estimation theory hold here as well. The estimator is the shift in the spectrum. Complex mathematics shows that its variance is given by

$$\sigma_{\omega_d}^2 = \frac{1}{\tau_o^2 2S/N_o} \tag{14.5.1}$$

where $2S/N_o$ is the SNR as before, and

$$\tau_o^2 = \frac{\int_{-\infty}^{+\infty} t^2 \, |A(t)|^2 \, dt}{\int_{-\infty}^{+\infty} |A(t)|^2 \, dt} \tag{14.5.2}$$

where $A(t)$ the complex envelope of the return signal, and the time origin of $|A(t)|^2$ is assumed at its centroid, that is, at the point,

$$\int_{-\infty}^{+\infty} t \, |A(t)|^2 \, dt \tag{14.5.3}$$

With the above, $|A(t)|^2$ divided by its energy has all the elements of a p.d.f. (everywhere positive, integrates to unity) and in view of (14.5.3), τ_o^2 behaves like a variance of some r.v. with this p.d.f. No surprise, then, if we announce that τ_o is called the **rms duration** of $A(t)$, as it gives the duration of $|A(t)|^2$.

So what conclusions can we draw from the above? Nothing really that we have not already guessed intuitively (it feels good, though, to see it mathematically). Aside from a high SNR, a radar's signal must be of long time duration, or narrow spectrum, if it is to measure Doppler shift well. More to the point, $|A(t)|^2$ must be of long duration. Now it can be shown (by an "inverse" Wiener–

Khintchine theorem) that $|A(t)|^2$ is the FT of $\mathfrak{IC}_A(v)$, the frequency ambiguity function of $A(t)$. Thus, what has to be narrow is this $\mathfrak{IC}_A(v)$, and this is what we have been saying all along in our analyses of radar systems.

So now the reader can sit back, relax, and ponder the issue of good radial velocity estimation. Which radars should be able to do it? CW radar? Pulsed radar? CW-FM radar? How about chirp radar? Assuming a given SNR, can you find the variance of the estimator?

Example 14.5.1

What is the error in measuring ω_d if the SNR is known and the radar signal is

$$m_r(t) = A \cos \omega_c t$$

Solution. For this signal

$$A(t) = A \qquad \text{(a constant)}$$

whose $\tau_o^2 = \infty$ and $\sigma_{\tau_o}^2 = 0$. We let the reader argue why this is only an academic result. Still it says something about CW radar that we had supported intuitively before.

So we know what it takes, in terms of signals, to measure range if we know or ignore velocity, and vice versa. But what should be the characteristics of a radar signal to measure both? A compromise?

14.6 SIMULTANEOUS ESTIMATION OF RANGE AND VELOCITY FOR A SINGLE TARGET

The problem at hand is the simultaneous estimation of two unknown parameters, t_o and ω_d (shift in time and frequency, respectively), from the same data. In estimation theory, this is a generalization of the basic problem of the estimation of a single parameter and a much more difficult one to carry out, as an error in one estimator could affect the other.

We ignore the formidable mathematics here as well and give the resulting variances of the estimators below. It turns out [see Rihaczek (1969) or Burdic (1968), for example] that

$$\sigma_{t_o}^2 = \frac{N_o}{\beta_o^2 2S} \frac{1}{(1 - p^2/\beta_o^2 \tau_o^2)} = \frac{N_o \tau_o^2}{2S(\beta_o^2 \tau_o^2 - p^2)} \qquad (14.6.1)$$

$$\sigma_{\omega_d}^2 = \frac{N_o}{\tau_o^2 2S} \frac{1}{(1 - p^2/\beta_o^2 \tau_o^2)} = \frac{N_s \beta_o^2}{2S(\beta_o^2 \tau_o^2 - p^2)} \qquad (14.6.2)$$

where τ_o^2 and β_o^2 are our previous rms duration and bandwidth of the complex envelope of the received signal [Eqs. (14.5.2) and (14.4.7), respectively] and S and $N_o/2$, as before. The new quantity is p and it needs some explanation before we get down to the conclusions.

This p is given by

$$p = 2\pi \frac{\displaystyle\int_{-\infty}^{+\infty} t\dot{\phi}(t)A^2(t)\,dt}{\displaystyle\int_{-\infty}^{+\infty} A^2(t)\,dt} \qquad (14.6.3)$$

where $\varphi(t)$ is the frequency modulation term in Eq. (14.4.4) of the signal received. It is quite obvious that if $\varphi(t) = 0$ (pulse radar, CW radar, etc.), then $p = 0$. In this case, the two variances become

$$\sigma_{t_o}^2 = \frac{N_o}{\beta_o^2 2S} \qquad (14.6.4)$$

$$\sigma_{\omega_d}^2 = \frac{N_o}{\tau_o^2 2S} \qquad (14.6.5)$$

the same as they were in the previous two estimation problems that we considered. Our dilemma here is obvious; if we increase the bandwidth, we get good range measurement but bad velocity measurement, and vice versa.

Can we increase both β_o^2 and τ_o^2, or to put it differently, can we find a signal (which has no phase modulation) for which both bandwidth and duration are large? We have something that can come to our aid here, the uncertainty principle of Eq. (2.9.5), which, as we stated in Section 2.9 holds for all definitions of bandwidth and duration. In terms of β_o^2 and τ_o^2, this principle is (see Burdic, 1968)

$$\beta_o^2 \tau_o^2 - p^2 \geq \pi^2 \qquad (14.6.6a)$$

or

$$\beta_o \tau_o \geq \pi \qquad \text{(if } p = 0\text{)} \qquad (14.6.6b)$$

so although for a *given* signal if you increase τ_o^2 you decrease β_o^2, it appears that the answer lies in picking a signal with high **bandwidth–duration (BD) product** to begin with. Still, with no phase modulation, there is not much you can do, as most such signals have low BD products, right around π or a little larger (see the Problems). So, if $p = 0$, you can do one of the two estimations well, but not both.

What if there is a phase modulation term $\varphi(t)$ in the radar signal, and $p \neq 0$? Here is where some remarkable things take place, as evidenced by the known popularity of radars employing such signals (pulse compression radars—mainly chirp). Indeed, it can be shown that signals with $\varphi(t) \neq 0$ can have high-BD products, in fact, as high as we want—theoretically, anyway.

So the answer *seems to be* to employ such signals and thus achieve high accuracy in both measurements. But things are never as they seem to be. If you add $\varphi(t)$, you also increase p, and high p is detrimental to the accuracy of measuring either of the two parameters, as Eqs. (14.6.1) and (14.6.2) show. So one must add $\varphi(t)$, but not too much; otherwise, p gets large—a situation that is difficult to balance. That is why the area is still under study and the search for good radar signals continues. Furthermore, accuracy in range and velocity are not the only things demanded of a radar; we have angle measurement and resolution problems, not to mention ambiguities. No type of signal yet on the horizon is a panacea. The ideas of this section are developed a bit more in the problems.

14.7 TARGET RESOLUTION: AMBIGUITY FUNCTIONS

Detecting and estimating the location of a target is, of course, bad enough. Add to it the problem of resolution and the situation becomes hopeless. No wonder, then, that it is rare that one radar is asked to perform all these functions and perform them well.

The problem of **resolution** is the ability of a radar to distinguish between two (or more) targets whose parameters (range, Doppler, etc.) are nearly equal in value. Multiple-target cases abound in practice, particularly in a war period when aircraft fly in formations or missiles are surrounded by decoys. When targets fly low, the radar receives echoes not only from the target, but from various other stationary objects (trees, mountains, buildings, etc.), and distinguishing the moving target in such an environment (clutter) is another resolution problem of immense importance.

The most important resolution problems involve range and velocity resolution and refer to the ability of a radar to distinguish targets with nearly equal ranges or velocities, respectively. In what follows we will consider only these two types of resolution problems, and we will discuss them only for the two-target case. Other resolution problems (angle, for example) or more than two targets, will be entirely omitted from the discussion and are left for further study by the reader.

We start our discussion with the simplest case of *range* resolution, that between two targets whose Doppler shifts are assumed zero. This will introduce us to the methodology and make the rest of the discussion easier to follow. Before we begin with the math, we refer the reader to Fig. 14.7.1, where the problem is illustrated in the absence of noise (noise obviously worsens it). In Fig. 14.7.1b the targets are close together and the two maxima start to merge into one. Problems! Note also, as a precursor to our discussion, that if the triangles are very narrow, resolution of the two maxima improves when the two targets get closer together.

Now the math. Let us start by assuming that the return signal has the form

$$m_r(t) = Am(t) + Am(t + t_o) \tag{14.7.1}$$

where $m(t)$ is the radar transmitted signal. This means that two echoes have returned from a single emitted signal, from two targets t_o time units apart. The return power of both echoes is assumed equal (that is, the targets have the same cross section). The signal $m(t)$ is

$$m(t) = a(t) \cos (\omega_c t + \varphi(t)) \tag{14.7.2}$$

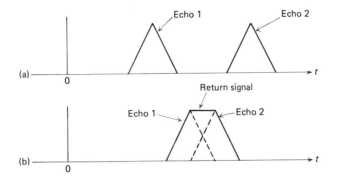

Figure 14.7.1 Illustration of range resolution: (a) easily resolved targets; (b) targets not easily resolved.

a general form, covering all radars. To simplify our analysis we have assumed that the time axis has been shifted so that one of the two echoes $[A_m(t)]$ has no time delay at all. Note also that at this point we are assuming that both targets are stationary (i.e., there is no Doppler shift in either of them).

To be able to distinguish between the two echoes, it is natural to expect that their signal "distance" is very large for all values of t_o (reread Section 2.10). Using our usual distance function of Eq. (2.4.18), the distance squared between $m(t)$ and $m(t + t_o)$ is of the form of Eq. (2.10.1). So, recalling Eq. (2.10.5), to maximize the distance between the two echoes, we must *minimize* the inner product of our two echoes for all values of t_o (denote it by τ), or the time auto-correlation function

$$R_m(\tau) = \int_{-\infty}^{+\infty} m(t)m(t + \tau)\, d\tau \qquad (14.7.3)$$

Now $m(t)$ of Eq. (14.7.2) can be written as

$$m(t) = \mathrm{Re}\,[a(t)e^{j\varphi(t)}e^{j\omega_c t}] = \mathrm{Re}\,[A(t)e^{j\omega_c t}] \qquad (14.7.4)$$

which means that

$$m(t + \tau) = \mathrm{Re}\,[A(t + \tau)e^{j\omega_c(t+\tau)}] \qquad (14.7.5)$$

In finding $R_m(\tau)$ for the $m(t)$ above, the exponential terms can be ignored because their functions correspond to target ranges smaller than the size of the target. Thus the time autocorrelation function $R_m(\tau)$ for radar range resolution can finally be written as

$$R_m(\tau) = \int_{-\infty}^{+\infty} \mathrm{Re}\,[A(t)]\,\mathrm{Re}\,[A(t + \tau)]\, dt \qquad (14.7.6)$$

$$= \mathrm{Re}\left[\int_{-\infty}^{+\infty} A(t)A^*(t + \tau)\, dt\right] \qquad (14.7.7)$$

where * denotes complex conjugate.

Now what does it mean to make $R_m(\tau)$ minimum for all τ? To begin with, at $\tau = 0$, $R_m(0)$ is the energy of the signal, and this had better be large for detection and estimation of the target parameters. So there is nothing much you can do there; $R_m(0)$ must be large. What about $\tau \neq 0$? A little thought [in conjunction with the properties of $R_m(\tau)$ and the type of signals and receivers used in radar] reveals that for $\tau \neq 0$ we want $R_m(\tau) = 0$. So the best choice for a signal $m(t)$ that will provide us with a good range resolution is one whose $R_m(\tau)$ is a narrow spike (an impulse) at $\tau = 0$, and this requirement is in line with the requirements for good detection, estimation, and range ambiguities. All of this we have actually seen before, but we repeat it here for the sake of completeness, throwing in a bit of mathematics to make it more convincing.

So much, then, for range resolution, if the Doppler shift ω_d of both targets is ignored. Now let us ignore range differences ($\tau = 0$) and try to see how we can resolve two targets in terms of *velocity*. This means that one of the echoes has the form

$$m(t) = \text{Re}\,[A(t)e^{j\omega_c t}] \tag{14.7.8}$$

and the other

$$m(t) = \text{Re}\,[A(t)e^{j\omega_c t}e^{j2\pi v t}] \tag{14.7.9}$$

where v is the Doppler shift. We assume that the first echo has $v = 0$ (axis shift) for the sake of simplicity here as well.

Running through the business of maximizing the distance between the two, we conclude that this time the answer lies with minimizing the integral

$$\int_{-\infty}^{+\infty} \text{Re}\,[A(t)]\,\text{Re}\,[A(t)e^{j2\pi v t}]\,dt = \text{Re}\left[\int_{-\infty}^{+\infty} A(t)A^*(t)e^{-j2\pi v t}\,dt\right] \tag{14.7.10}$$

Recalling at this point the generalized Parseval identity,

$$\int f(t)g^*(t)\,dt = \frac{1}{2\pi}\int F(\omega)G^*(\omega)\,d\omega \tag{14.7.11}$$

the integral above becomes

$$\frac{1}{2\pi}\,\text{Re}\left[\int_{-\infty}^{+\infty} A(\omega)A^*(\omega - v)\,d\omega\right] \tag{14.7.12}$$

which is nothing but the real part of the frequency autocorrelation function $\mathcal{K}_A(v)$ of $A(t)$. Thus we see that in order to have good velocity resolution, we must make this

$$\mathcal{K}_A(v) = \frac{1}{2\pi}\,\text{Re}\left[\int_{-\infty}^{+\infty} A(\omega)A^*(\omega - v)\,d\omega\right] \tag{14.7.13}$$

minimum, something we also knew, having argued it in the frequency domain before. In other words, to detect and measure shifts in the spectra of the two echoes, one must have (ideally) a $\mathcal{K}_A(v)$ that is high at $\omega = 0$ (an impulse) and zero elsewhere, in analogy with resolving shifts in time.

The stage is finally set for the case of combined *range and velocity* resolution. In this case, the two echoes are

$$m(t) = \text{Re}\,[A(t)e^{j\omega_c t}] \tag{14.7.14}$$

and

$$m(t) = \text{Re}\left[A(t + \tau)e^{j\omega_c(t+\tau)}e^{j2\pi vt}\right] \qquad (14.7.15)$$

and the maximization of the distance leads to the minimization of

$$\text{Re}\left[\chi(\tau, v)\right] = \text{Re}\left[\int_{-\infty}^{+\infty} A(t)A^*(t + \tau)e^{-j2\pi vt}\, dt\right] \qquad (14.7.16)$$

where

$$\chi(\tau, v) = \int_{-\infty}^{+\infty} A(t)A^*(t + \tau)e^{-j2\pi vt}\, dt \qquad (14.7.17)$$

is a combined correlation in both time and frequency shifts, and it is called the **radar ambiguity function**, although the name is not universally accepted. Often, the term $|\chi(\tau, v)|^2$ is called the radar ambiguity function. Other names floating around for both are **time-frequency correlation function**, **signal ambiguity diagram or surface**, **matched-filter-response function**, and so on. No matter what it is called, though, this $\chi(\tau, v)$ [or $|\chi(\tau, v)|^2$ which can be plotted] is of paramount importance in radar signal design. Whole volumes have been written on it since its inception (Woodward, 1953), as it seems to hold the key to all the problems that appear in radar signal design.

Needless to say, a complete discussion of this function is beyond our scope. We will simply mention and discuss some of its properties below, hoping to mobilize the reader's interest for further study.

The first thing to note is that

$$\chi(0, 0) = \int_{-\infty}^{+\infty} |A(t)|^2\, dt \qquad (14.7.18)$$

which is the energy of the envelope of the signal. The point $(0, 0)$ is the maximum value of $|\chi(\tau, v)|^2$, for it can be easily shown by the Cauchy–Schwartz inequality that (show it)

$$|\chi(\tau, v)|^2 \leq |\chi(0, 0)|^2 \qquad (14.7.19)$$

Note also that

$$\chi(\tau, 0) = \int_{-\infty}^{+\infty} A(t)A^*(t + \tau)\, dt = R_A(\tau) \qquad (14.7.20)$$

$$\chi(0, v) = \int_{-\infty}^{+\infty} A(t)A^*(t)e^{-j2\pi vt}\, dt = \mathcal{K}_A(v) \qquad (14.7.21)$$

whose real parts are the time and frequency autocorrelations we encountered earlier, Eqs. (14.7.7) and (14.7.10), respectively. So we see that the ambiguity function is more general and reduces to the other two autocorrelations, as we had originally indicated in Chapter 2 [below Eq. (2.10.20)].

Another useful expression involving $\chi(\tau, v)$ is

$$\int_{-\infty}^{+\infty}\int_{-\infty}^{+\infty} |\chi(\tau, v)|^2\, d\tau\, dv = |\chi(0, 0)|^2 \qquad (14.7.22)$$

which we leave for the reader to prove. This property tells us that the volume under $|\chi(\tau, v)|^2$ is a constant, and therefore if it rises somewhere, it must fall somewhere else. Since radar designers usually want low ambiguity at a region (τ, v), they must understand that if they get it, the ambiguity will by necessity be higher somewhere else. You squeeze it here, in other words, and it pops up somewhere else, like a rubber mattress. It is a sort of a "conservation-of-ambiguity" situation.

The ideal case would be to find a signal whose $|\chi(\tau, v)|$ is an impulse at $\tau = 0$, $v = 0$, and zero elsewhere. Such a signal does not exist, so we are stuck with compromises. To get a feeling for how this $\chi(\tau, v)$ encompasses in it the various abilities of a radar in resolution problems, we take up a couple of examples below. Finding $|\chi(\tau, v)|^2$ and plotting it is, incidentally, quite messy, and in the examples that follow we will not bother with the actual algebra involved.

Example 14.7.1

Find and sketch the ambiguity surface for the signal

$$m(t) = \left(\frac{2}{\pi T^2}\right)^{1/4} e^{-t^2/T^2} \cos \omega_c t \qquad (14.7.23)$$

that is, for a cosine AM-modulated by a Gaussian signal.

Solution.   This is probably the only one that can be done rather easily in closed form. Recall, however, that it has a low-BD product (no phase function), and it is not much good for estimating range and velocity at the same time.

Well, if you insert the envelope (it is real here) into Eq. (14.7.17), bring out the τ term, complete the square, and so on, you end up with

$$\chi(\tau, v) = e^{-1/2[t^2/T^2 + v^2 T^2/4]} \qquad (14.7.24)$$

which is real, so we do not need to bother with $|\chi(\tau, v)|^2$. This ambiguity function is sketched (for a typical value of T^2) in Fig. 14.7.2.

Sometimes it is convenient to draw a top view of the ambiguity surface and to note the regions where $|\chi(\tau, v)|^2$ (or $\chi(\tau, v)$) is nonzero. This is done for

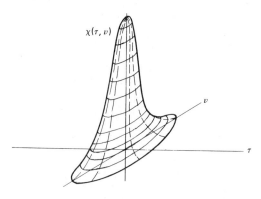

Figure 14.7.2   Ambiguity surface for the signal of Eq. (14.7.24).

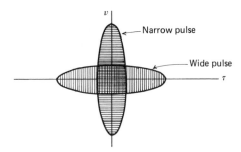

Figure 14.7.3 Ambiguity cross sections
for a Gaussian pulse.

this ambiguity function in Fig. 14.7.3 for two cases of T^2. Note and discuss the
difference, in these two cases, as an exercise.

Example 14.7.2

Find and discuss the ambiguity function $\chi(\tau, \nu)$ for the pulse radar signal.

Solution.   The signal emitted by the radar in this case is

$$m(t) = \text{rect}\left(\frac{t}{T_p}\right) \cos \omega_c t \tag{14.7.25}$$

and this is the form of the echoes as well, except for the fact that they might have time
and frequency shifts in them.

Now for this signal

$$A(t) = \text{rect}\left(\frac{t}{T_p}\right) \tag{14.7.26}$$

and

$$\chi(\tau, \nu) = \int_{-\infty}^{+\infty} \text{rect}\left(\frac{t}{T_p}\right) \text{rect}\left(\frac{t+T}{T_p}\right) e^{-j2\pi\nu t} \, dt \tag{14.7.27}$$

Of course, to get an idea of what is happening, one must plot $|\chi(\tau, \nu)|$ or
$|\chi(\tau, \nu)|^2$ as a function of τ and ν, a very complex business, as it is a surface in two
dimensions. Anyway, in Fig. 14.7.4 we have given a few plots of $|\chi(\tau, \nu)|^2$ as a function
of ν, for a few specific values of τ. It can be seen there (with a little imagination) how
the ambiguities stand for various values of τ and ν.

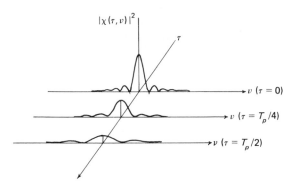

Figure 14.7.4   Plots of $|\chi(\tau, \nu)|^2$ for a pulse radar signal for $\tau = 0, T_p/4, T_p/2$.
(From Franks, 1969.)

Example 14.7.3

Discuss the ambiguity function for a chirp signal.

Solution. We omit the mathematics except for recalling that our signal is

$$m(t) = \text{rect}\left(\frac{t}{T_p}\right) \cos\left(\omega_c t + \frac{\mu}{2}t^2\right) \tag{14.7.28}$$

We go directly to the top-view type of diagram in Fig. 14.7.5. It is obvious that although the chirp signal looks good, it, too, has problems in certain regions of (τ, v), as we noted in the section on chirp radar. These sections depend on the amount of modulation $\mu/2$.

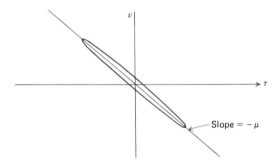

Figure 14.7.5   Crosssection of $|\chi(\tau, v)|^2$ for a chirp radar.

Our purpose in this section is only to introduce the subject, and not to discuss it to its full potential. So we let it go at this point. The interested reader is referred to Skolnik (1970), Vakman (1968), or Barton (1975) for further study of the role of $\chi(\tau, v)$ in waveform selection in radar systems. We should note in closing that the resolution problem was considered entirely in the absence of noise. Adding noise of any kind does not affect the conclusions that we have reached here.

Velocity resolution is of outmost importance in picking a moving target in the presence of clutter (noise caused by echoes due to trees, mountains, etc.) as such noise comes from stationary targets that produce no Doppler. All in all, this $\chi(\tau, v)$ seems to embody most that is useful in radar analysis. It tells us about resolution in both range and velocity. It reduces to $R(\tau)$ and $\mathcal{H}(v)$, which have a lot to say about accuracy in range and velocity measurements via their bandwidth or duration. It even has in it the energy of the signal $R_m(0)$, so important in SNR, which figures in detection as well as parameter estimation. It may not be a panacea, but it does not turn out to be a chimera either.

PROBLEMS, QUESTIONS, AND EXTENSIONS

14.1. Consider a detection problem like the one in Section 14.2, but in which the r.v. $O(T)$ has a Laplacian p.d.f. with mean zero (no target) or with mean S (target). Sketch the receiver operating characteristics for SNR $= 0$, 2, or 5.

14.2. Our discussion on detection was based on a pulse radar. Discuss detection for the case of a chirp radar.

14.3. Find the variance of the range estimator of a radar if the signal emitted has the form of an isosceles triangle, and the S/N is 20. Choose your own triangle.

14.4. Find the variance of the velocity estimator if the signal emitted has the form of an isosceles triangle (choose your own), and the S/N is 10.

14.5. State and prove an inverse Wiener–Khintchine theorem that can justify the statement made in Section 14.5 that $|A(t)|^2$ is the FT of $\mathcal{K}_A(v)$.

14.6. Show that ρ of Eq. (14.6.3) can also be given in terms of the FT of $A(t)$ and $\varphi(t)$.

14.7. A Gaussian pulse of the form

$$f(t) = e^{-t^2/2\tau_o^2}$$

appears to have the lowest BD product (equality in the uncertainty principle). Show it.

14.8. Consider our known chirp signal

$$m(t) = A \, \text{rect}\left(\frac{t}{T_p}\right) \cos\left(\omega_c t + \frac{\mu}{2} t^2\right)$$

and show that $\rho = (\mu/2\pi)\tau_o^2$. Now find τ_o^2 and β_o^2, and examine the accuracies in range and velocity measurements. (*Warning:* To find a "reasonable" β_o^2, you must approximate the rect with a trapezoid. This is not easy to carry through. Check the literature.)

14.9. Find the BD product:
(a) For a Gaussian chirp pulse

$$m(t) = e^{-t^2/2\tau_o^2} \cos\left(\omega_c t + \frac{\mu}{2} t^2\right)$$

(b) For a signal $f(t)$ with FT

$$F(\omega) = \left(\frac{\sin 10^2\omega}{10^2\omega}\right)^2$$

14.10. The fact that ρ effects the accuracy of both measurements can be seen a bit better by the following. Take the product of $\sigma_{t_o}^2$ and $\sigma_{\omega_d}^2$, which expresses the accuracy of both measurements. Note that for $\rho = 0$, this product stresses the fact that high BD is desirable. Now let $\rho \neq 0$, and use the (radar) uncertainty principle, Eq. (14.6.6a), on this product, to get an upper bound on it. Do you see that high BD increases the upper bound of the accuracy of the simultaneous measurement?

The rest of the problems deal with ambiguity functions and are meant to extend the theory of Section 14.7.

14.11. Prove expression (14.7.22).

14.12. Show and discuss the meaning of

(a) $\chi(\tau, 0) = \dfrac{1}{2\pi} \displaystyle\int_{-\infty}^{+\infty} |A(\omega)|^2 e^{-j\omega\tau} \, d\omega$

(b) $\chi(0, v) = \displaystyle\int_{-\infty}^{+\infty} |A(t)|^2 e^{-j2\pi vt} \, dt$

14.13. Define a cross-ambiguity function for two signals and discuss what it reduces to for $\tau = 0$ and $\nu = 0$.

14.14. Carry out the mathematics and arrive at the result of Eq. (14.7.24).

14.15. Although we discussed the ambiguity function of a rect (t/T_p) signal, we did not really find it. Vakman (1968) claims that it can be written as

$$\chi(\tau, \nu) = \begin{cases} \dfrac{\sin^2 (\nu T/2)(1 - |\tau|/T)}{(\nu/2)^2} & \text{for } |\tau| < T \\ 0 & \text{elsewhere} \end{cases}$$

Check this reference (and the reference it refers to) and see if you can work this result out.

14.16. It is shown in Berkowitz (1965) that $|\chi(\tau, \nu)|^2$ for a chirp pulse is given by

$$|\chi(\tau, \nu)| = \begin{cases} \left| \dfrac{\sin (1/2)(\mu\tau - \nu)(1 - |\tau|/T_p)}{(\mu\tau - \nu)/2} \right|, & |\tau| < T_p \\ 0, & \text{elsewhere} \end{cases}$$

Check this reference and learn the derivation.

14.17. The examples on the section on ambiguity functions were based on the assumption that the radar emits only one "pulse" and not a few similar ones, as the case may be due to "illumination" time. The ambiguity function of a burst of "pulses" is more useful since *range ambiguities* and *blind speeds* are evident in its diagram. Try to come up with a top-view ambiguity diagram for three pulses, and discuss what this diagram shows. (You will find help in Berkowitz, 1965, p. 209.)

=== 15 ===

ELECTRONIC WARFARE

Si vis pacem, para bellum.
(*If you desire peace, prepare for war.*)

Latin saying

15.1 INTRODUCTION

Now that we know how the various communication and radar systems work
and what we must do to improve their performance, the time has come to see
what we can do to jam them up and "render them inoperative"—the enemy's,
of course, not ours.

Electronic warfare is officially[1] defined as the employment of electronic
devices and techniques for the purposes of:

1. Determining the existence and disposition of the enemy's electronic aids
   to warfare
2. Destroying or degrading the effectiveness of the enemy's electronic aids to
   warfare
3. Preventing the destruction of friendly electronic aids to warfare

These pompous statements—the whole field abounds with euphemisms—
mean nothing more than (1) spy (electronically) on the enemy, (2) do his devices
in, and (3) avoid having your devices done in by him. If you are wondering what
this kind of thing is doing in a book that poses as a university textbook, we
cannot really blame you. But then again, this field *is* a scientific endeavor, probably the best endowed one of all. Governments give it top priority and the

---

[1] See The *International Countermeasures Handbook*, or Boyd et al. (1978).

taxpayers' money flows out like Niagara Falls. So why not introduce it early in a student's education? Splendid careers can be made in this area, and the opportunities should be open to all who desire action. As Thucydides once remarked: "To those who call yourselves men of peace, I say: You are not safe unless you have men of action at your side."

The above three goals have evolved to three distinct subfields which are named and discussed below.

Electronic reconnaissance. **Reconnaissance**, in general, is a military euphemism for "spying," that is, the collection and analysis of information for pinpointing everything about the enemy: troop numbers, plans of attack, weapons, even political developments. **Electronic reconnaissance** (ER), which is of interest to us here, pertains to gathering and analyzing information about the enemy by *electronic means* (not hiring an agent, for example). Its chief purpose is the development of methods (and devices) for receiving and analyzing signals emitted by the enemy, so that his capabilities in the use of the electromagnetic spectrum (radar and communication systems, mainly) are ascertained. Its ultimate form would be a huge receiving station, an "ear" so to speak, that hears every signal that the enemy emits and then passes it on to a giant computer which analyzes it. The end result would be a full picture of the present and future enemy capabilities in communications and radar systems (including weapons systems) as well as their locations, movements, and the like. Big task!

Electronic countermeasures. The field of electronic countermeasures (ECM) has been developed to fulfill goal 2 above. Once reconnaissance has informed us of the enemy's capabilities, our next task is to develop techniques for destroying them, or at least for degrading them. A typical achievement in the area is the ability to jam an enemy's communication and radar system. If our planes are on a mission deep in enemy territory (we assume a war situation), they had better carry some devices that will make their detection by enemy radar difficult. It is quite obvious that this field is closely related to electronic reconnaissance, described above. It is very difficult to degrade a radar's performance appreciably unless you know a lot about what kind of radar it is. Your efforts to jam some radars (monopulse, for instance) with noise may, in fact, be worse than doing nothing at all.

Electronic counter-countermeasures. If we are engaged in ER and ECM, it is foolish not to assume that the enemy is also. ER is supposed to tell us that, anyway, and, in fact, to specify precisely what the enemy is capable of doing to our own abilities to make use of the electromagnetic spectrum. Knowing that, we must develop methods to safeguard these abilities, and all such methods fall in the electronic counter-countermeasures (EC^2M) subfield.

It is quite obvious that the three subfields described above are coordinated in a dynamic, continuously changing interdevelopment. ER, for example, may detect the enemy capability on a new enemy radar for missile guidance. ECM is immediately alerted, to develop a technique for jamming it. Now the enemy's ER discovers these efforts and alerts its own EC^2M section to improve the system, or jam the jammer that is meant to jam it. This gives rise to EC^3M, as a new method must be found for jamming the radar, or for jamming the jammer that jams the jammer. . . . Well, the thing never ends, really. It is like a poker game between two players where an unlimited number of raises are allowed and both players are holding a straight flush. Each player raises the pot, unwilling to stop the game. Except that what we are talking about is not a game—it is a deadly business. In case you are entertaining some doubts about the deadliness, read the example below, taken directly from Schlesinger (1961). It is a real story, and not the only one, of course.

During early 1942, the RAF Coastal Command started using L-band radars for locating and sinking German U-boats as they were surfacing to recharge their batteries. The Germans soon produced a countermeasure to this, an L-band search receiver, which detected the radar signals early enough for U-boats to crash-dive before radar detection. Now it was the turn of the Allies to produce an EC^2M device, after they realized, of course, what was happening (reconnaissance). Their "move" was a new S-band search radar installed early in 1943. U-boats started to sink regularly, and the German L-band receivers were unable to produce a warning. The Germans believed at first that the allies had developed a new infrared detection device (bad reconnaissance) and spent a lot of time and effort trying to combat (EC^3M) this new threat. They lost a lot of submarines before they figured out what was happening.

This chapter is devoted to a brief presentation of the fundamental ideas and techniques of the three subfields of electronic warfare. The whole area is vast, and increasing with phenomenal speed, unmatched by any other field of scientific endeavor. Some of the thickest books available are in the field of electronic warfare (see, e.g., Boyd et al., 1978), and this is quite remarkable in view of the fact that only a small portion of what is known can be found in the literature; the rest is secret.

15.2 ELECTRONIC RECONNAISSANCE

15.2.1 Introduction

If a country wishes to develop its defenses, it must do so in the presence of great uncertainties about the offensive capabilities of the enemy. General reconnaissance has the purpose of reducing these uncertainties, by gathering and analyzing information about the enemy. Electronic reconnaissance (ER) helps in this general effort by collecting and analyzing signals emitted by the enemy's electronic devices. There are other means, also electronic, which can help in gathering information, such as satellite photography and radar, but such means are "active," so to speak—they emit their own signals and analyze their return—and will not be considered here. The ER we shall introduce deals only with accepting (intercepting) and analyzing the signals emitted by the enemy devices.

A typical ER scenario is the following. An aircraft stocked with receivers, analyzers, and the like flies over enemy territory, or next to the enemy's border and intercepts signals that are caught by its antenna. Such signals may be taped for later analysis, or may be analyzed on the spot if the means are available on the aircraft. The direction of the signals as they hit the antenna helps to estimate the location of the devices that emitted them. The type of signal that is received identifies to a great extent the type of device that emitted it, and therefore the capabilities of the enemy in waging electronic warfare. This basic scenario can vary a lot, of course. The receiver may be in a submarine (or ship) that cruises inside or outside the territorial waters of the enemy country. The interceptor-receiver may be in an aircraft that is flying a mission, and in direct communication with a countermeasure device (a noise jammer, say) for immediate action. But whatever the variations may be, the *basic* functions of the ER device is *location* (or at least determination of direction) of the enemy device, and *identification* of its function. Typically, for example, the device may warn that a "conical scan radar is range-tracking us from such and such direction," assuming, of course, that it can do so, and is not simply a recorder of intercepted signals for analysis later.

More specifically, a sophisticated **interceptor-analyzer** should be able to decide the following.

1. *Number and location of enemy devices.* A good ER system should be able to determine the number of enemy devices that are emitting signals, as well as their precise location. Such information is of utmost importance in ways that are hard to imagine unless you have experience in the area. Knowing, for example, that the enemy country has five different radar systems (search, conical scan, monopulse, etc.) is certainly important. But knowing that all five are concentrated in *one* location is even more important. It probably means that this location has installations that must be protected, and such a location makes a top-priority target. In any case, identifying the number and location of many devices is very hard, particularly if they are emitting at the same time. It is a complex problem, and such problems are hard to solve even if you designed the signal and know what to expect at the receiver.

2. *The type of system (device) that emitted the signal.* Most signals received by an interceptor-analyzer are a sort of "identity card" of the device that emits them, and proper analysis can determine this device. A received signal with AM modulation from pulse to pulse designates a possible conical-scan radar, particularly if its spectrum is analyzed and found to be as expected. So it is with most devices that are known to exist a priori, and

the receiver is set up to receive them and analyze them. If the emitting enemy device is something new (remember the S-band Allied radar), the interceptor may miss it entirely, or if it receives it, it may not recognize it. Secrecy of all new developments is of utmost importance in this area, the closest one can come to a *supreme weapon* at this stage of the game. If the enemy can find out what you have, he can always find (or buy) a way to countermeasure it. It is only a matter of time and money, nothing more.

Aside from the above, ER systems perform other tasks as well, some of them matters of life and death. A good ER system (or the operators that manage them) are always on the alert for high signal "traffic," for example, as it indicates many possible types of planned action by the enemy. High traffic of one type of signal may indicate testing of a new weapons system. High traffic of all sorts of signals (radar, communication, etc.) may indicate troop movements for a possible attack. The enemy knows that you are keeping track of these "traffic indicators" and that is why he often announces his military maneuvers in advance, if they are not offensive in nature, lest he be misunderstood and find himself the target of a preemptive primary attack. It is a risky business, and mistakes can cost lives.

15.2.2 A General ER System

To get a better overall picture of ER, we present and discuss a typical interceptor-analyzer in block diagram form, shown in Fig. 15.2.1. The one shown is, of course, quite general, and we do not mean to imply that all ER systems have all the boxes (functions) portrayed, as many of them are designed for special purposes and are of limited scope.

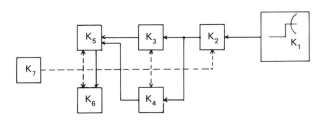

Figure 15.2.1   General ER system.

The first box, K_1, represents the antenna system. This system is asked to perform a lot of diverse functions—receive signals in a huge frequency band, help in estimation of location, and so on—and often these functions cannot be performed with a single antenna. So it is not unusual to have more than one antenna there.

The second box, K_2, is the receiver. Its basic function is to receive the signal emitted by the enemy device and send it to boxes K_3 and K_4 for further analysis. Its characteristics are those of a good overall receiver (high S/N for example, wide bandwidth, etc.), except that now it is asked to detect signals of unknown form, so it is not exactly ready for all eventualities as it would be if it served as a receiver in a communication or radar system. Its most important characteristic is *very wide bandwidth* (in regions where signals are emitted now or *may* be emitted in the future), so that a signal is not missed entirely.

Box K_3 is the signal analyzer of the ER system and it may be combined with K_2 as a single box. It is here that key characteristics (parameters) of the received signal are measured (type of modulation, bandwidth, carrier frequency, any PRF, etc.) so that the *identity* of the device can be ascertained in another box. Thus K_2 and K_3 make up a combination receiver-analyzer for all possible signals that might be received. If the signal received comes from a communication system, K_3 might be asked to demodulate it and ship it to another box for storing or transmitting elsewhere.

The next box, K_4, is the subsystem whose purpose is to pinpoint the location of the enemy device (or devices). Naturally, it works in cooperation with the receiver-analyzer (especially in problems of resolution) and the antenna system, as well. It should be fast and accurate, particularly if the situation calls for immediate action by a ECM device (as in a mission, say).

K_5, the next box, accepts the results of the various measurements done in K_3 and K_4 and stores them or decides the identity of the emitting device as best it can on the spot. It all depends on the type of mission that the whole system is on. If it is a strictly ER mission, the results are stored for eventual analysis at an analysis center. If the ER system accompanies a fighter plane on a mission to protect it, the decision must be made on the spot, as radar tracking by the enemy may result in a hit and downing of the plane. ER systems in such situations must have an a priori idea what to expect, they must be small in size, and so on; otherwise, they could become a detriment rather than a help.

Box K_6 is a communications transmitter for sending the information obtained elsewhere for analysis. Again, whether this box is included or not depends on the situation. If the ER device is on a satellite, for example, this box is a *must*.

The final box K_7 is a sort of control center for the rest of the boxes, coordinating their various performances. It can be a complicated device. Many decisions must be made here that affect the operation of the other boxes, as, for example, the time required for an analysis of a received signal when there are many around, and so on.

This whole analysis is not to be taken literally, of course. These boxes are entirely pictorial, meant to organize our thinking on the various functions of the overall ER system. In a typical ER practical system, these boxes do not exist at all, but the various functions must somehow be performed.

15.2.3 Receivers

The most fundamental function of an ER receiver is to intercept the enemy signal (i.e., not to miss it). The second function of importance is to measure the center frequency of the signal's spectrum, and, naturally, to estimate the entire spectrum as well. After all, demodulation of the signal received is usually necessary for further analysis in both communication and radar signals, so knowledge of carrier frequency is important. Besides that, the simplest kind of ECM against an enemy device is noise, which jams it. Knowledge of the center frequency and bandwidth is necessary if the noise is to be in the proper frequency region, and therefore effective.

Let us describe briefly some of the general types of receivers used in ER (called sometimes **Ferret** receivers), without much detail or mathematical adornment—the available space does not permit it. Those interested can pursue the matter further in Torrieri (1981).

1. *Scanning receivers.* Scanning receivers are, typically, superheterodyne receivers such as those we discussed in our study of AM communication systems. The scanning is done automatically over a very wide band of frequencies, and if an output is obtained, the scanning stops for recording. If the enemy is using a frequency-hopping SS system, you have problems trying to hop around to find his signal.

2. *Nonscanning receivers.* The most basic type here is the **wide-open receiver**, which can be thought of as a BPF with a huge bandwidth (see Fig. 15.2.2a) that accepts everything that can possibly be transmitted, but cannot really tell you where its spectrum is located, or its center frequency.

The wide-open receiver can be modified to improve the frequency resolution by using many parallel filters (see Fig. 15.2.2b) and by observing which of them gives an output. The ability of this receiver to estimate center frequency and spectral bandwidth depends on the number of parallel filters that are used to split up the available region. To increase the accuracy of measurement, without a lot of complexity, the **channelized** (or **matrix**) **receiver** of Fig. 15.2.2c is often used, a clever mechanism that is explained below by means of an example.

Let us assume that the receiver covers the frequency region 10 to 13 GHz. If a signal enters this receiver, the first thing it faces is three wideband filters, one at 10 to 11, one at 11 to 12, and one at 12 to 13 GHz, filters that split up the region into three subregions. No matter which of the three filters the signal passes through, it will be multiplied by a local cosine, which will shift its spectrum down to the region covered by the second bank of filters, the region 1 to 2 GHz, for example. Let us say that there are three filters there, covering the regions 1 to 1.33, 1.33 to 1.66, and 1.66 to 2 GHz, respectively. The signal will

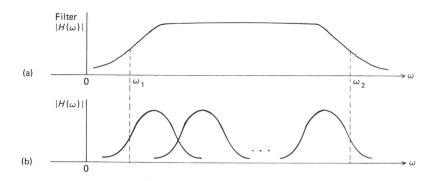

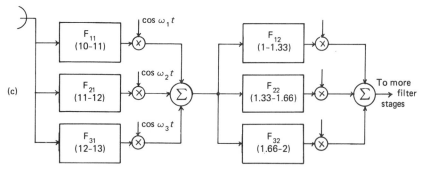

Figure 15.2.2 (a) Wide-open Ferret receiver; (b) parallel filter type; (c) channelized receiver.

pass through one of them and the story repeats until the final stage, which can be as "fine" as we want and as we can design. By keeping track of the filters the signal passes, its bandwidth and center frequency can be estimated. If, for example, the signal enters filter F_{11}, we know that its spectrum lies in the region 10 to 11 GHz. If next it passes the filter F_{32}, we know that its spectrum lies in the upper part of the region 10 to 11 GHz, that is in the region 10.66 to 11 GHz, and so on. Needless to say, logic circuits keep track of the specific filters that the signal traverses, and announce the final result. We might add that this receiver has problems if more than one signal is received simultaneously.

That is all that we are going to say about interceptor-receivers. We hope that it is enough to stimulate the reader to pursue the matter further in the literature.

15.2.4 Estimation of Direction and Location

Another primary function of an ER system is the measurement of *direction*, and then the precise *location* of the enemy device that emits a received signal. Let us consider these two things separately, even though in most cases they are dependent on each other. Generally speaking, once direction is established, measurement of the location coordinates follows. There are cases, however, when only the direction is desired, as for example, when the enemy device is on a moving aircraft. In such cases, the ER system is working with ECM and the direction is often enough for appropriate interference action against the enemy device.

Measurement of direction.   Quite obviously, the antenna plays the key role here, with direction being established when its position gives maximum output. Pencil-beam antennas are needed for precise direction, and they can consist of a lot of stationary ones, or a single scanning one to cover the entire geographical area of interest.

1. *Stationary antennas.* Such a system is shown schematically in Fig. 15.2.3a, presumably covering a horizontal area of 2π radians. Each antenna is connected to a different receiver, and the one with the biggest output marks the direction of the enemy device. The system is obviously bulky and unsuitable for installation on fighter planes, where space is at a premium.

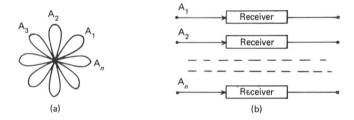

Figure 15.2.3   Stationary antenna system for direction finding.

2. *Scanning antennas.* In these systems the space is covered with a pencil-beam antenna which performs a scanning movement (what kind?) similar to those used in acquisition. Of course, whenever we have a scanning situation, there is always a probability that an intermittent signal will be missed. High-powered analyses exist for finding the probability of interception for various signals and scanning movements, but most are only of theoretical interest.

Estimation of location.   Long-range strategic planning requires knowledge of the location of all enemy installations that emit signals. Flights to accomplish this are of two types, and are explained below.

1. *Overflight.* When the ER device is located on an aircraft or satellite that flies over enemy territory, pencil-beam antennas are directed downward and (depending on the height) cover a specific area, as shown in Fig. 15.2.4.

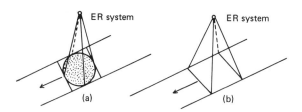

Figure 15.2.4   Overflight area covering: (a) circle, (b) square.

If a signal is received during such a flight, it is recorded together with the other parameters that can help in the estimation of location, such as height and location of aircraft. Needless to say, there is uncertainty as to the exact location of the enemy device, which is equal to the surface covered by the antenna beam. In other words, we can determine the location of a device only within the limits of the area covered. Now we can try many flights with overlapping areas to reduce the uncertainty, but this is not always successful, as the device may be "silent" on the second trip around.

2. *Horizontal measurement.* In times of peace it is difficult to violate another country's airspace, and barring the use of a satellite, the only thing left is what is called *horizontal* measurement of location. This is done by having 'the aircraft fly along the border of the enemy country, directing its antenna horizontally toward it. It is obvious that the reception of a signal does not specify the location of the device in this case, only its direction. What is needed is reception of the same signal from two (or more) different locations, as shown in Fig. 15.2.5. The area of uncertainty is also shown in the figure, and it depends on the beam's width, the distance L between points, and other factors. A detailed analysis of this method can be found in Schlesinger (1961).

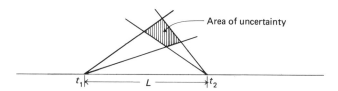

Figure 15.2.5   Horizontal reception from two points (triangularization).

With this we close our very brief introduction to the field of ER. We must stress that our approach was quite shallow. Although we mentioned a few things about *locating* enemy devices and *receiving* enemy signals, we said nothing about the methods of recognizing the nature of these signals and the identification of the device that emitted them. In any event, we will assume that all these tasks can be accomplished and proceed to introductory discussions of ECM and ECCM for communication and radar systems.

15.3 ECM AND ECCM FOR COMMUNICATION SYSTEMS

Let us now consider some methods for disrupting enemy communication systems, methods used mainly during wartime. During periods of peace, our usual efforts are to intercept, "listen" and record enemy conversations, and then analyze them for information that will aid in overall strategic planning for the future.

Disruption (or jamming) is done, of course, with noise, noise that can be produced by amplifying the output of a noise source (thermal, for example) and then modulating a carrier for transmission, if needed. In general, the mathematical form of the emitted noise is

$$N(t) = a(t) \cos (\omega_c t + \varphi(t)) \qquad (15.3.1)$$

where $a(t)$ and $\varphi(t)$ are both random processes. That such noise will degrade the enemy's communication receiver performance is known by our previous analyses.

Some very important questions in this area are the following. What should the probabilistic description of the noise be (in terms of p.d.f. and/or spectrum) for it to be most effective? Can the form (15.3.1) be reduced to a simpler form $[a(t) = K, \varphi(t) = K,$ or both] and still provide satisfactory disruption in enemy communications?

The answer depends, of course, on the type of enemy communication system that we are attempting to jam, and therefore on the effectiveness of our ER. Let us consider next some special cases that illustrate the thinking and methods of analysis in the design of effective ECM for communication systems.

Enemy system unknown.   In the absence of complete identification of the enemy system, the only thing we know is its approximate transmission bandwidth (after all, we did intercept and listen in some frequency band). Methods such as the maximum entropy principle discussed in Section 11.6 conclude that the best noise in such cases is Gaussian (complete description) white (partial description) noise, with a spectrum that covers entirely the suspected enemy band. If the band is narrow, the noise jammer packs and emits all its power in this band (**spot noise**). If the band is very wide (suspected FM, SS systems, etc.), there are two options, illustrated in Fig. 15.3.1.

1. *Barrage noise.* Spread out the noise power over the entire band. Naturally, such a technique spreads out the power over a large frequency area, and

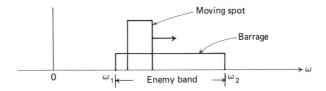

Figure 15.3.1   Barrage and moving spot noise.

if the enemy band is narrow (unknown to you, however), the jamming becomes ineffective.

2. *Moving (swept) spot noise.* Pack all the power in a narrow band and sweep the suspected wide band periodically. Optimum sweeping speeds have been studied and related to performance of various systems, but such analyses are beyond the scope of this book.

Both methods have their advantages and disadvantages. The best choice would be a barrage noise jammer with so much power that it covers the entire band with plenty of power, but that is not possible in all situations.

Enemy system known.   Analyses (and applications) can get a boost if the enemy system is known. Of course, the conclusions are not always in line with practical results, but they do provide good insight into the problems. In what follows we give a sample of an analysis when the enemy system is of analog type (AM, PM, FM). Our goal is to derive the form of optimum jamming noise if the enemy's system is known a priori.

Let us assume that the communication system's signal has the form

$$m(t) = a(t) \cos (\omega_1 t + \varphi_1(t)) \tag{15.3.2}$$

a form that includes all analog systems (AM, FM, PM) with a sinusoidal carrier. We will start our analysis with this general form and then specialize it later for specific systems.

The form of the noise is

$$N(t) = b(t) \cos (\omega_2 t + \varphi_2(t)) \tag{15.3.3}$$

the same form but with different symbols, so our analysis can be clear.

Ignoring all other noises (thermal, etc.), the output of the usual BPF in the receiver of the enemy communication system will be

$$X(t) = m(t) + N(t) \tag{15.3.4}$$

that is, we are assuming that all of $N(t)$ made it through without distortion.

Inserting Eqs. (15.3.2) and (15.3.3) into Eq. (15.3.4) and using known trigonometric formulas, we are led to

$$X(t) = R(t) \cos (\omega_1 t + \varphi_1(t) + \theta(t)) \tag{15.3.5}$$

where

$$R(t) = [a^2(t) + b^2(t) + 2a(t)b(t) \cos (\omega_3 t + \varphi_3(t))]^{1/2} \tag{15.3.6}$$

and,

$$\theta(t) = \tan^{-1} \frac{b(t) \sin (\omega_3 t + \varphi_3(t))}{a(t) + b(t) \cos (\omega_3 t + \varphi_3(t))} \tag{15.3.7}$$

where

$$\omega_3 = \omega_2 - \omega_1, \qquad \varphi_3(t) = \varphi_2(t) - \varphi_1(t) \tag{15.3.8}$$

Now we are ready to look at a couple of specific systems that involve ECM. Both cases involve AM and are meant to be illustrative of the ideas.

If the system is AM of any kind, $\varphi_1(t) = 0$ and therefore $\varphi_3(t) = \varphi_2(t)$. Let us assume that the AM system is of the large-carrier type, where the demodulation is done with an envelope detector. The final output of the system $o(t)$ will obviously be $R(t)$, which is rewritten as

$$o(t) = a(t)\left[1 + \frac{b^2(t)}{a^2(t)} + 2\frac{b(t)}{a(t)}\cos(\omega_3 t + \varphi_2(t))\right]^{1/2} \qquad (15.3.9)$$

so we see that instead of $a(t)$, the receiver gets the above.

To do something about this we assume that $a(t) \geq 2b(t)$, a case not too unreasonable in most situations. With this condition, if $o(t)$ is expanded into a Taylor series in the parameter $b(t)/a(t)$ around the origin, and only the first three terms are kept as significant,

$$o(t) = a(t) + b(t)\cos(\omega_3 t + \varphi_2(t))$$
$$+ b(t)\left[\frac{b(t)}{4a(t)} - \frac{b(t)}{4a(t)}\cos(2\omega_3 t + 2\varphi_2(t))\right] \qquad (15.3.10)$$

Let us see if we can draw now some conclusions from Eq. (15.3.10). If $b(t)$ and $\varphi_2(t)$ are constants—the jammer emits only a tone—and $\omega_3 = 0$, most of the power of the jamming terms can be blocked by a capacitor, so *this is not an effective ECM signal*. If $\varphi_2(t)$ is a constant—the jammer emits AM noise—the effectiveness is low only if $\omega_3 \approx 0$ and $\varphi_2 \approx \pi/2$. Taking into account that if the jammer emits AM noise, its power is concentrated into a narrow band, we have to conclude that this (AM noise) is a *good way to jam a large-carrier AM system*.

If the AM system is DSB and uses a synchronous detector, $X(t)$ is multiplied by $2\cos\omega_1 t$, and $o(t)$ becomes (check it)

$$o(t) = a(t) + b(t)\cos(\omega_3 t + \varphi_2(t)) \qquad (15.3.11)$$

quite effective with or without $\varphi_2(t)$. In fact, the noise here effects synchronization. If the transmitter tries to overcome this by sending a little of the carrier, it can be detected by the jammer's ER system, and more effective jamming can be produced with $\omega_3 = 0$.

Similar analyses can be done for FM and PM systems, and can be found in Torrieri (1981). The overall results are that an AM type of noise is effective against AM, PM, and FM systems, whereas a simple tone can be effective only on jamming PM and FM systems. This is very interesting in view of the wide spectra possessed by the last two systems.

When the systems are of the PCM or mixed type, analyses zero in on the degradation of their ability to detect bits without errors. Spread-spectrum systems appear to be the hardest to jam in view of their *very* wide spectra. Again, interestingly enough, tone jamming seems to be effective against such systems, particularly if their center frequency is known. More details on these issues and others (cryptography, for example) can be obtained in Torrieri (1981) or the special issues on military communications of the *IEEE Transactions on Communications* (1980).

What about ECCM in this area? Well, from the point of view of enemy "listening," the best appear to be fiber optics systems, which cannot be tapped. Barring those (as in mobile communications in the field), the best system depends on the situation. Sometimes the "best" may be the "worst" if you change the situation. Communications between aircraft flying in close proximity may be better done with a poor system than with a very good one, because the latter may be more easily intercepted by the enemy for "listening."

The challenging situations in these areas are unending. Only questions of morality seem to limit the number of people engaged in the development of the field.

15.4 ECM AND ECCM IN RADAR SYSTEMS

15.4.1 Introduction

ECM and ECCM for radar systems is highly developed, for such systems play a key role in wartime situations, a role that can tip the balance and change the eventual outcome of a war.

ECM methods here are classified into two broad general categories: (1) **noise jamming**, which attempts to mask the enemy radar's echoes in heavy noise and degrade the radar's performance; and (2) **deception jamming**, which attempts to lead the enemy radar to an erroneous estimation of parameters. Within each category, ECM systems are again broken down into two classes: (a) **active systems**, those that emit energy to accomplish their goals; and (b) **passive systems**, those that emit nothing but, instead, reflect the enemy's signal to accomplish their goals.

All in all, then, we have four different types of ECM to discuss here, and we shall do so in a logical order.

15.4.2 Noise Jamming

Noise, as we know, degrades the performance of radar systems; it affects the R_{max} for good detection, the range and velocity estimation, and so on, and thus is judged as a good, basic, and relatively inexpensive way to realize an ECM system.

Active noise countermeasures.   The emission of some type of noise toward the enemy radar is considered active ECM noise jamming. We mentioned in the preceding section that such noise is usually white in spectrum, and that it can be barrage, spot, or sweeping spot, as the occasion demands. The overall SNR decreases as a result of noise jamming, and the effects can be estimated and incorporated into the receiver operating characteristics (see Fig. 14.2.1), where the degradation in performance becomes clear.

The effects of noise jamming on R_{max} (i.e., the radar equation) are obvious. There are a number of new radar equations that one can derive (see Nathanson, 1969), depending on the location of the jammer relative to the radar and target, and so on, but we will let the reader look them up in the literature if interested.

Jamming an enemy radar with noise is not a job for an amateur. It must be done with great care, or it could be detrimental rather than beneficial. One must realize that the echo power of radar varies inversely with R^4 (two-way transmission), whereas the jammer's signal arrives at the radar with power inversely related only to R^2. This has consequences which are explained below.

A jammer usually travels on an aircraft accompanying a mission, and co-operates closely with an ER device. If during the trip the ER device alerts the jammer, it does not follow that the jammer should start operating right away. *It may be too early.* The fact that the ER device detects the radar's signal (one-way transmission) does not mean that an echo is detectable by the radar. Early jamming may arrive stronger than a weak echo and cause the radar to suspect a target's presence.

Another problem is the *duration* of jamming. Quite obviously, after the decision is made to start jamming, it cannot continue indefinitely. It must stop now and then, to give the ER device time to assess the jamming damage (if any), to see if other radars (and what kind) have entered the picture, and so on. The interval of time that the jammer interupts transmission is called **look-through**, and it is a critical quantity. If it is too short, the jamming may be ineffective. If it is too long, the aircraft may be the target of a "home-on-jam" missile.

As the jammer approaches a radar's location, the radar's echoes increase in power in relation to the jammer's power. This is due to the fact that the radar has a lot more power at its disposal than does the jammer. There comes a distance beyond which the jammer becomes ineffective. This distance, say R_s, is called the **self-screening range** (or **crossover range**) and can be taken to be the distance at which the powers are approximately equal. It can be shown (see Skolnik, 1962) that it is given approximately by

$$R_s = \left[\frac{P_T G_0 \sigma B_j J}{P_j G_j 4\pi B_r S} \right]^{1/2} \tag{15.4.1}$$

where P_T, G_0, and σ are as defined in Section 4.4; J/S is the jammer-to-signal ratio; B_j the jammer bandwidth; B_r the radar bandwidth; P_j the jammer's transmitted power; and G_j the jammer's antenna gain. It is quite dangerous for the jammer to come closer than this distance, and in view of the fact that one cannot estimate it during a mission, another problem in created. There are thousands of scenarios with thousands of possible solutions, enough to occupy those in the "smoke-filled" rooms for generations to come.

What about ECCM against noise jamming? Well, good radar design is, of course, good ECCM against *any* jamming. The type of radar used is also a factor. Monopulse radar, for example, can track a target using the jamming noise as a signal for tracking. The enemy can still invalidate a monopulse with two jammers at two different locations (**buddy mode**, it is called), so there is no end to this EC"M game. See also Problem 15.1 for another radar case for which active noise jamming is theoretically ineffective.

Passive noise jamming.   The whole idea behind noise jamming is to decrease the S/N at the enemy's receiver system. Emitting noise toward the radar (active) increases the total N, and this reduces the S/N term. There are, however, other methods of decreasing S/N without emitting noise, methods that either *increase N* or *decrease S*. Let us discuss briefly the principal methods.

1. *Chaff.* The most typical example of passive noise jamming is the use of chaff. Chaff is made up of many (thousands) small dipole reflectors constructed of aluminum or aluminum-covered fiberglass in the form of strips wrapped up in bundles. These dipoles are released (or rocketed out) from an aircraft and scattered by the wind, forming a reflecting cloud. Though small, they can be designed so that they reflect a sizable amount of a radar's signal (i.e., they appear to have a high radar cross section). A small bundle can have an effective cross section equivalent to that of a large aircraft.

In passive noise jamming, chaff is dropped in such a way that it creates a cloud corridor through which aircraft can fly undetected; the chaff reflections amount to too much noise for the radar to sense a return echo. In the old days when airplanes were slow-flying, a chaff bundle could even create a "false" target—the radar could start tracking it by mistake. Nowadays, high-speed aircraft can easily be differentiated from slow-moving chaff, so it is used primarily as a smoke screen.

Figuring out the right chaff for the specific application takes some calculation. To get an idea of what is happening with chaff, consider the following. The radar cross section of N pieces of chaff has been found to obey approximately (see Schlesinger, 1961)

$$\sigma = 0.18 \ \lambda^2 N \tag{15.4.2}$$

where λ is the radar signal's wavelength, and the chaff pieces are scattered uniformly in all directions. Let us assume that the pieces are cut from aluminum foil, and that they are $\lambda/2$ long (note the dependence of length to the radar), 0.01 in. wide, and 0.001 in. thick. With these dimensions σ reduces to

$$\sigma = 30{,}000 \frac{W_c}{f} \tag{15.4.3}$$

where σ is in square feet; W_c, the total chaff weight, is in pounds; and f, the radar signal carrier frequency, is in kilocycles. This last equation determines the amount of chaff needed to effect a desired σ. Actually, since the dipoles are resonant, they cover only a limited frequency range. So, many assorted sizes must be cut and thrown together if a wide spectrum is to be covered.

Chaff is an easy, inexpensive way to create noise ECM, and is still popular despite modern advances in the field. ECCM techniques against them are based on their primary disadvantage—their low speed of movement. CW radars, pulse-Doppler, or radars with MTI are relatively immune from the effects of chaff and similar methods of creating noise cloud screens. For besides chaff, there are other ways of doing the same thing—aerosols of solid particles, fog oil, and so on—shown to be particularly effective against laser-guided weapons.

Recent suggestions in the area of "cloud" formation include *space ionization* techniques made possible by the burning of certain elements (sodium, cesium, etc.), or even by the use of *nuclear explosions* in the atmosphere. Such clouds of ions create problems in reflection, refraction, and attenuation in the radar signals, and thus targets that fly through them are lost. More on these issues can be found in the *International Countermeasures Handbook*, and Vadim et al. (1969, Chap. 9).

2. *Reduction of reradiated energy.* As we mentioned earlier, another way to decrease the enemy S/N is by decreasing S, the energy of the received echo. There are various methods for doing that, methods that include changing the shape of the aircraft or covering the aircraft with energy-absorbing materials, and we shall discuss the main ones below. It should be stressed that such passive methods are considered very important. They can be combined with active methods of noise jamming, effectively reducing the performance requirements of the latter.

(a) *Shape of aircraft (or target).* Good aerodynamic design can reduce σ (the target cross-sectional area) and therefore S, but despite good achievements in this area, S still remains high enough to require further reduction by some of the other methods.

(b) *Absorbing coverings.* If a radar signal hits a target, the energy of the reflected echo depends not only on the shape of the target, but also on the electric and magnetic properties of its reflecting surfaces. If you could construct an aircraft out of material that completely *absorbs* the incident wave, it would be invisible to the radar. Materials that absorb waves are usually soft, and cannot be a part of the structure of the aircraft; they can only be used as coverings.

Absorbing coverings are quite popular and of various types. They are usually applied in layers, the outer layer matched (in dielectric constant) to free space, and the other layer with smoothly increasing dielectric constant, so that no energy is reflected back toward the radar. They are usually made (the actual materials are secret) of polysterene, with pieces of graphite (or carbon), whose density changes from layer to layer. The English covering type AF 33, for example, is a mixture of porous rubber and coal dust. Many layers of this covering can reduce reflectivity to 1 %, but only in certain ranges of wavelength.

(c) *Interference coverings.* Such coverings are made of special materials so that the reflections from them and the reflections from the metal of the aircraft neutralize each other. They are usually thinner than absorbing coverings, and their characteristics are similar to the absorbing coverings, so the two methods can be combined. A typical example is the MXI type, made of rubber and iron carbonyl. Its thickness is around 2 mm, its weight 7 kg/m², and its reflectivity below 5% in the region $\lambda = 3.15$ to 3.25 cm.

(d) *Scattering coverings.* The idea behind such coverings is to reflect the incident radar waves toward many directions so that the amount of energy redirected back to the radar is small. Such coverings usually consist of a layer of insulating material, on which metallic strips are glued or sprayed, oriented in different directions and connected to each other in different ways. Such coverings turn the surface of the aircraft into small equivalent reradiating oscillators which scatter the incident energy into many directions. The return energy toward the radar usually depends on the angle of incidence, but it can become quite small—a portion of what it would be without them.

All of these coverings have, of course, the disadvantage that they cannot cover moving parts very effectively, or the "tail" of ionized particles that a supersonic aircraft leaves behind it. They are also bulky, they add weight, and they are effective in only certain ranges of radar wavelength (a good ECCM is variation in the frequency of the radar). Often they are used to cover only sharp edges of high reflectivity, not the entire aircraft. Such coverings are also used on the ground, buildings, bridges, and so on, as they can be effective against enemy radar during raids.

Great amounts of time and money continue to be spent in this area, as—theoretically at least—it can lead to the *ultimate weapon*, a plane invisible to enemy radar. The news media often announce that the perfect covering has been discovered by this or that country. It could be true, or it could be the other big weapon in electronic warfare—the bluff.

In closing we remark again that passive noise jamming is usually coordinated with active jamming. A well-planned mission includes all the available means of ECM (and ECCM), passive and active, and all various submethods within each type. And good reconnaissance is an integral part of the overall mission, both in the planning stage and during its actual course.

15.4.3 Deception Jamming

Here we are dealing with refined, sophisticated methods of leading the enemy radar astray. Deception ECM is the product of the more dissembling side of human nature.

Passive deception ECM.   These are methods of *distorting* the enemy's radar return signal (so that its conclusions are erroneous), without emitting any electromagnetic energy in the process.

Chaff, as we noted, can be used as a false target, at least at its initial stages, before it scatters and becomes a cloud. The use of decoys is also quite popular. **Decoys** are usually small aircraft designed to appear as large ones, by clever use of signal-enhancement devices such as corner reflectors, Luneburg reflectors, and the like. They can be carried aboard bombers and launched at the proper moment to confuse enemy radar. Sometimes they are even outfitted with small noise jammers to make the deception even more complete. A good ECCM against them is the decision to shoot down everything in sight. That takes care of the real target together with the decoys, but it takes a lot of missiles to carry out.

Other passive deception methods exist and can be quite effective. A rotating corner reflector mounted on the aircraft can be quite detrimental to the operation of a conical-scan radar, as the return echoes have AM on their envelope. If the rotation speed is similar to the rotation (around the axis) of the enemy conical-scan, angle tracking can be broken and the radar sent to a searching mode. Two or more reflectors mounted on separate parts of the aircraft degrade even the monopulse radar's performance in certain cases. These and other passive methods are inexpensive ECM deception techniques, quite popular with small countries that cannot afford the active devices described next.

Active deception ECM.   Human ingenuity reaches its zenith here, where the possibilities are endless. It is, of course, impossible to cover all the available ECM and EC"M techniques in this area; they depend a lot on the type of radar versus aircraft "cat and mouse" game being played. We will simply outline some of the most usual methods below and let the reader pursue the matter in the literature. Their principal characteristic is *small power*. Ingenuity and sophistication seldom depend on heavy force reserves.

1. *False targets*. A jammer that uses this general method of deception emits a signal which is similar to that used by the enemy radar. The radar receives the signal, mistakenly takes it for an echo of its own emitted signal, and assumes the existence of a target that is not really there. Occupying the enemy radar with a false target is a great way to ensure the safety of your own aircraft and the eventual success of a mission. Needless to say, this deception maneuver must cooperate closely with a reconnaissance device for good identification of the emitted enemy radar signal.

False target jammers range from the very simple to the highly complex. The simplest case is against a pulse (noncoherent) radar, where knowledge of the carrier's frequency and pulse width are the only things needed to duplicate the signal. Once these two things are measured, the jammer can emit similar pulses a short time after the radar's pulses are received, giving the impression that the target is *farther* than it actually is (typically 12 μs of delay correspond to 1 nautical mile).

Beyond this, false target methods become more complicated and require better analysis of the radar's signal and greater abilities at duplicating it. For example, misleading the radar to think that the target is *closer* than it really is requires measurement of the radar's antenna rotation and the PRF, so that the enemy pulses can be anticipated and false ones sent slightly before the real ones arrive.

Knowledge of the radar's antenna rotation speed can also help the jammer create false targets at other directions, by emitting pulses when the radar is looking "elsewhere." Such pulses must be of stronger energy, as they enter the enemy antenna through the sidelobes. They are easier to accomplish in shorter radar-jammer distances.

False target jammers (called **repeaters**) can be even more sophisticated. They can be programmed to emit pulses that simulate a whole flock of false targets flying in formations, with specified "nonexistent" paths leading to nowhere at all. Others emit many false targets in randomly changing positions, creating psychological problems for the radar operator, who tries to ascertain the existence of a target—having "seen" it twice—but never being able to do so. To make the deception more believable, a little noise is sometimes emitted with the signal, so that the radar is fooled into assuming that the (false) target carries a noise jammer aboard.

All these things work best with simple pulse radars, whose signals are easy to duplicate or delay and retransmit (**transponders**). As soon as the radars become more complicated (chirp, frequency agile ones, etc.), false target methods lose their effectiveness. Still they have been able to keep up in complexity with their corresponding radars and they continue to be an effective ECM method that is not about to be abandoned.

2. *Range gate pull-off.* False targets are mainly effective when the enemy radar is in a searching mode. If the radar is in a tracking mode, they are usually ineffective, and other means are needed to make it "break the track."

The range gate pull-off (RGPO) deception method is intended to break the range tracking of a pulse radar, and it works as explained below. It is assumed that the reader recalls the gate method (actually double gate) of range tracking discussed in Section 10.4.

Let us assume that a PR is tracking an aircraft with an RGPO jammer aboard. As soon as a radar pulse is received, an identical but stronger pulse is emitted, and the enemy radar is automatically adjusted to follow it. This pulse is then gradually delayed more and more, until the radar is "walked-off" into a new and wrong gate. Such **gate stealing**, as it is called, is quite an effective way to break a range track of a simple pulse radar, but not so effective against radars with complex signals that are hard to duplicate. It must be borne in mind that complicated radar signals require complicated reconnaissance devices to identify them, and this adds to the complexity of the overall RGPO method.

A similar method exists for CW radars, the idea being to walk them off the *velocity gate.* The jammer emits a strong sinusoid whose frequency is at first equal to that of the radar, and then is gradually increased or decreased until the radar puts it on a new and wrong gate. In keeping with the terminology this deception mechanism is called **velocity** (or **speed**) **gate pull-off** (VGPO). In actual practice both RGPO and VGPO work only temporarily, and the radar can usually reestablish tracking of the real target. The jammer then repeats the routine above, keeping the radar continuously busy (on–off target), effectively degrading its overall performance. You cannot really ask for more than that.

3. *False* (or *inverse*) *modulation.* Angle tracking is another "target" of deception jammers, as such tracking is often quite dangerous. False modulation is meant to break the track of a conical-scan angle-tracking radar, and as the name implies, it takes advantage of the fact that such a radar uses AM pulse-to-pulse modulation to zero in the right direction of the target.

Let us recall that when the CS radar is on target (its antenna rotating in a conical scan), the return echoes all have equal amplitudes. At that point, the jammer can start emitting pulses with AM on them, simulating echoes coming from a shifted target. The end result is to cause the radar to break track, even though it is usually only for short periods. Nevertheless, the jammer repeats the procedure and the CS radar's performance is degraded.

Such a method has no effect on monopulse tracking radars, as the reader can guess from knowing their principles of operation.

4. *Other countermeasures: combinations.* The three methods described above (actually four with VGPO) constitute the most basic functions used in deception jamming. There are many others, of course, designed for specific radars, or even "circuits" that the radars use in processing their signals. One example of this variety is the **AGC countermeasure**, a method applicable to radars with AGC circuits. It basically uses a strong pulse (identical to the radar's signal) to decrease the threshold of the AGC and then drops it, causing the radar to lose the real but weak pulse. Sometimes, when the parameters of the enemy radar's signal (frequence, PRF, etc.) are not known, a strong pulse of long duration is effective, as it masks all the return echoes within the illumination time and confuses the radar (**broad pulse jamming**). Even noise itself can be used as a deception device when combined with other parameters that cause deception. One such case is **blinking**, in which two jammers flying at about the same azimuth angle with respect to the radar emit spot noise interchangeably, rendering the enemy radar incapable of concentrating on one of the two targets.

As we mentioned earlier, all the methods described in this section (and many others) can be used in combinations. A good overall jammer should be capable of both noise and deception. Furthermore, if the mission is well planned —with good a priori reconnaissance of the threat environment—it is obvious that many countermeasures are needed to carry it through, including passive methods such as chaff, aerosols, and the like. Modern jammers can even be programmed a priori for a certain mission, so that all the operator has to do is watch for the eventuality of something unexpected.

Specific examples of ECM combinations against weapon systems, such as antiaircraft fire control radars and surface-to-air missiles, can be found in the *International Countermeasures Handbook*. These combinations involve both noise and deception active types of countermeasures, and they may have been proven effective in actual experiments in the field.

15.4.4 ECCM in Radar Systems

Although we have indicated here and there in our discussion of ECM possible ECCM action, let us summarize the idea so that the reader will feel a bit more knowledgeable about it. After all, it represents one of the three subfields of the overall area of electronic warfare. In what follows the assumption is that the enemy is engaged in ECM, and we, in ECCM.

Generally speaking, all radars can be jammed *if the enemy is willing to pay the money for it*. With this in mind, the ECCM policy is to design such radars so that the price becomes prohibitive for the enemy.

The best ECCM, then, is sound radar design, that is, receivers with high S/N, good dynamic range, shielding, and so on. The antenna should also be well designed, with low sidelobes, so that false targets are minimized in various azimuth angles. A good MTI subsystem is always a necessity as a protection against chaff or other such slow-moving cloud-forming countermeasures.

Signal design is probably the most important counter-countermeasure of all. We saw in Chapter 14 how important signal design is if the radar is to accomplish its basic functions of measuring target parameters and resolving more than one target. Now we see that the signal design must take into consideration an added requirement, *immunity to countermeasures*. Interestingly enough, wide-spectrum (chirp, pulse compression, etc.) signals seem to have the edge again, in both noise and deception jamming. If the spectrum of the radar signal is very wide, a noise jammer must spread its power thin if it is barrage, or spend less time per frequency band if it is swept-spot. Similarly, a signal with a wide spectrum is effective against deception, for such signals are hard to process and immitate for the production of false targets, RGPO, and so on.

Nothing we say here, of course, is going to be close to "the last word" on the issue, as the field is growing in leaps and bounds. For every "toad there is an antitoad" as the standard joke goes in this and other competitive areas.

PROBLEMS, QUESTIONS, AND EXTENSIONS

15.1. Consider a CW radar system whose emitted signal is

$$m(t) = A \cos \omega_c t$$

Next assume that the radar receives

$$X(t) = m(t) + N_j(t)$$

where $N_j(t)$ is a realization of a jamming noise process. Show that time cross-correlation of $X(t)$ with a locally generated periodic signal (with period $2\pi/\omega_c$) helps in detecting $m(t)$ if $m(t)$ and $N_j(t)$ are time uncorrelated. This shows why a noise jammer is relatively ineffective for CW radar.

15.2. Write a paper on the fundamental ideas of cryptography. A good start is Roden, 1982, Chap. 10 or Torrieri, 1981, Chap. 6.

15.3. Find the radar cross section of a bundle of chaff strips weighing 0.3 lb. The strips are cut for a radar of carrier frequency $f_c = 5$ kHz, and their dimensions are $\lambda/2 \times 0.01$ in. $\times$ 0.001 in.

15.4. Look up in the literature and discuss the notions of (**a**) burnthrough, (**b**) asynchronous pulsed jamming, (**c**) coherent repeater jammer, and (**d**) identification friend or foe (IFF).

15.5. Study the article "What Jams What" in *The Handbook of Electronic Countermeasures* (1977–78, p. 449), and keep it in mind for future reference.

APPENDIX *A*
FREQUENCY ALLOCATIONS

1. *AM Broadcast Band.* From 540 to 1600 kHz. Bandwidth 10 kHz. Carriers 30 kHz (or more) apart.
2. *FM Broadcast Band.* From 88 to 108 MHz. Carriers 200 kHz apart.
3. *VHF Television.* From 54 to 88 MHz and from 174 to 216 MHz.
4. *UHF Television.* From 470 to 638 MHz.
5. *Radar Bands*

Band Name	Frequency
P	230–1000 MHz
L	1–2 GHz
S	2–4 GHz
C	4–8 GHz
X	8–12.5 GHz
Ku	12.5–18 GHZ
K	18–26.5 GHz
Ka	26.5–40 GHz
Millimeter	>40 GHz

APPENDIX B

GAUSSIAN P.D.F. TABLE

$$F(x) = \frac{1}{\sqrt{2\pi}} \int_{-\infty}^{x} e^{-y^2}\, dy$$

x	.00	.01	.02	.03	.04	.05	.06	.07	.08	.09
0.0	.5000	.5040	.5080	.5120	.5160	.5199	.5239	.5279	.5319	.5359
0.1	.5398	.5438	.5478	.5517	.5557	.5596	.5636	.5675	.5714	.5753
0.2	.5793	.5832	.5871	.5910	.5948	.5987	.6026	.6064	.6103	.6141
0.3	.6179	.6217	.6255	.6293	.6331	.6368	.6406	.6443	.6480	.6517
0.4	.6554	.6591	.6628	.6664	.6700	.6736	.6772	.6808	.6844	.6879
0.5	.6915	.6950	.6985	.7019	.7054	.7088	.7123	.7157	.7190	.7224
0.6	.7257	.7291	.7324	.7357	.7389	.7422	.7454	.7486	.7517	.7549
0.7	.7580	.7611	.7642	.7673	.7704	.7734	.7764	.7794	.7823	.7852
0.8	.7881	.7910	.7939	.7967	.7995	.8023	.8051	.8078	.8106	.8133
0.9	.8159	.8186	.8212	.8238	.8264	.8289	.8315	.8340	.8365	.8389
1.0	.8413	.8438	.8461	.8485	.8508	.8531	.8554	.8577	.8599	.8621
1.1	.8643	.8665	.8686	.8708	.8729	.8749	.8770	.8790	.8810	.8830
1.2	.8849	.8869	.8888	.8907	.8925	.8944	.8962	.8980	.8997	.9015
1.3	.9032	.9049	.9066	.9082	.9099	.9115	.9131	.9147	.9162	.9177
1.4	.9192	.9207	.9222	.9236	.9251	.9265	.9279	.9292	.9306	.9319
1.5	.9332	.9345	.9357	.9370	.9382	.9394	.9406	.9418	.9429	.9441
1.6	.9452	.9463	.9474	.9484	.9495	.9505	.9515	.9525	.9535	.9545
1.7	.9554	.9564	.9573	.9582	.9591	.9599	.9608	.9616	.9625	.9633
1.8	.9641	.9649	.9656	.9664	.9671	.9678	.9686	.9693	.9699	.9706
1.9	.9713	.9719	.9726	.9732	.9738	.9744	.9750	.9756	.9761	.9767
2.0	.9772	.9778	.9783	.9788	.9793	.9798	.9803	.9808	.9812	.9817
2.1	.9821	.9826	.9830	.9834	.9838	.9842	.9846	.9850	.9854	.9857
2.2	.9861	.9864	.9868	.9871	.9875	.9878	.9881	.9884	.9887	.9890
2.3	.9893	.9896	.9898	.9901	.9904	.9906	.9909	.9911	.9913	.9916
2.4	.9918	.9920	.9922	.9925	.9927	.9929	.9931	.9932	.9934	.9936
2.5	.9938	.9940	.9941	.9943	.9945	.9946	.9948	.9949	.9951	.9952
2.6	.9953	.9955	.9956	.9957	.9959	.9960	.9961	.9962	.9963	.9964
2.7	.9965	.9966	.9967	.9968	.9969	.9970	.9971	.9972	.9973	.9974
2.8	.9974	.9975	.9976	.9977	.9977	.9978	.9979	.9979	.9980	.9981
2.9	.9981	.9982	.9982	.9983	.9984	.9984	.9985	.9985	.9986	.9986
3.0	.9987	.9987	.9987	.9988	.9988	.9989	.9989	.9989	.9990	.9990
3.1	.9990	.9991	.9991	.9991	.9992	.9992	.9992	.9992	.9993	.9993
3.2	.9993	.9993	.9994	.9994	.9994	.9994	.9994	.9995	.9995	.9995
3.3	.9995	.9995	.9996	.9996	.9996	.9996	.9996	.9996	.9996	.9997
3.4	.9997	.9997	.9997	.9997	.9997	.9997	.9997	.9997	.9998	.9998
3.5	.9998	.9998	.9998	.9998	.9998	.9998	.9998	.9998	.9998	.9998
3.6	.9998	.9999	.9999	.9999	.9999	.9999	.9999	.9999	.9999	.9999
3.7	.9999	.9999	.9999	.9999	.9999	.9999	.9999	1.0000	1.0000	1.0000
3.8	.9999	.9999	.9999	.9999	.9999	.9999	.9999	1.0000	1.0000	1.0000

Note: The table holds if the r.v. X is standard normal, that is, $E(X) = 0$, $\sigma_x^2 = 1$. If it is not, it must be "standardized":

$$Z = \frac{X - E(X)}{\sigma_x}$$

BIBLIOGRAPHY

ABRAMSON, N., *Information Theory and Coding.* New York: McGraw-Hill Book Company, 1963.

ARKOUMANEAS, E., "Effectiveness of a Ground Jammer," *IEE Proc.*, Vol. 129, Pt. F, No. 3, June 1982.

ASH, R., *Information Theory.* New York: John Wiley & Sons, Inc., 1965.

AVGERIS, T. G., AND N. S. TZANNES, "On Reconstituting a Process from Its Quantized Form in the Presence of Noise," *Inf. Sci.*, Vol. 15, 1978, pp. 127–141.

———, E. LITHOPOULOS, AND N. S. TZANNES, "Application of the Mutual Information Principle to Spectral Density Estimation," *IEEE Trans. Inf. Theory*, Vol. IT-26, No. 2, Mar. 1980, pp. 184–189.

BALAKRISHNAN, A. V., *Communication Theory.* New York: McGraw-Hill Book Company, 1968.

BARTON, D. K., *Radars*, Vols. I–V. Dedham, Mass.: Artech House, Inc., 1975.

———, *Radar Systems Analysis.* Dedham, Mass.: Artech House, Inc., 1976.

BAYLESS, J. W., S. J. CAMPANELLA, AND A. J. GOLDBERG, "Voice Signals: Bit-by-Bit," *IEEE Spectrum*, Vol. 10, No. 10, Oct. 1973, pp. 28–34.

BEAN, B. R., AND G. D. THAYER, *CRPL Exponential Reference Atmosphere*, National Bureau of Standards, Monograph No. 4. Washington, D.C.: U.S. Government Printing Office, Oct. 29, 1959.

BENDAT, J. S., AND A. S. PIERSOL, *Random Data: Analysis and Measurement Procedures.* New York: John Wiley & Sons, Inc., 1971.

BENNET, W., AND J. R. DAVEY, *Data Transmission.* New York: McGraw-Hill Book Company, 1965.

BERGER, T., *Rate Distortion Theory: A Mathematical Basis for Data Compression.* Englewood Cliffs, N.J.: Prentice-Hall, Inc., 1971.

BERKOWITZ, R. S., *Modern Radar. Analysis, Evaluation and System Design.* New York: John Wiley & Sons, Inc., 1965.

BLACKMAN, R. B., AND J. W. TUKEY, *The Measurement of Power Spectra.* New York: Dover Publications, Inc., 1958.

BOYD, J. A., ET AL., EDS., *Electronic Countermeasures.* Los Altos, Calif.: Peninsula Publishing, 1978.

BURDIC, W. S., *Radar Signal Analysis.* Englewood Cliffs, N.J.: Prentice-Hall, Inc., 1968.

BURG, J. P., "Maximum Entropy Spectral Analysis." Presented at the *37th SEG Annual Meeting*, Oklahoma City, Okla., 1967. In Childers (1978).

CARLSON, A. B., *Communication Systems*, 2nd ed. New York: McGraw-Hill Book Company, 1975.

CATTERMOLE, K. W., *Principles of Pulse Code Modulation.* London: Illife Books Ltd,, 1969.

CHILDERS, G., ED., *Modern Spectrum Analysis.* New York: IEEE Press, 1978.

CHOU, W., ED., *Computer Communications*, Vol. I: *Principles.* Englewood Cliffs, N.J.: Prentice-Hall, Inc., 1982.

COOK, C. E., "Pulse Compression—Key to More Efficient Radar Transmission," *Proc. IRE*, Vol. 48, No. 3, Mar. 1960, pp. 310–316. Also in Barton (1975).

———, AND M. BERNFELD, *Radar Signals: An Introduction to Theory and Application.* New York: Academic Press, Inc., 1967.

———, F. W. ELLERSICK, L. B. MILSTEIN AND D. L. SCHILLING, GUEST EDITORS, "Special Issue on Spread Spectrum Communications," Two Parts, *IEEE Trans. Commun.* Vol. COM-30, No. 5, May 1982.

———, J. PAOLILLO, M. BERNFELD AND C. A. PALMIERI, "Matched Filtering, Pulse Compression and Waveform Design," Part II, *Microwave J.*, Jan. 1965. Also in Barton (1975), Vol. 3.

COOPER, G. R., AND C. D. McGILLEM, *Methods of Signal and Systems Analysis.* New York: Holt, Rinehart and Winston, 1967.

———, AND C. D. McGILLEM, *Probabilistic Methods of Signal and Systems Analysis.* New York: Holt, Rinehart and Winston, 1971.

COSTAS, J. P., "Synchronous Communication," *Proc. IRE*, Vol. 44, Dec. 1956, pp. 1713–1718.

DEUTSCH, R., *Nonlinear Transformations of Random Processes.* Englewood Cliffs, N.J.: Prentice-Hall, Inc., 1965.

DIXON, R. C., *Spread Spectrum Systems.* New York: John Wiley & Sons, Inc., 1975.

———, ED., *Spread Spectrum Techniques.* New York: IEEE Press, 1976.

FRANKS, L. E., *Signal Theory.* Englewood Cliffs, N.J.: Prentice-Hall, Inc., 1969.

FROEHLICH, J. P., *Information Transmittal and Communicating Systems.* New York: Holt, Rinehart and Winston, 1969.

GAGLIARDI, R., *Introduction to Communication Engineering.* New York: John Wiley & Sons, Inc., 1978.

GALLAGER, R. G., *Information Theory and Reliable Communication*. New York: John Wiley & Sons, Inc., 1968.

GIBBY, R. A., AND J. W. SMITH, "Some Extensions of Nyquist Telegraph Transmission Theory," *Bell Syst. Tech. J.*, Vol. 44., Sept. 1965, pp. 1487–1510.

GOLD, B., AND C. M. RADER, *Digital Processing of Signals*. Lincoln Laboratory Publication. New York: McGraw-Hill Book Company, 1969.

GOLDMAN, S., *Information Theory*. Englewood Cliffs, N.J.: Prentice-Hall, Inc., 1953.

GREGG, W. D., *Analog and Digital Communication*. New York: John Wiley & Sons, Inc., 1977.

HANCOCK, J. C., AND P. A. WINTZ, *Signal Detection Theory*. New York: McGraw-Hill Book Company, 1966.

HAYKIN, S., *Communication Systems*. New York: John Wiley & Sons, Inc., 1978.

HELSTROM, C. W., *Statistical Theory of Signal Detection*. Elmsford, N.Y.: Pergamon Press, Inc., 1968.

HONOLD, P., *Secondary Radar: Fundamentals and Instrumentation*. Erlangen, West Germany: Siemens Aktiengesellschaft/London: Heyden & Sons Ltd., 1976.

INGELS, F. M., *Information and Coding Theory*. Scranton, Pa.: International Textbook Company, 1971.

INOSE, H., AND Y. YASUDA, "A Unity Bit Encoding Method by Negative Feedback," *Proc. IEEE*, Vol. 51, Nov. 1963, pp. 1524–1535.

JAHNKE, E., AND F. EMDE, *Tables of Functions*. New York: Dover Publications, Inc., 1945.

JAVID, M., AND E. BRENNER, *Analysis, Transmission and Filtering of Signals*. New York: McGraw-Hill Book Company, 1963.

JAYANT, N. S., "Digital Coding of Speech Waveforms: PCM, DPCM and DM Quantizers," *Proc. IEEE*, Vol. 62, No. 5, May 1974, pp. 611–632.

JELINEK, F., *Probabilistic Information Theory: Discrete and Memoryless Models*. New York: McGraw-Hill Book Company, 1968.

JOHNS, P. B., AND T. R. ROWBOTHAM, *Communication Systems Analysis*. London: Butterworth & Company (Publishers) Ltd., 1972.

KERR, D. E., ED., *Propagation of Short Radio Waves*, Chap. 5. New York: McGraw-Hill Book Company, 1951.

KHINCHIN, A. I., *Mathematical Foundations of Information Theory*. New York: Dover Publications, Inc., 1957.

KLAUDER, J. R., ET AL., "The Theory and Design of Chirp Radars," *Bell Syst. Tech. J.*, Vol. 39, No. 4, July 1960, pp. 745–808.

KOENIG, W., "Coordinate Data Sets for Military Use," *Bell Lab. Rec.*, Vol. 36, May 1958.

KOLMOGOROFF, A., "Grundbegriffe der Wahrscheinlichkeitsrechnung," *Ergeb. Mat. Grenz.*, Vol. 2, No. 3, 1933.

LACY, E. A., *Fiber Optics*. Englewood Cliffs, N.J.: Prentice-Hall, Inc., 1982.

LATHI, B. P., *Signals, Systems and Communications*. New York: John Wiley & Sons, Inc., 1965.

——, *Communication Systems*. New York: John Wiley & Sons, Inc., 1968.

LEE, Y. W., *Statistical Theory of Communication*. New York: John Wiley & Sons, Inc., 1960.

LIGHTHILL, M. J., *An Introduction to Fourier Analysis and Generalized Functions*. New York: Cambridge University Press, 1959.

LINDSEY, W. C., AND M. K. SIMON, *Telecommunications Systems Engineering*. Englewood Cliffs, N.J.: Prentice-Hall, Inc., 1973.

LOGAN, B. F., AND M. R. SCHROEDER, "A Solution to the Problem of Compatible Single-Sideband Transmission," *IRE Trans. Inf. Theory*, Vol. IT-8, Sept. 1962.

LUCK, D. G. C., *Frequency Modulated Radar*. New York: McGraw-Hill Book Company, 1949.

LUCKY, R. W., J. SALZ, AND E. J. WELDON, JR., *Principles of Data Communication*. New York: McGraw-Hill Book Company, 1968.

MAKSIMOV, M. V., ET AL., *Radar Anti-jamming Techniques*. Dedham, Mass.: Artech House, Inc., 1979.

MANDL, M., *Principles of Electronic Communications*. Englewood Cliffs, N.J.: Prentice-Hall, Inc., 1973.

MAX, J., "Quantizing for Minimum Distortion," *IEEE Trans. Inf. Theory*, Vol. IT-6, Mar. 1960, pp. 7–12.

McGILLEM, C. D., AND G. R. COOPER, *Continuous and Discrete Signal and Systems Analysis*. New York: Holt, Rinehart and Winston, 1974.

MIDDLETON, D., *An Introduction to Statistical Communication Theory*. New York: McGraw-Hill Book Company, 1960.

NATHANSON, F. E., *Radar Design Principles*. New York: McGraw-Hill Book Company, 1969.

NETTLETON, R. W., AND G. R. COOPER, "Performance of a Frequency Hopped Differentially Modulated Spread Spectrum Receiver in a Rayleigh Fading Channel," *IEEE Trans. Veh. Technol.*, Vol. I, VT-30, Feb. 1981, pp. 14–19.

NEWMAN, W. I., "Extension to Maximum Entropy Method," *IEEE Trans. Inf. Theory*, Jan. 1977, pp. 89–93.

NOVICK, A., "Echolocation in Bats—Some Aspects of Pulse Design," *Am. Sci.*, Vol. 59, No. 2, Mar. 1971, pp. 198–209.

NYQUIST, I. H., "Certain Topics in Telegraph Transmission Theory," *Trans. AIEE*, Vol. 47, Feb. 1928, pp. 617–644.

PANTER, P. F., *Modulation, Noise and Spectral Analysis*. New York: McGraw-Hill Book Company, 1965.

———, *Communication Systems Design: Line-of-Sight and Tropo-Scatter Systems*. New York: McGraw-Hill, Book Company, 1972.

PAPOULIS, A., *The Fourier Integral and Its Applications*. New York: McGraw-Hill Book Company, 1962.

———, *Probability, Random Variables, and Stochastic Processes*. New York: McGraw-Hill Book Company, 1965.

———, *Circuits and Systems*. New York: Holt, Rinehart and Winston, 1965.

PEEBLES, P. Z., JR., *Communication Systems Principles*, Reading, Mass.: Addison-Wesley Publishing Company, 1976.

PICKHOLTZ, R. L., D. L. SCHILLING, AND L. B. MILSTEIN, "Theory of SS Systems—A Tutorial," *IEEE Trans. Commun.*, Vol. COM-30, No. 5, May 1982, pp. 855–884.

PIERCE, J. R., AND E. C. POSNER, *Introduction to Communication Science and Systems.* New York: Plenum Press, 1980.

PINSKER, M. S., *Information and Information Stability of Random Variables and Processes.* San Francisco: Holden-Day, Inc., 1964.

RABINER, L. R., AND B. GOLD, *Theory and Application of Digital Signal Processing.* Englewood Cliffs, N.J.: Prentice-Hall, Inc., 1975.

RAEMER, H. R., *Statistical Communication Theory and Applications.* Englewood Cliffs, N.J.: Prentice-Hall, Inc., 1969.

REZA, F. M., *An Introduction to Information Theory.* New York: McGraw-Hill Book Company, 1961.

RHODES, D. R., *Introduction to Monopulse.* Dedham, Mass.: Artech House, Inc., 1980.

RIHACZEK, A. W., *Principles of High Resolution Radar.* New York: McGraw-Hill Book Company, 1969.

RODEN, M. S., *Introduction to Communication Theory.* Elmsford, N.Y.: Pergamon Press, Inc., 1972.

———, *Digital and Data Communication Systems.* Englewood Cliffs, N.J.: Prentice-Hall, Inc., 1982.

ROSIE, A. M., *Information and Communication Theory.* New York: Van Nostrand Reinhold Company, Inc., 1973.

SAGE, A. P., AND J. L. MELSA, *Estimation Theory with Applications to Communication and Control.* New York: McGraw-Hill Book Company, 1971.

SARWATE, D. V., AND M. B. PURSLEY, "Cross-Correlation Properties of Pseudo-Random and Related Sequences," *Proc. IEEE*, Vol. 68, May 1980, pp. 598–619.

SCHLESINGER, R. J., *Principles of Electronic Warfare.* Englewood Cliffs, N.J.: Prentice-Hall, Inc., 1961.

SCHMIDT, H., *Analog Digital Conversion.* New York: Van Nostrand Reinhold Company, Inc., 1970.

SCHWARTZ, M., *Information, Transmission, Modulation and Noise.* New York: McGraw-Hill Book Company, 1959.

———, W. R. BENNETT, AND S. STEIN, *Communication Systems and Techniques.* New York: McGraw-Hill Book Company, 1966.

SHANMUGAM, K. S., *Digital and Analog Communication Systems.* New York: John Wiley & Sons, Inc., 1979.

SILVER, S., ED., *Microwave Antenna Theory and Design.* MIT Radiation Laboratory Series, Vol. 12. New York: McGraw-Hill Book Company, 1949.

SINNEMA, W., *Digital, Analog, and Data Communication.* Englewood Cliffs, N.J.: Prentice-Hall, Inc., 1982.

SKOLNIK, M. I., *Introduction to Radar Systems.* Tokyo: McGraw-Hill Kogakusha Ltd., 1962.

———, EDITOR-IN-CHIEF, *Radar Handbook.* New York: McGraw-Hill Book Company, 1970.

SMITH, B., "Instantaneous Companding of Quantized Signals," *Bell Sys. Tech. J.*, Vol. 36, May 1957, pp. 653–709.

SOLOMON, G., "Optimal Frequency Hopping Sequences for Multiple Access," *Proc. 1973 Symp. Spread Spectrum Commun.*, Vol. I, AD915852, pp. 33–35.

SPILKER, J. J., JR., *Digital Communications by Satellite*. Englewood Cliffs, N.J.: Prentice-Hall, Inc., 1977.

STANLEY, W., *Electronic Communication Systems*. Englewood Cliffs, N.J.: Prentice-Hall, Inc., 1982.

STARK, H., AND F. B. TUTEUR, *Modern Electrical Communications*. Englewood Cliffs, N.J.: Prentice-Hall, Inc., 1979.

STIFFLER, J. J., *Theory of Synchronous Communications*. Englewood Cliffs, N.J.: Prentice-Hall, Inc., 1971.

STILL, A., *Communication through the Ages*. New York: Murray Hill Books, Inc., 1946.

STONE, R. R., JR., AND H. F. HASTINGS, "A Novel Approach to Frequency Synthesis," *Frequency*, Sept–Oct. 1963.

STREMLER, F. G., *Introduction to Communication Systems*. Reading, Mass.: Addison-Wesley Publishing Co., 1977.

TAUB, H., AND D. L. SCHILLING, *Principles of Communication Systems*. New York: McGraw-Hill Book Company, 1971.

The International Countermeasures Handbook, (Comes out every two years), Palo Alto, Cal.: EW Communications, Inc.

TORRIERI, D. J., *Principles of Military Communication Systems*. Dedham, Mass.: Artech House, Inc., 1981.

TRIBUS, M., *Rational Descriptions, Decisions and Designs*, Chap. VII. Elmsford, N.Y.: Pergamon Press, 1969.

TZANNES, M. A., D. POLITIS, AND N. S. TZANNES, "A General Method of Minimum Cross-Entropy Spectral Estimation," *Proceedings of MECO* (Measurement and Control) *Conference*, Aug. 28—Sept. 1, 1983, Athens, Greece.

TZANNES, N. S., AND T. G. AVGERIS, "A New Approach to the Estimation of Continuous Spectra," *Kybernetes*, Vol. 10, 1981, pp. 123–133.

———, AND J. P. NOONAN, "The MIP and Applications," *Information and Control*, Vol. 22, Feb. 1973, pp. 1–12.

VAN TREES, H. L., *Detection, Estimation and Modulation Theory, Parts I, II, III*. New York: John Wiley & Sons, Inc., 1971.

VAKMAN, D. E., *Sophisticated Signals and the Uncertainty Principle in Radar*. New York: Springer-Verlag, 1968.

VITERBI, A. J., *Principles in Coherent Communication*. New York: McGraw-Hill, Inc., 1966.

———, AND J. K. OMURA, *Principles of Digital Communication and Coding*. Tokyo: McGraw-Hill Kogakusha, Ltd., 1979.

VOELKER, H., "Demodulation of Single Sideband Signals Via Envelope Detection," *IEEE Trans. of Communication Technology*, Feb. 1966, pp. 22–30.

WAX, N., ED., *Selected Papers on Noise and Stochastic Processes*. New York: Dover Publ., Inc., 1954.

WOODWARD, PM., *Probability and Information Theory, with Applications to Radar*. New York: McGraw-Hill, Inc., 1953. Also Pergamon Press, Inc., 1955.

INDEX

DATE DUE

DEMCO 38-297